International Symposium and School of Young Scientists on Interfacial Phenomena and Heat Transfer (IPHT 2016)

MATEC Web of Conferences Volume 84 (2016)

Novosibirsk, Russia
2 - 4 March 2016

Editors:

D.V. Zaitsev
O.A. Kabov

ISBN: 978-1-5108-3278-7

Printed from e-media with permission by:

Curran Associates, Inc.
57 Morehouse Lane
Red Hook, NY 12571

Some format issues inherent in the e-media version may also appear in this print version.

Printed by Curran Associates, Inc. (2016)

For additional information, please contact EDP Sciences – Web of Conferences at the address below.

EDP Sciences – Web of Conferences
17, Avenue du Hoggar
Parc d'Activité de Courtabœuf
BP 112
F-91944 Les Ulis Cedex A
France

Phone: +33 (0) 1 69 18 75 75
Fax: +33 (0) 1 69 28 84 91

contact-edps@webofconferences.org

Additional copies of this publication are available from:

Curran Associates, Inc.
57 Morehouse Lane
Red Hook, NY 12571 USA
Phone: 845-758-0400
Fax: 845-758-2633
Email: curran@proceedings.com
Web: www.proceedings.com

TABLE OF CONTENTS

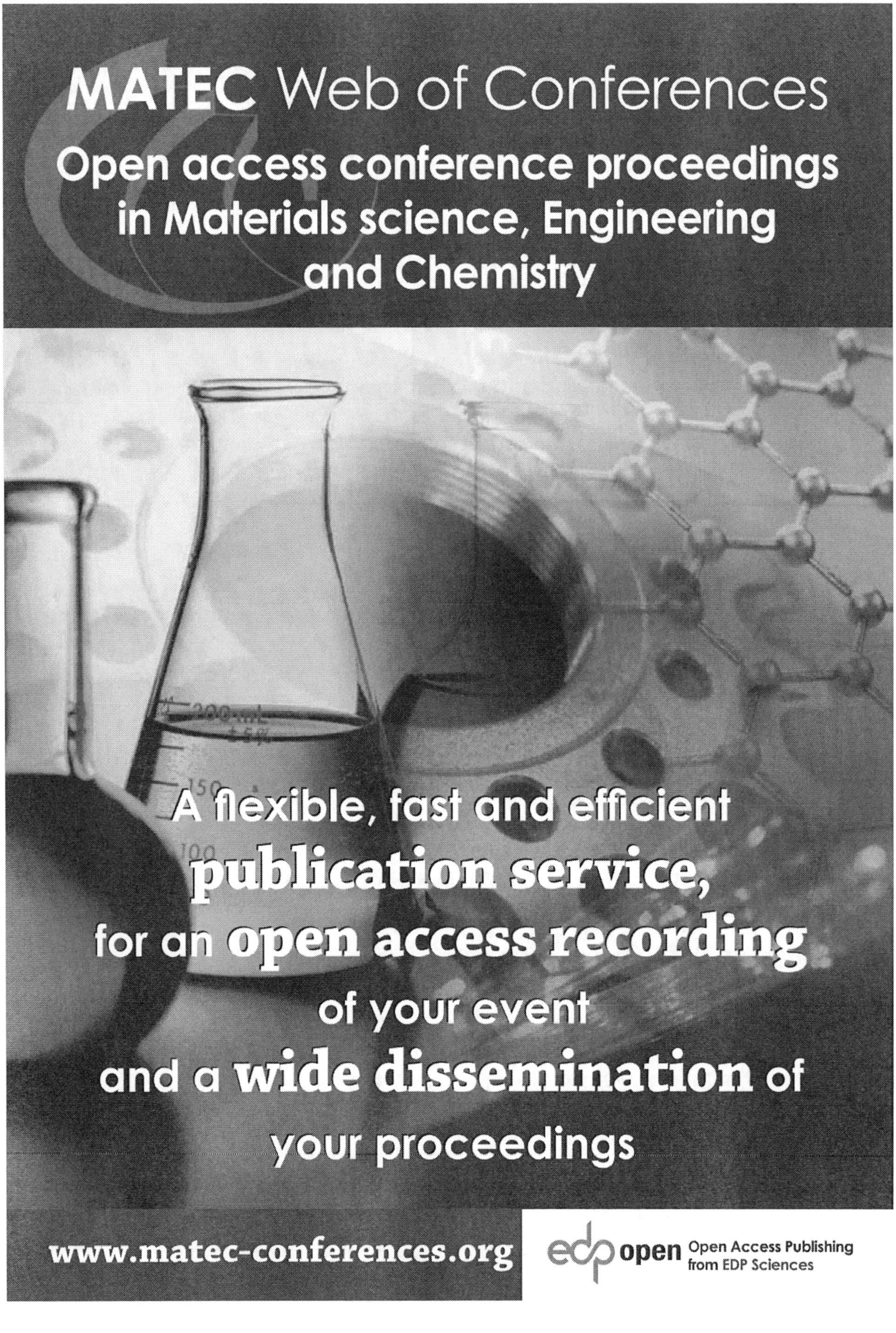
MATEC Web of Conferences
Open access conference proceedings in Materials science, Engineering and Chemistry
A flexible, fast and efficient publication service, for an open access recording of your event and a wide dissemination of your proceedings
www.matec-conferences.org
edp open Open Access Publishing from EDP Sciences

edp sciences

MATEC Web of Conferences

Statement of Peer review

In submitting conference proceedings to MATEC Web of Conferences, I certify to the Publisher that I adhere to the **Policy on Publishing Integrity** of the journal in order to safeguard good scientific practice in publishing.

1. All articles have been subjected to peer review administered by the proceedings editors.
2. Reviews have been conducted by expert referees, who have been requested to provide unbiased and constructive comments aimed, whenever possible, at improving the work.
3. Proceedings editors have taken all reasonable steps to ensure the quality of the materials they publish and their decision to accept or reject a paper for publication has been based only on the merits of the work and the relevance to the journal.

Title, date and place of the conference

International Symposium and School of Young Scientists "Interfacial Phenomena and Heat Transfer" ("IPHT-2016")
March 2-4, 2016, Novosibirsk, Russia

Proceedings editor(s):

Zaitsev D.V., PhD

Kabov O.A., DcS, Professor

Date and editor's signature

September 28, 2016

MATEC Web of Conferences **84**, 00001 (2016) DOI: 10.1051/matecconf/20168400001
International Symposium IPHT 2016

Three-dimensional investigation of liquid film structure at the initial area of annular-dispersed flow

Sergey Alekseenko[1,2], Andrey Cherdantsev[1,2], Mikhail Cherdantsev[1], Sergey Isaenkov[1] and Dmitriy Markovich[1,2]

[1]*Kutateladze Institute of Thermophysics SB RAS, Lavrentiev ave.,1 , Novosibirsk, 630090, Russia*
[2]*Novosibirsk State University, Pirogov str., 2, Novosibirsk, 630090, Russia*

Abstract. Initial stage of downward flow of gas-sheared liquid film in a vertical rectangular duct was studied using brightness-based laser-induced fluorescence technique. Measurements were resolved along both longitudinal and transverse coordinates and time. The initial high-frequency waves which are formed at the inlet were found to be two-dimensional. These waves are promptly broken into localised horseshoe-shaped waves which merge downstream to form large-scale quasi-2D disturbance waves. Peculiarities of three-dimensional evolution of waves of different types were studied in a wide range of flow parameters.

1 Introduction

In annular-dispersed flow liquid film flows on the inner wall of the pipe sheared by high-velocity gas stream flowing in the center of the pipe. In such a flow large-scale disturbance waves are formed on film surface. Presence of disturbance waves is necessary for liquid entrainment into the gas core. These waves cause enhancement of pressure drop and heat transfer. Modeling of wavy structure of liquid film is complicated by multi-scale nonlinear wavy processes and strong turbulence in both phases. To verify the existing models and to create new models detailed experimental information describing the interactions of waves is required.

Experiments by local conductivity probes showed that near the inlet film surface is covered by high-frequency small-amplitude waves [1]. In recent paper [2] spatiotemporal evolution of waves in one longitudinal section of 15 mm pipe was studied near the inlet using laser-induced fluorescence technique. It was concluded that disturbance waves are formed due to multiple coalescence of initial high-frequency waves that appear due to Kelvin-Helmholtz instability.

The goal of the present paper is to investigate the process of formation of disturbance waves in three-dimensional approach, i.e., with measurements of local film thickness resolved in both longitudinal and transverse coordinates and time.

2 Experimental setup and measurement technique

In pipe geometry obtaining 3D film thickness profile over the whole wetted perimeter requires an experimental system consisting of several cameras, which is expensive and complicated. Besides, it would be difficult to completely eliminate optical distortions appearing due to the curvature of the pipe [3]. It was recently shown [4] that the wavy structure of gas-sheared films is qualitatively the same in channels of different shape, size and orientation. Thus, it was considered that investigation of 3D-evolution of waves near the inlet would be more comprehensive in a rectangular duct which provides clear optical access to the whole width of the channel (Fig. 1). The duct's dimensions are 700 mm (length) by 50 mm (width) by 5 mm (height). Liquid film is formed by 0.5 mm width tangential slot between one of the long walls and a thin stainless steel plate.

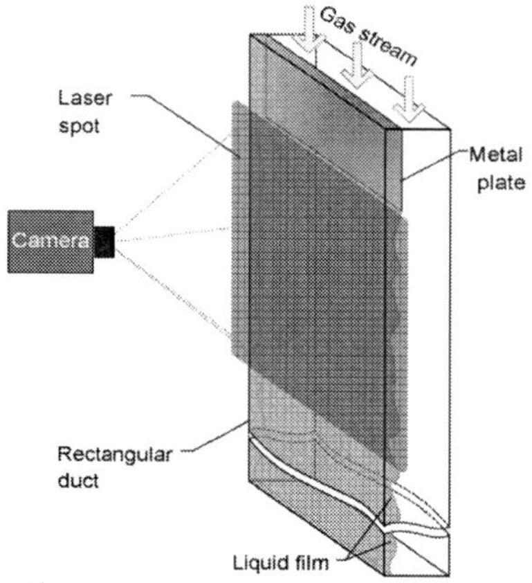

Figure 1. A sketch of measurement section.

Brightness-based laser-induced fluorescence (LIF) method was used for film thickness measurements. Fluorescent dye (Rhodamine-6G) was dissolved in working liquid. The area of measurements was illuminated by continuous 2W laser with wavelength of 532 nm. Reemitted fluorescent light was recorded by high-speed camera and recalculated into local film thickness. Detailed description of this method can be found in [3]. In present experiments, frame rate of high speed camera is 1 KHz; spatial resolution is 0.2 mm/pixel. The size of the investigated area is 50 mm in transverse direction, y, and 200 mm in longitudinal direction, x.

3 Results and discussion

According to x-t-representation (Fig. 2a), the disturbance waves are formed due to coalescence of initial waves which appear just after the inlet. This process turns out to be more sophisticated in 3D-approach. Fig. 2b shows instantaneous shape of film surface for time moment marked by the dashed line in Fig. 2a. The initial waves are two-dimensional near the inlet, but farther downstream they are broken into small-scale transversely localised horseshoe-shaped waves. Further downstream avalanche-like merging of large number of these 3D waves occurs to form the disturbance waves. The latter are characterised by much larger transverse size which grows downstream approaching to that of the duct.

Thus, the degree of two-dimensionality - or transverse coherence - of film surface can be described as follows. It is very high near the inlet due to two-dimensional character of Kelvin-Helmholtz instability; when the initial waves develop, increasing in amplitude and nonlinearity, transverse instability of the waves leads to complete loss of two-dimensionality. But, due to action of gas shear, coalescence of small waves produces wider waves which are surprisingly stable to transverse perturbations and continue to absorb the smaller waves and grow in transverse size. Supposedly, this stability is explained by much longer wavelength in longitudinal direction in comparison to that of the initial waves. As a result, the transverse coherence is regained gradually. This is in agreement to the observations of the paper [1], though further downstream this increase may also occur due to coalescence of narrower disturbance waves to form wider waves, as it was noticed in [4].

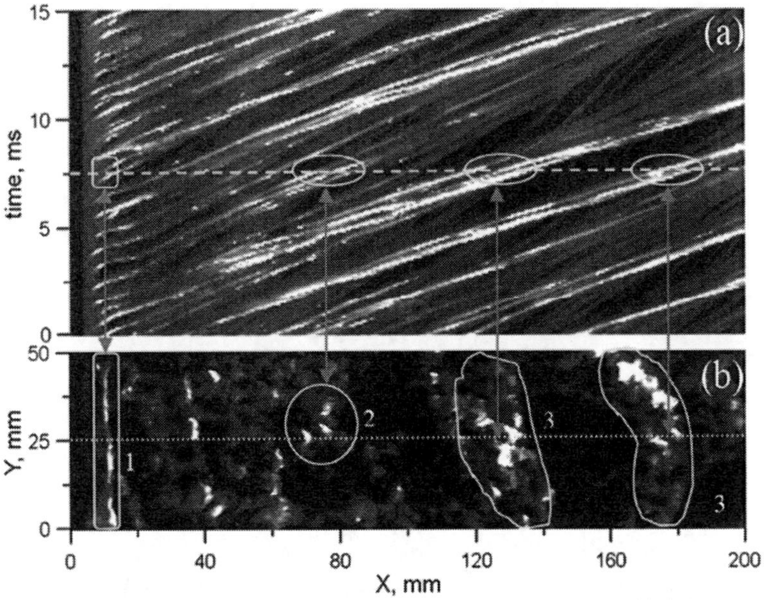

Figure 2. Formation of disturbance waves in *x-t* (a) and *x-y* (b) representation. Local brightness of the image is directly proportional to local film thickness. $Re_L = 300$, $V_g = 26$ m/s. 1 – initial 2D waves, 2 – 3D waves, 3 – disturbance waves.

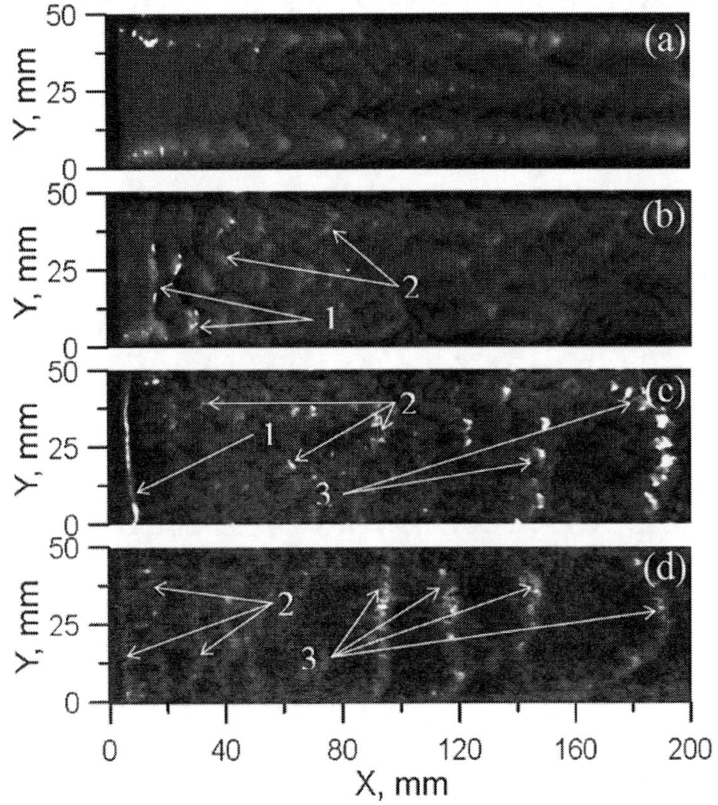

Figure 3. Instantaneous three-dimensional film thickness profiles for $Re_L = 220$. $V_g = 14$ m/s (a), $V_g = 18$ m/s (b), $V_g = 26$ m/s (c), Vg=43 m/s (d). 1 – initial 2D waves, 2 – 3D waves, 3 – disturbance waves.

Fig. 3 shows examples of instantaneous shapes of film surface at different gas velocities, V_g. At low V_g (Fig. 3a) the film surface is covered by longitudinally-oriented rivulets. It was recently observed [5] that this structure is also typical for adiabatic falling films. Increasing gas speed makes the film thickness more uniform which is inherent to annular flow. Length over which the initial waves are two-dimensional shortens as V_g increases (Fig. 3b-d). The increase leads to stronger rates of amplitude growth of the initial waves; hence, the transverse instability becomes significant earlier. At the highest gas speeds (Fig. 3d), the 2D waves can hardly be noticed; they are broken into 3D waves immediately after the inlet. Transition from the 3D waves to disturbance waves also occurs closer to the inlet at larger V_g. At low V_g (Fig 3a-b) the disturbance waves are not observed within the area of interrogation, whilst at the largest V_g the disturbance waves appear at the distance of less than 100 mm from the inlet. This effect can be explained by larger number of 3D waves with larger scatter in velocity and time separation, which increases probability of coalescence.

4 Conclusions

The mechanism of formation of disturbance waves, discovered recently in frames of 2D-approach, was confirmed during the present 3D experiments in a duct of a different shape. It was found that during the initial stages of flow of gas-sheared liquid films the transverse coherence of waves shows non-monotonic behaviour due to cascade of events dominating at certain stages, namely, two-dimensional Kelvin-Helmholtz instability, three-dimensional transverse instability and formation of large-scale quasi-2D disturbance waves. Range of distances from the inlet over which each stage is achieved depends on flow parameters, primarily, on the gas velocity. The gas flow also straightens the film in transverse direction, suppressing the formation of longitudinally oriented rivulets, but enhances formation of transversely oriented disturbance waves.

Acknowledgements

The work was supported by Russian Scientific Foundation (project 16-19-10449).

References

1. Y. Zhao, C.N. Markides, O.K. Matar, G.F. Hewitt, Int. J. Multiphase Flow, **55**, 111–129 (2013).
2. S.V. Alekseenko, A.V. Cherdantsev, M.V. Cherdantsev, S.V. Isaenkov, D.M. Markovich, Int. J. Multiphase Flow, **77**, 65–75 (2015).
3. S. Alekseenko, A. Cherdantsev, M. Cherdantsev, S. Isaenkov, S. Kharlamov, D. Markovich, Exp. Fluids, **53**, 77-89 (2012).
4. A.V. Cherdantsev, D.B. Hann, B.J. Azzopardi, Int. J. Multiphase Flow, **67**, 52-64 (2014).
5. S.M. Kharlamov, V.V. Guzanov, A.V. Bobylev, S.V. Alekseenko, D.M. Markovich, Phys. Fluids, **27**, 114106 (2015).

DOI: 10.1051/matecconf/20168400002

3D instability of liquid films and rivulets

Sergey Alekseenko[1,2], Sergey Aktershev[1,2], Aleksey Bobylev [1,2], Vladimir Guzanov [1,2] Sergey Kharlamov[1,2] and Dmitriy Markovich[1,2]

[1]*Kutateladze Institute of Thermophysics SB RAS, Lavrentiev ave., 1 , Novosibirsk, 630090, Russia*
[2]*Novosibirsk State University, Pirogov str., 2, Novosibirsk, 630090, Russia*

Abstract. This article presents the results of experimental and theoretical study of 3D instability of waves in liquid films and rivulets, falling down a vertical plate. The method of laser induced fluorescence was used to obtain instant shapes of three dimensional waves in a liquid films and rivulets and to investigate the regularities of formation of 3D wave patterns. In particular, during 2D-3D wave transition in the film flow, we observed a strong transverse redistribution of liquid leading to the formation of rivulets on the surface of isothermal liquid film, which is a phenomenon not described previously. The results of simulation of 2 D and 3D regular waves in the single rivulet are compared with the authors' previous experimental data.

1 Introduction

Gravity-driven film flow of a viscous liquid down a flat plane is a well known example of convectively unstable free surface flow with instabilities of different kinds leading to the appearance and interaction of surface waves with a great variety of characteristics. Specific type of film flow whereby the fluid flows in the form of many streamlets is typically called rivulet flow, whereas an individual streamlet, bounded by two contact lines, is called a rivulet. Special attention has been paid to rivulet flows because of their practical value when used in a variety of devices in power engineering and chemical technology.

This study presents the results of experimental and theoretical investigations of instabilities in the liquid film and rivulet flows. In the first part of study, the experimental results for transition from two-dimensional (2D) to three-dimensional (3D) waves in falling liquid films are presented. The method of laser induced fluorescence was used to obtain instant shapes of three dimensional waves and to investigate the regularities of formation of 3D wave patterns arising due to transverse instability of 2D waves. The regular 2D waves with prescribed frequency f were forced by harmonic modulation of liquid flow rate. Distilled water, water-glycerin solution (WGS) and water-ethanol solution (WES) were used as working fluids. The obtained results were compared to the results from the published literature on the modeling of 3D wave regimes of film flow.

The second part of the study is devoted to numerical simulation of 3D waves on the surface of a rivulet flowing down a vertical plate. The model [1] is used for numerical simulation of the forced waves in the rivulet. The results of the simulations are compared with the authors' previous experimental data [2]. The calculations are performed only for the conditions implemented in the authors' previous experiments. The comparison shows that the applied model adequately describes the shape of the wave surface of a rivulet, although the wave propagation velocity and wavelength are underestimated.

2 Experimental setup and measurement technique

The sketch of experimental setup is shown in Fig. 1. Experiments were carried out on the vertical glass plate with dimensions 20 cm in transversal and 25 cm in vertical directions. Liquid film was formed by a slot distributor with adjustable slot width. The method of laser induced fluorescence (LIF) was used to measure instantaneous local film thickness on the area of 12 × 12 cm with spatial resolution of 0.125 mm.

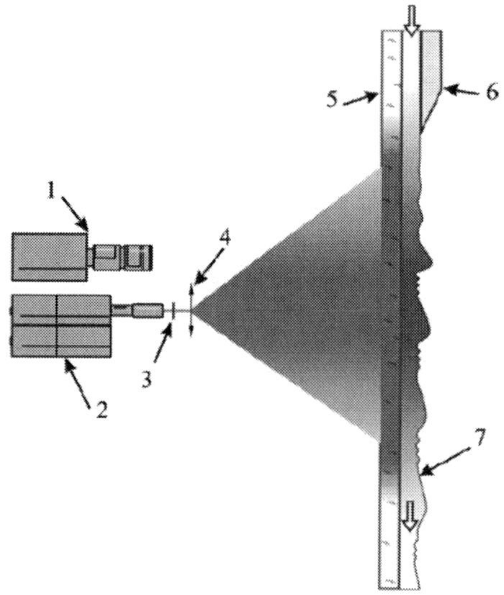

Figure 1. Experimental setup. (1) CCD camera; (2) laser; (3) scattering fluorocarbon film; (4) expanding lens; (5) glass plate; (6) slit distributor; (7) liquid film. White arrows show flow direction.

The LIF method is based on the reconstruction of local film thickness in accordance with local light intensity emitted by a small amount of fluorescent dye dissolved in working liquid and is quite similar to the method of "fluorescent imaging" originally introduced in [3] to study wave dynamics on inclined films. The main difference between the LIF and fluorescent imaging methods is that in the former coherent laser light is used for fluorescence excitation and additional measures should be taken to suppress speckles in illuminated area. As a simple remedy from the speckles, we placed scattering film made of porous fluorocarbon between the laser and expanding lens (Fig. 1). In our experiments working liquids were doped with small additions, about 30–40 mg/l, of fluorescent dye Rhodamine 6 G. To minimize optical distortions arising due to wavy free surface of the film, we arranged the light source and the camera at the dry side of the glass plate, i.e., excitation and registration were carried out through glass. A more detailed description of the method can be found in [4].

3 Results and discussion

At moderate Reynolds numbers *Re*, two scenarios of the patterns evolution were observed. For low frequencies of 2D waves excitation *f*, the leading humps initially run away from the trailing humps. Then their amplitudes and velocities decrease so that the trailing humps which move all the time with approximately constant celerity begin to overtake the leading humps. As the trailing humps approach the leading ones, the latter begin to restore their amplitudes and the pattern structure almost repeats itself through several wave lengths. An example of such oscillatory scenario obtained previously in modeling [5] and observed in experiments [6] is shown in Fig. 2(*a*). For high excitation frequencies,

 DOI: 10.1051/matecconf/20168400002

the site between trailing humps, which was left by running downstream leading hump, is filled by another leading hump that arrives from above so that the pattern structure also almost repeats itself through several wave lengths as shown in Fig. 2(*b*). We will refer to such scenario as "passing through" one.

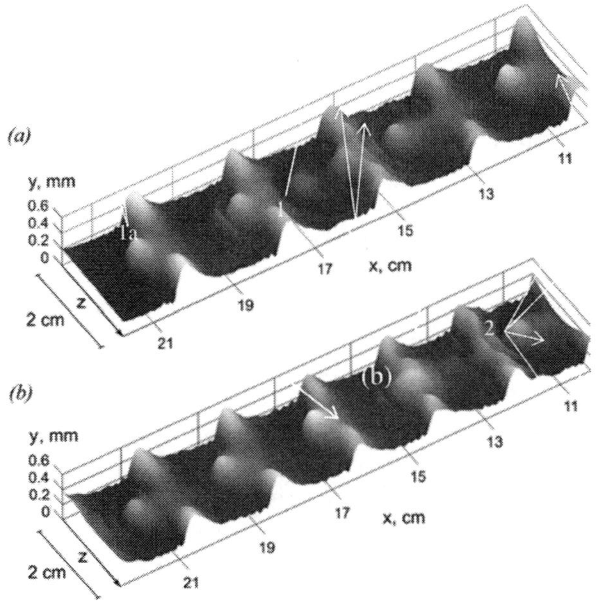

Figure 2. Scenarios of 3D wave patterns evolution. $Re = 50$ and $L = 2$ cm. (*a*) $f = 17$ Hz, oscillatory pattern. (*b*) $f = 20$ Hz, passing through pattern.

Figure 3 demonstrates the calculated shape of the developed waves on a WGS rivulet in comparison with experimental data for sufficiently high values of Reynolds number and forcing frequency.

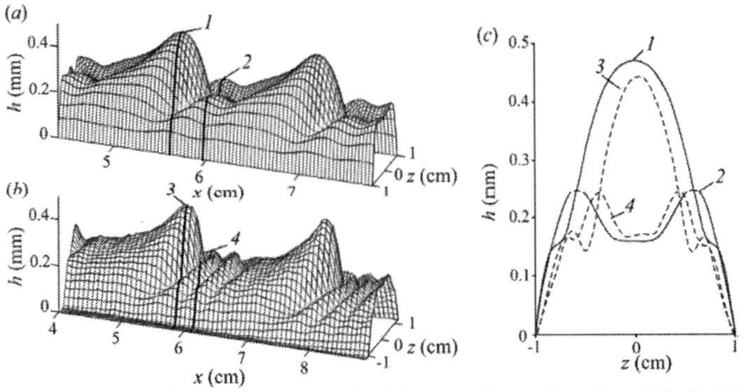

Figure 3. Shape of the wave in the WGS rivulet compared with experimental at $Re = 36$, $f = 23$ Hz. (*a*) calculated 3D surface, (*b*) experimental 3D surface, (*c*) rivulet cross sections, (1, 2) calculation, (3, 4) experiment.

As it is clear from the figure, the calculation well describes the shape of the peak with capillary ripples in front of the wave, as well as the rivulet profiles in different cross-sections. Significant discrepancy with the experimental data is observed in the wavelength. The calculated distance between peaks is 1.5 cm, while experiment gives 2 cm.

4 Conclusions

We have investigated experimentally the process of transition from 2D to 3D waves in liquid films falling down a vertical plate with a primary goal to examine to which extent the modern modeling approaches [5, 7] adequately describe real film flows. The method of laser induced fluorescence is used to obtain instant shapes of three dimensional waves and to investigate the regularities of formation of 3D wave patterns arising due to transverse instability of 2D waves. Although many details of 3D wave patterns correspond well, there are a few significant distinctions between our experiments and modeling. In particular, during 2D-3D wave transition, we observed a strong transverse redistribution of liquid leading to the formation of rivulets on the surface of isothermal liquid film, which is a phenomenon not described previously. Our results show that forming 3D waves always resemble streak-like waves as in [8] but not Λ-solitons as in [9, 10].

A numerical simulation of nonlinear three-dimensional forced waves in a straight rivulet, flowing down a vertical plate, was carried out. The calculations were performed for the two liquids (WES and WGS) with different physical properties, used in experiments [2]. Compared with the film flow, the waves in rivulets have significant differences. In particular, the wave patterns for the cases of "small" and "large" contact wetting angles are significantly different. The calculations have shown that for WES rivulet ("large" contact wetting angle) the waves are 2D in nature, while in WGS rivulet ("small" contact wetting angle) 3D horseshoe-waves develop. Comparison of experimental data on nonlinear waves with the calculations, made for different values of the Reynolds number within a wide range of the forcing frequency leads to the following conclusions. The applied model [1] is appropriate to describe in general the shape and amplitude of the waves on the rivulet surface at moderate Reynolds numbers. In particular, for WGS rivulet the model predicts waves of step-wise shape, which were observed in the experiments at low forcing frequencies. For moderate forcing frequencies the model predicts waves with capillary ripples in front of the wave peak. The shape of these waves is in good agreement with the experimental data. For WES rivulet the model predicts well the amplitude of developed waves and the shape of the rivulet wave surface. Besides, for WES rivulet the model predicts no capillary ripple in front of the main peak that is consistent with the experimental data.

Acknowledgements

The work was supported by Russian Scientific Foundation (projects 15-01-06702 and 15-01-04320).

References

1. E.A. Demekhin, V.Y. Shkadov, Fluid Dynamics. **19**, 689 (1984).
2. S.V. Alekseenko, A.V. Bobylev, V.V. Guzanov, D.M. Markovich, S.M. Kharlamov, Thermophys. Aeromech., **17**, 371 (2010).
3. J. Liu, J. D. Paul, and J. P. Gollub, J. Fluid Mech. **250**, 69 (1993).
4. S.M. Kharlamov, V.V. Guzanov, A.V. Bobylev, S.V. Alekseenko, D.M. Markovich, Phys. Fluids, **27**, 114106 (2015).
5. G.F. Dietze, W. Rohlfs, K. Nährich, R. Kneer, B. Scheid, J. Fluid Mech. **743**, 75 (2014).
6. S.V. Alekseenko, V.V. Guzanov, D.M. Markovich, S.M. Kharlamov, Tech. Phys. Lett. **38**, 739 (2012)
7. B. Scheid, C. Ruyer-Quill, P. Manneville, J. Fluid Mech. **562**, 183 (2006).
8. P. Adomeit and U. Renz, Int. J. Multiphase Flow, **26**, 1183 (2000).
9. E.A. Demekhin, E.N. Kalaidin, S.Kalliadasis, S.Yu. Vlaskin, Phys. Rev. E, **82**, 036322 (2010).
10. E.A. Demekhin, E.N. Kalaidin, A.S. Selin, Phys. Fluids, **22**, 092103 (2010).

Numerical modelling of thermocapillary deformation in a locally heated thin horizontal volatile liquid layer

Ella Barakhovskaya[1], Igor Marchuk[1,2,a] and Jean Claude Legros[2]

[1]*Novosibirsk State University, Pirogova Str. 2, Novosiborsk, 630090, Russia*
[2]*Kutateladze Institute of Thermophysics, Russian Academy of Sciences, Prosp. Lavrentyev 1, Novosibirsk, 630090, Russia*

Abstract. The thermocapillary flow in a thin horizontal layer of viscous incompressible liquid with free surface is considered. The deformable liquid layer is locally heated. The problem of thermocapillary deformation of the locally heated horizontal liquid layer has been solved numerically for two-dimensional unsteady case. The lubrication approximation theory is used. Capillary pressure, viscosity and gravity are taken into account. Evaporating rate is supposed to be proportional to the temperature difference between the liquid surface and ambient. Heat transfer in the substrate is also simulated. The deformation of the free surface has been calculated for different values of the heating power and thickness of the liquid layer. Initially the liquid layer has flat surface and uniform temperature. The model predicts the thermocapillary deformation of the liquid surface and the formation of the thin residual layer of the liquid.

1 Problem statement

Heat transfer in thin liquid layers with local heating is an important challenge for thermal stabilization technique of electronic equipment [1-3]. The problem consider a thin horizontal volatile liquid layer on the substrate with local heater, Figure 1. Liquid is viscous and incompressible. Heater is thin and uniform. Two-dimensional unsteady problem is considered. Initially the liquid layer has flat surface and uniform constant temperature T_0. At the time $t_0=0$ the heater is activated and starts heating the substrate and liquid. Shear stress occurs on the liquid surface caused by the heterogeneity of its temperature. Thermocapillary flow and deformation of the liquid surface is developing. The mechanism of convective heat transfer in the liquid and deformation of surface in the heat problem is also taken into account. Gravity, surface tension, thermocapillary effect, viscosity are included in the evolution equation for the liquid layer thickness.

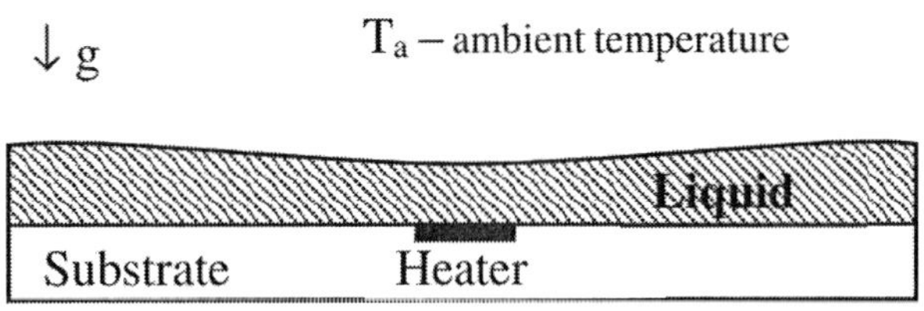

Figure 1. Scheme of the process: liquid layer on the substrate with local heater.

[a] Corresponding author: marchuk@itp.nsc.ru

2 Mathematical formulation

2.1 Basic equations

2.1.1 Motion equation of the liquid in the approximation of the lubrication theory

The mathematical formulation of the problem includes the Navier-Stokes equations, continuity and thermal conductivity [4]:

$$\frac{d\boldsymbol{u}}{dt} = \boldsymbol{F} - \frac{1}{\rho}\nabla p + \nu\Delta\boldsymbol{u}$$
$$\nabla\boldsymbol{u} = 0 \tag{1}$$
$$\frac{\partial T}{\partial t} + \boldsymbol{u}\nabla T = a\Delta T + \boldsymbol{F}$$

The dimensional variables that are used into equations: $\boldsymbol{u}$ - velocity, $\boldsymbol{F}$- vector of external forces, ρ - density, p - capillary pressure, ν - kinematic viscosity, T - temperature, a - thermal diffusivity.

2.1.2 The evolution equation for liquid film thickness

The evolution equation for liquid film thickness in the case of horizontal substrate, in thin layer approximation [4-6], is as follows:

$$h_t + divq + J_{ev} = 0 \tag{2}$$

Where, h_t - velocity of the surface, $q = \frac{h^3}{3\mu} f + \frac{h^2}{2\mu}\boldsymbol{\tau}$ - liquid flow rate along the x direction, $f = grad\left(\rho g h + \sigma H\right)$ - gradient of the modified pressure, $\boldsymbol{\tau} = \sigma_T gradT$ - thermocapillary tangent stress, μ - dynamic viscosity, ρ - density of the liquid, $\boldsymbol{g}$ - gravity vector, h - film thickness, σ - surface tension, σ_T - linear coefficient, which determines the dependence of the surface tension on temperature, T - temperature, $H = h_{xx} / \left(1 + h_x^2\right)^{3/2}$ - double mean curvature. $J_{ev} = J_{ev}^*(T_s(x) - T_a) / \Delta T$ - evaporation velocity, $T_s(x)$ - temperature of the surface, $J_{ev}^* = \alpha_{ev} / r_{lv}\rho$, α_{ev} - heat transfer coefficient for evaporation, r_{lv} - latent heat of vaporization.

2.1.3 The energy equation

The law of temperature distribution in the liquid layer is determined by the energy equation:

$$\frac{\partial T}{\partial t} + u\nabla T = a\Delta T + \frac{Q}{\rho C_p} \tag{3}$$

Where $Q = Q(t,x,z) = \begin{cases} Q = 0, \text{ outside the heater} \\ Q = const \neq 0, \text{ on the heater} \end{cases}$ - bulk density of the heat sources, the function $Q(t,x,z)$ has a constant nonzero value in the area of the heater which is known from the experimental conditions, $a = \dfrac{\lambda}{\rho \cdot C_p}$ - thermal diffusivity; λ - coefficient of thermal conductivity; C_p - specific heat; ρ - density of the liquid.

2.2 The boundary and initial conditions

2.2.1 The boundary conditions

Since it is made the replacement of the variables for the rectification of the liquid surface, then move on to the conjugation conditions at the liquid-substrate interface.

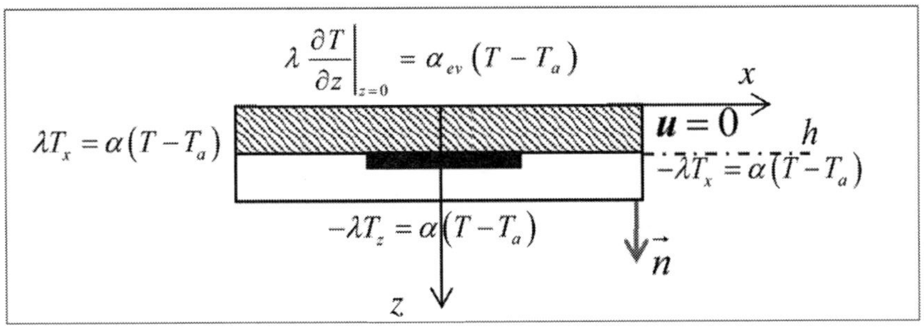

Figure 2. Boundary conditions.

$$\begin{cases} \lambda \dfrac{\partial T}{\partial x}\Big|_{x=-L} = \alpha(T - T_a), \text{ left boundary} \\ -\lambda \dfrac{\partial T}{\partial x}\Big|_{x=L} = \alpha(T - T_a), \text{ right boundary} \\ -\lambda \dfrac{\partial T}{\partial z}_{z=h+h_w} = \alpha(T - T_a), \text{ bottom boundary} \\ \lambda \dfrac{\partial T}{\partial z}\Big|_{z=0} = \alpha_{ev}(T - T_a), \text{ top boundary} \end{cases} \quad (4)$$

In the condition of the top boundary $\alpha_{ev} = J^*_{ev} r_{lv} \rho$, α_{ev} is the heat transfer coefficient for evaporation, r_{lv} - latent heat of vaporization, J^*_{ev} - evaporation rate coefficient. The coefficients for evaporation rates have been calculated using the data published in [7], $J^*_{ev} = 3.8\,\mu\text{m/s}\cdot\text{K}$, $\alpha_{ev} = 9.72\,\text{kW/m}^2\text{K}$.

2.2.2 The initial conditions

Initially the liquid layer has flat surface and uniform temperature:

$$h(0,x,z) = h_0 = const,\ T(0,x,z) = T_0 = const.$$

3 Numerical modelling

The evolution equation of the liquid layer thickness (2) is approximated at grid's nodes with finite volume method [6]. System of nonlinear algebraic equations obtained by the approximation is solved with Newton's method. The Jacobians are calculated using numerical linearization [6]. The grid with uniform steps in the space variables in the liquid and solid phases is set: $x_i = i \cdot dx$, $i = 0,..,N$. The time step τ is also uniform. $h_i^k = h(x_i, t_k)$ is the value of the liquid layer thickness in the time t and i node.

$$\frac{h_i^{k+1} - h_i^k}{\tau} dx^2 + q_{i-\frac{1}{2}} dx + q_{i+\frac{1}{2}} dx = 0 \tag{5}$$

Geometric parameters and parameters of the liquid, substrate and heater are specified.

Table 1: parameters of the system in 20 ^{0}C.

	Ethanol, 20 ^{0}C	Water, 20 ^{0}C	Substrate (sapphire), 20 ^{0}C
ρ density, kg/m^3	802,9	998.3	3980
Cp specific heat, $J/kg \cdot m$	2390	4182	761
λ coefficient of thermal conductivity, $W/m \cdot K$	0.1692	0.6092	25
μ dynamic viscosity, $kg/m \cdot s$	0,001145	0.001018	
r latent heat of vaporization, (J/kg)	938200	2563000	
σ surface tension, N/m	0.02203	0.07275	
σ_T - linear coefficient of the dependence of the surface tension on temperature, $(N/m \cdot K)$	−0.0000936	−0.0001575	
Heater			
Length $L_h = 40mm$, width $h_h = 0.5mm$, thickness $h_{heater} = 0,5\mu m$, Heating power P_h=3 W			

The temperature in the liquid is calculated without allowance of the deformations, since the heater is thin, it is considered that $Q = Q(t, r, z)$ is concentrated only in heater units. The difference analog of the energy equation (3) is solved by the method of fractional steps [8].

$$\left.\begin{aligned}\frac{T^{n+\frac{1}{2}}-T^{n}}{\tau}=\frac{1}{2}a\left(\Lambda_1 T^{n+\frac{1}{2}}+\Lambda_2 T^{n}\right)+\frac{Q^n}{2C_p\rho}\\ \frac{T^{n+1}-T^{n+\frac{1}{2}}}{\tau}=\frac{1}{2}a\left(\Lambda_1 T^{n+\frac{1}{2}}+\Lambda_2 T^{n+1}\right)+\frac{Q^n}{2C_p\rho}\end{aligned}\right\},\quad \begin{aligned}\Lambda_1=\frac{T_{i-1,j}-2T_{i,j}+T_{i+1,j}}{dx^2}\\ \Lambda_2=\frac{T_{i,j-1}-2T_{i,j}+T_{i,j+1}}{dz^2}\end{aligned} \tag{6}$$

Boundary and initial conditions are set in the finite differences equations:

$$\begin{aligned}&T(j,0)=T(j,2),\ j=1,..,M,-\ \text{symmetry}\\ &T(j,N+1)=T(j,N-1),\ j=1,..,M-\ \text{adiabatic right wall}\\ &T(0,i)=T(2,i)-\frac{\alpha}{\lambda}(T(1,i)-T_a)2dh,\ i=1,..,N\\ &T(M+1,i)=T(M-1,i)-\frac{\alpha_w}{\lambda_w}(T(M,i)-T_a)2dh_w,\ i=1,..,N\end{aligned} \tag{7}$$

Initial condition: $T^0_{i,j}=0,\ i=0,..,N,\ j=1,..,M-1$.

The resulting system of linear algebraic equations is solved at each step with the tridiagonal matrix algorithm. Calculations are performed successively. The time step for the evolution equation is done after the time step for the energy equation.

4 Numerical results

The deformation of the free surface has been calculated for different values of the heating power and thickness of the liquid layer. The model predicts the formation of the thin residual layer of the liquid and film breakdown at sufficiently intensive heating. Calculated deformations of the liquid layer with the time are shown in, Fig. 3. The thin residual layer is formed in the center, over the heater. It predicts the formation of dry spots. A local decreasing of the liquid surface temperature is shown in Fig. 4, curve 4. It is caused by the intense evaporation and decreasing of the thermal resistance of the liquid layer.

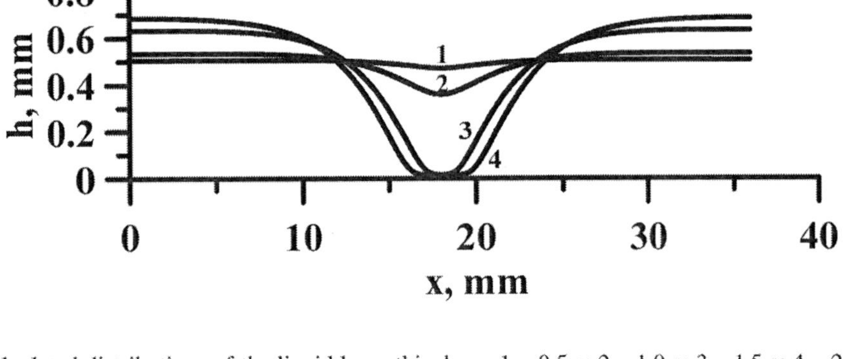

Fiqure 3. Calculated distributions of the liquid layer thicnkess. 1 – 0.5 s; 2 – 1.0 s; 3 – 1.5 s; 4 – 2.0 s. Heating power 3 W, water.

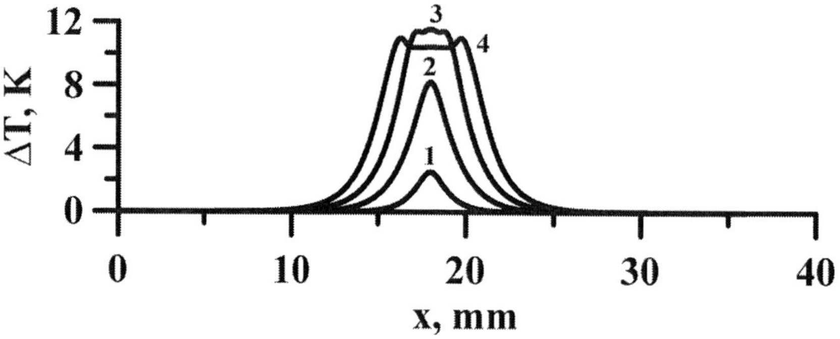

Figure 4. Calculated distributions of the surface temperature. 1 – 0.5 s; 2 – 1.0 s; 3 – 1.5 s; 4 – 2.0 s. Heating power 3 W, water.

Conclusions

The numerical calculations with given model of thin horizontal liquid layer with local heating have shown that local heating of a horizontal liquid layer causes deformation of the liquid surface and thermocapillary flow. Significant reduction of liquid film thickness occurs in the greatest value of the temperature gradient zone. The model predicts the formation of the thin residual layer of the liquid. A local decreasing of the liquid surface temperature is caused by the intense evaporation and the decreasing of the thermal resistance of the liquid layer.

Acknowledgements

The study was financially supported by the Russian Science Foundation (Project 15-19-20049).

References

1. I. V. Marchuk, Journal of Engineering Thermophysics, **18** (3), 227-237 (2009).
2. O. A. Kabov, Proc. of the ASME International Heat Transfer Conference **8** 229 (2010)
3. Zaitsev D.V. and Kabov O.A. Experiments in Fluids, **39**, 712 (2005).
4. Oron, A., Davis, S.H., Bankoff, S.G., Reviews of Modern Physics, **69**(3), 931-980 (1997).
5. Gatapova, E.Ya., Kabov, O.A. Microgravity Science and Technology, **19** (3-4), 132-134 (2007).
6. I.V. Marchuk, Journal of Engineering Thermophysics, **18** (3), 227-237 (2015).
7. A. A. Fedorets, O. A. Kabov, I. V. Marchuk, Technical Physics Letters, **37** (2), 116-118 (2011).
8. N.N. Yanenko, *The Method of Fractional Steps* (1971).

Influence of thin fluoropolymer film deposition on wettability of the silicon oxide nanowires array

Evgeniy Baranov[1,a], Sergey Khmel[1], Alexandr Zamchiy[1], Ekaterina Kirichenko[1], Aleksey Safonov[1], Nikolay Timoshenko[1] and Veronica Sulyaeva[2]

[1]*Kutateladze Institute of Thermophysics SB RAS, 630090 Novosibirsk Ac. Lavrentiev ave. 1, Russia*
[2]*Nikolaev Institute of Inorganic Chemistry SB RAS, 630090 Novosibirsk Ac. Lavrentiev ave. 3, Russia*

Abstract. In this work, we studied influence of fluoropolymer thin film deposition on wettability of the silicon oxide nanowires array. Deposition of fluoropolymer coating on the silicon oxide nanowires array changes the surface properties from hydrophilic to hydrophobic, and micro and nano scale surface roughness does not change significantly. In addition, it was shown that the deposition of fluoropolymer coating on nanowires protects the nanostructures from bundling together as a result of the attractive capillary forces that arise during evaporative drying.

1 Introduction

The surfaces with different wettability has wide industrial applications [1]. Recently increased interest in such surfaces in bio [2] and nano [3] technologies. Especially interesting are cases when the wetting takes limits – surface to become super hydrophilic or super hydrophobic. To change the wettability in various ways, such as nano structuring of surfaces, surface functionalization by chemical methods or by applying thin films of hydrophobic materials [4].

In this work, we studied influence of fluoropolymer thin film deposition on wettability of the silicon oxide nanowires array. If the obtained silicon oxide nanowires array were hydrophilic nature, the fluoropolymer thin film after deposition, they became hydrophobic. Untreated arrays are typically found to bundle together under surface tension forces during drying. Fluoropolymer treatment allows to avoid of undesirable liquid surface tension effects that can alter the morphology of nanowire arrays irreversibly and detrimentally [5].

2 Experiment

2.1 Synthesis of the silicon oxide nanowires array

Silicon oxide nanowires were synthesized from mixture monosilane-argon (40 sccm) with gas diluent hydrogen (600 sccm) by gas-jet electron beam plasma CVD method [6, 7]. Oxygen (3 sccm) was supplied directly into the vacuum chamber. The synthesis was carried out on c-Si substrate with tin film mass thickness about 60 nm. The process of nanowire synthesis on the substrate with the catalyst

[a] Corresponding author : itpbaranov@gmail.com

consists of three stages: heating up to operating temperature, treatment of hydrogen plasma and the actual growth of nanowires by vapor–liquid–solid (VLS) mechanism. The synthesis temperature was 335°C.

2.2 Deposition of thin fluoropolymer film

In this study, the thin fluoropolymer coating is deposited by the HWCVD method [8-10]. Hexafluoropropylene oxide (C_3F_6O) was used as the precursor gas of the fluoropolymer coating. The cooled substrate holder with the diameter of 90 mm was in the center of the chamber, and substrate with the silicon oxide nanowires array was fixed there. Above the substrates at a distance of 50 mm, there is the catalytic activator in the form of a mesh with pitch of 8 mm. The temperature of the wire was constant. The precursor gas pressure in the deposition chamber was 0.5 Torr. The precursor gas flow rate is 100 sccm. The temperature of activating wire mesh is 580°C, the target temperature was about 30°C. The thickness of the fluoropolymer film is about 100 nm.

2.3 Measuring techniques

The surface morphology was determined by the methods of scanning electron microscopy (SEM) using the JEOL JSM-6700F microscope.

The contact angle of the sample with water was measured by DSA-100 KRUSS device with high-precision system of liquid supply with a minimal dosing step of 0.1 mcl. optic system, and diffusion halogen light source (50 W).

3 Results and discussion

SEM image on Fig. 1a shows of the silicon oxide nanowires array. Each microrope consists of numerous nanowires with average diameter about 15 nm. Nanowires bunch is on the bottom of the catalyst particle, whereas the nanowires are absent at the top of it.

Analysis of the photograph (Fig. 1b) of the water droplet on the silicon oxide nanowires grown on c-Si substrate shows the measured contact angle was <20°. This confirms that the surface of the silicon oxide nanowires array, synthesized in this work, is hydrophilic in nature. The measured contact angle was different for the left and right side of the drop. We can explain this phenomenon by the fact that microropes are not perpendicular to the surface, and have a different angle of inclination. In addition, it was observed that the droplets on the surface have a very large imbibition layer [11]. We assume that water penetrate easily into grooves constituting the relief in case of Cassie impregnating wetting [12]. SEM image of the silicon oxide nanowires array after exposure to water is shown on Fig. 1c. It is evident that microropes stuck together in bundles of several ones. We assume that this is the influence of the water capillary forces that arise during evaporation process from the nanostructured surface.

SEM image on Fig. 1d shows of the silicon oxide nanowires array covered with the thin fluoropolymer film. The surface of the microropes is covered with a thin layer of the film, so it is not visible the structure of individual nanowires. Herewith on the top of microrope, on the catalyst particle, structure with high roughness can be observed.

The wettability of the surface of the silicon oxide nanowires array covered with the thin fluoropolymer film is shown on Fig 1e. The observed static water contact angle was >107°. This confirms that the surface of the silicon oxide nanowires array covered with the thin fluoropolymer film is hydrophobic in nature. This means that the covering with the thin fluoropolymer film has significant impact on the properties of the surface such as wettability. In this case, the measured contact angle was approximately equal for the left and right side of the drop.

DOI: 10.1051/matecconf/20168400004

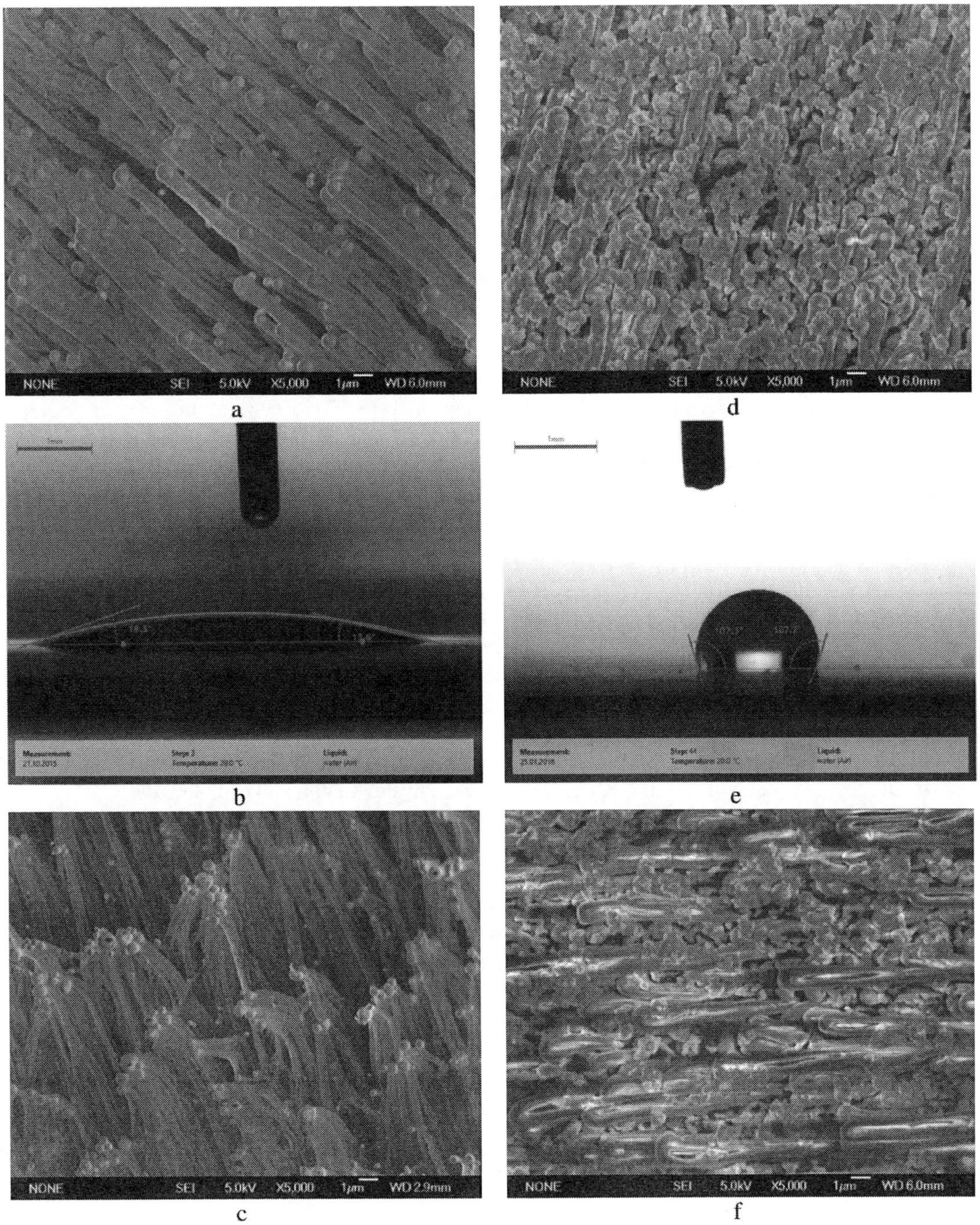

Figure 1. a – SEM image of the silicon oxide nanowires array: b – the photograph of the water droplet on the silicon oxide nanowires; c – SEM image of the silicon oxide nanowires array after exposure to water; d – SEM image of the silicon oxide nanowires array covered with the thin fluoropolymer film; e – the photograph of the water droplet on the silicon oxide nanowires array covered with the thin fluoropolymer film; f – SEM image of the silicon oxide nanowires array covered with the thin fluoropolymer film after exposure to water.

SEM image of the silicon oxide nanowires array covered with the thin fluoropolymer film after exposure to water is shown on Fig. 1f. Obtaining of the clear picture of the structure with help of the electron microscope is a big problem, because of strong charging of the surface. But it is safe to say that microropes didn't stuck in bunches after water evaporation. The thin fluoropolymer film deposition of the silicon oxide nanowires array shows protective properties for the nanostructured surface.

4 Conclusion

In this work, we studied influence of fluoropolymer thin film deposition on wettability of the silicon oxide nanowires array. If the obtained silicon oxide nanowires array were hydrophilic nature, the fluoropolymer thin film after deposition, they became hydrophobic. In addition, it was shown that the coating of polymer nanowires protects the nanostructure from water.

Acknowledgements

The work was supported by the grant of the President of the Russian Federation for young scientists MK 5052.2016.08 (deposition of thin fluoropolymer film) and by the grant of the Russian Science Foundation #15-19-30038 (synthesis of the silicon oxide nanowires array).

References

1. N. Verplanck, Y. Coffinier, V. Thomy, R. Boukherroub, Nanoscale Res. Lett., **2**, 577 (2007)
2. G. Piret, Y. Coffinier, C. Roux, O. Melnyk, R. Boukherroub, Langmuir, **24**, 1670 (2008)
3. C. Kuo, C. Gau, Journal of The Electrochemical Society, **157**, K201 (2010)
4. L. B. Boinovich, A. M. Emelyanenko, Russ. Chem. Rev., **77**, 583 (2008)
5. K. Lau, J. Bico, K. Teo, M. Chhowalla, G. Amaratunga, W. Milne, G. McKinley, K. Gleason, Nano Lett., **3**, 1702 (2003)
6. A. Zamchiy, E. Baranov, S. Khmel, Physica status solidi (c), **11**, 1397 (2014)
7. E. Baranov, A. Zamchiy, and S. Khmel, Technical Physics Letters, **39**, 1023 (2013)
8. K. Lau, H. Lewis, S. Limb, M. Kwan, K. Gleason, Thin Solid Films, **395**, 288 (2001)
9. K. Kenneth, Y. Mao, H. Lewis, S. Murthy, B. Olsen, L. Loo, K. Gleason, Thin Solid Films, **501**, 211 (2006)
10. A. Rebrov, A. Safonov, N. Timoshenko, Technical Physics Letters, **35**, 395 (2009)
11. N. Dhillon, J. Buongiorno, K. Varanasi, Nat. Commun. **6**, 8247 (2015)
12. A. Mohamed, A. Abdullah, N. Younan, Arabian Journal of Chemistry, **8**, 749 (2015)

DOI: 10.1051/matecconf/20168400005

Film thickness measurement for elongated bubble flow in microchannel using LIF

German V. Bartkus[1,2a] and Vladimir V. Kuznetsov[1,2]

[1]*Kutateladze Institute of Thermophysics SB RAS, Novosibirsk, Russia*
[2]*Novosibirsk State University, Novosibirsk, Russia*

Abstract. Two-phase gas-liquid flow in microchannels occurs in a variety of modern industrial applications. Using high-speed video recording, laser scanning and method LIF the characteristics of upward gas-liquid flow in rectangular microchannels with inlet T-mixer are considered. Statistical characteristics of gas-liquid flow and distributions of the phases in cross-section of a microchannel, with hydraulic diameter less than the capillary constant, were identified in a wide range of liquid and gas flow rates. Experiments were carried out for the flow of water-nitrogen in microchannels with cross section 420x280 μm and 720x1500 μm.

1 Introduction

Multiphase microsystems based on microchannels are of great interest for a variety of applications. Typical dimensions of the channels in these microsystems are in the range from hundreds of nanometers to hundreds of micrometers, which ensures the excellent heat and mass transfer properties and continuous operation mode. Studies of gas-liquid flows in microchannels have shown that it is possible to identify the basic flow patterns: isolated bubbles, Taylor bubbles, slug-annular, annular and churn flow which properties have been well studied [2, 3]. Important parameters for optimizing the heat and mass transfer processes in microchannels are the length of the bubble, the liquid film thickness and their distribution. Laser scanning was used in [4, 5] for determination of the statistical characteristics of the gas-liquid flows. The liquid film thickness for the Taylor bubbles in square and rectangular capillaries was measured in [6, 7]. It was shown that the dimensionless film thickness in round and rectangular microchannel has different dependence on the capillary number.

The main objective of the present work is experimental study of the multi-scale upward gas-liquid flow characteristics in rectangular microchannel of the microfluidic devices. The statistical parameters of the flow were defined and local film thickness was measured in microchannels with cross section 420x280 μm and 720x1500 μm using dual laser scanning and laser-induced fluorescence method. Distilled water and nitrogen were used as the working fluids.

2 Experimental facility and procedure

The experimental setup for the study of the upward two phase gas-liquid flow in a rectangular microchannel is shown in Figure 1 (a). The nitrogen is used as gas phase which is fed from a tank (1)

[a] Corresponding author : germanbartkus@gmail.com

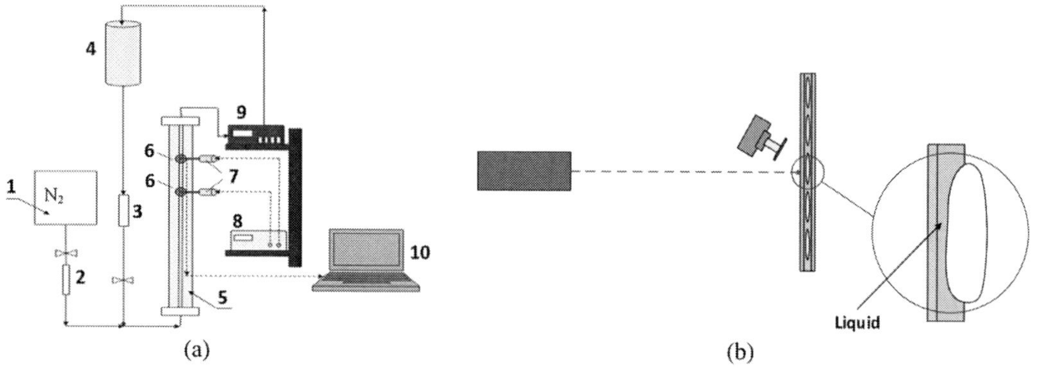

Figure 1. Experimental setup and scheme of the LIF method application.

via the control valve and gas flow meter (2) to the test section (5). Water flows from the tank (4) through flow meter to the control valve (3) and further into T-shaped mixer at the inlet of the experimental section. Experimental section is a microchannel, consisted from glued glass plates of length L = 50 cm with gap, which size is much smaller than the capillary constant (420x280 μm, 720x1500 μm). From the exit of experimental section, the gas-liquid mixture is moved by a peristaltic pump (9) to the water tank (4) where the gas goes to the atmosphere, and saturated water is returned to the experimental section. All experiments were carried out at temperatures between 19 and 21° C.

The study of the gas-liquid flow patterns was conducted using the method of dual laser scanning, see Figure 1 (a). Two lasers (7) are positioned for illumination of one side of the rectangular channel and the apertures are used to reduce the diameter of the laser beams. The lasers are powered using voltage source (8). The intensity of transmitted light is measured by the photodiodes (6) installed at the opposite side of the channel. Electric voltage from the photodiode is analyzed using high-speed ADC and processed on the computer (10) over time frame from one to three minutes. Flow patterns are registered using digital video camera. Using laser-induced fluorescence (LIF) method (see Fig. 1 (b)), which provides high spatial resolution without making hydrodynamic disturbances, the distribution of local thickness of liquid film along the long side of the channel was measured for elongated bubbles flow and transition flow. Rhodamine 6G was used as a fluorescent dye.

3 Flow patterns for upward gas-liquid flow

Gas-liquid flow in microchannels is accompanied by wide variety of phenomena that are not typical for the large scale. High speed video shows that the periodic flow with elongated gas bubbles (Taylor bubbles), the transition flow and annular flow are the main patterns of the flow for microchannels with the gap substantially less than the capillary constant. In the microchannel with cross section 720x1500 μm, the flow with elongated bubbles and stable liquid plugs is characterized by periodicity and observed in the range of liquid superficial velocities J_{liq} = 0.07 ÷ 0.407 m/s and gas superficial velocities J_{gas} = 0.04 ÷ 1.0 m/s. Superficial velocity was calculated as if the liquid or gas were the only one flowing in a channel cross sectional area. Increasing of the superficial gas velocity above the specified values leads to destruction of the elongated bubbles and the flow loses its periodicity. Figure 2 shows the photograph of the flow at the beginning of elongated bubbles destruction (b) and the corresponding signal from the photodiode (a), in which the existence of the gas phase corresponds to the lowest electric voltage. Such flow is the transition between periodic flow with elongated bubbles and annular flow.

Statistical analysis of the duration of gas and liquid phase existence at photodiode, conducted in the assumption of lognormal distribution of measured values, allows us to propose new method for determination of the flow pattern. Figure 2 (c) shows the square root from overall variance for the duration of gas and liquid phase existence at second photodiode depending on superficial gas velocity. The liquid superficial velocity equals to 0.17 m/s. The data shows that the periodic flow with elongated bubbles correspond to low values of the variance. After reaching the threshold value of the

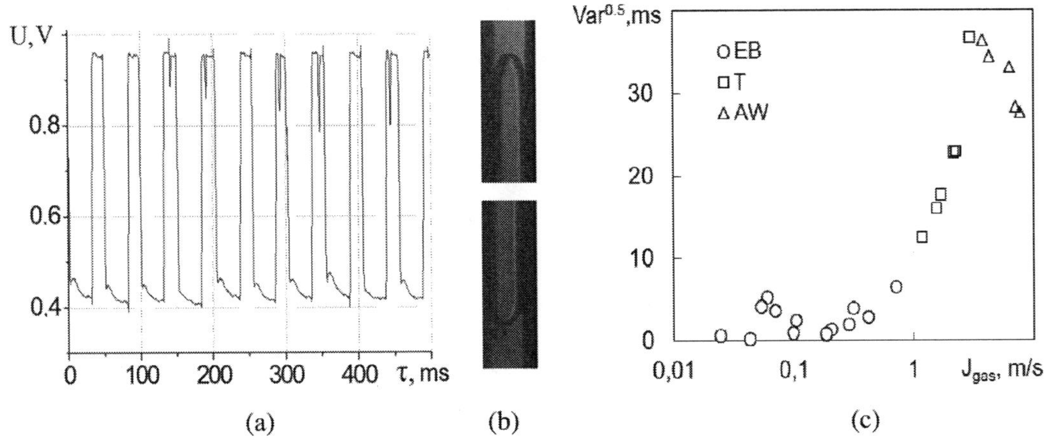

Figure 2. Dynamics of voltage on the photodiode – (a) and corresponding photo of the flow – (b) for J_{liq} = 0.27m/s, J_{gas} = 0.61 m/s; overall variance of the signal from second photodiode for J_{liq} = 0.17 m/s – (c).

superficial gas velocity equals to 1.55 m/s, the overall variance increases significantly. It shows change in the flow pattern and transformation to transition flow. The highest values of the overall variance correspond to the superficial gas velocities at which complete overlap of the channel cross section by liquid plugs and partial overlap of the channel cross section by disturbance waves on the liquid film surface exist simultaneously. Therefore, the transformation to annular flow corresponds to the maximum in the overall variance because in this flow it decreases due to disappearing of the liquid plugs and reducing the intensity of disturbance waves.

4 Local liquid film thickness measurements

The amount of liquid which left on channel wall behind the elongated bubble head gives the important information for analysis of the heat and mass transfer in microchannels. It is known that for capillary tube the film thickness δ is a monotonic increasing function of the capillary number $Ca = U_b\mu/\gamma$ and experimental data were fitted in [8] to obtain the empirical equation named the Taylor law:

$$\delta/D_h = 0.67Ca^{2/3}/(1+3.35Ca^{2/3}). \quad (1)$$

Here D_h is the hydraulic diameter of the channel, U_b is the bubble velocity, μ is the dynamic liquid viscosity and γ is the surface tension. The liquid film in a rectangular microchannel consists of a thin liquid layer at the channel wall and liquid filled the channel corners. Using laser-induced fluorescence (LIF) method, which provides high spatial resolution without making hydrodynamic disturbances, the distribution of local film thickness along a long side of the channel was measured for elongated bubbles flow and transition flow in the vertical rectangular microchannel with cross section 420x280 μm. The distribution of a liquid film thickness along the long side of microchannel measured by LIF is shown on Figure 3 (a). The data are presented for three gas superficial velocities and fixed liquid superficial velocity 0.059 m/s. One can see that capillary forces deform the interface and maximal film thickness is observed in the middle of long side of microchannel as it was observed in [6]. Accurate film thickness measurements for whole meniscus in the channel's corner using this technique are impossible. Using high-speed video recording and method LIF, the velocities of elongated bubble were measured and compared with mixture superficial velocities.

The dependence of dimensionless film thickness on Capillary number was determined and compared with existing correlations. Figure 3 (b) shows the dependence of dimensionless liquid film thickness in the middle of a long side of the channel's cross section with length a=420 μm (scaled by microchannel hydraulic diameter) on the Capillary number. Here the experimental data of [6, 7] are presented also. The experiments were done for elongated bubble flow and transition flow. The film thickness is presented for two dimensionless distances from the bubble head L/a. Figure 3 (b) shows

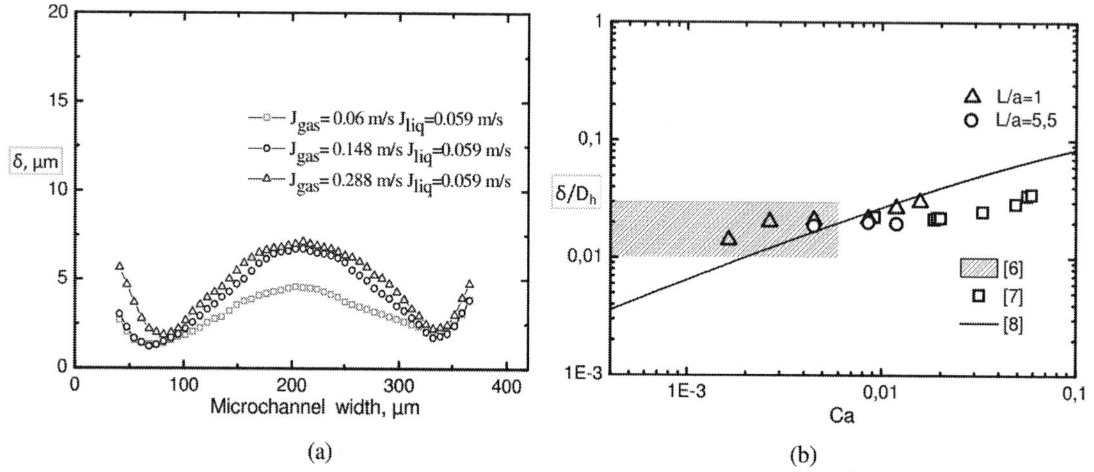

Figure 3. Distribution of liquid film thickness along long side of a microchannel measured by LIF – (a); dependence of dimensionless film thickness on the capillary number – (b).

that film thickness can be predicted by Eq. 1 only for small distance from bubble head. Increasing the distance from bubble head leads to reducing the film thickness due to capillarity induced liquid flow to the channel corners. It is the reason of the failure to comply with Taylor law in rectangular microchannel. At distance L/a=5.5 the experimental data show good coincidence with data [7].

5 Conclusions

It was observed that the elongated bubbles flow, transition flow and annular flow are the main flow patterns for rectangular microchannel with hydraulic diameter substantially less than the capillary constant. It was shown that the perspective method to define the boundary of corresponding flow patterns is statistical processing of the data obtained by laser scanning. Liquid film thickness was measured for elongated bubble flow and transition flow at two dimensionless distances from the bubble head. The LIF measurements show that the capillary forces deform the interface and maximal film thickness is observed in the middle of a long side of the microchannel's cross section. The obtained data on local film thickness confirm the failure of the Taylor law at long distance from the bubble head due to transverse capillary flows in a rectangular microchannel.

Acknowledgements

The materials of section 3 were obtained at the Kutateladze Institute of Thermophysics SB RAS by the grant from Russian Science Foundation (project 14-49-00010) and the materials of section 4 were partially supported by RFBR grant 15-08-07506-a.

References

1. M.N. Kashid, A. Renken, L. Kiwi-Minsker, Chem. Eng. Sci. **66** (2011)
2. J.W. Coleman, S. Garimella, Int. J. Heat Mass Trans. **42** (1999)
3. F.A. Kawahara, P.M.Y. Chung, M. Kawaji, Int. J. Multiphase Flow **28** (2002)
4. R. Revellin, J.R. Thome, J. Micromech. Microeng. **17** (2007)
5. I.A. Kozulin, V.V. Kuznetsov, J. Appl. Mech. Tech. Phys. **52** (2011)
6. D.M. Fries, F. Trachsel, P.R. Von Rohr, Int. J. Multiphase Flow **34** (2008)
7. W.B. Kolb, R.L. Cerro, Chem. Eng. Sci. **46** (1991)
8. P. Aussillous, D. Quere, Phys. Fluids **12** (2000)

MATEC Web of Conferences **84**, 00006 (2016) DOI: 10.1051/matecconf/20168400006
International Symposium IPHT 2016

Influence of surface roughness and porosity of inclusion in water droplet on heat transfer enhancement

Anastasia G. Borisova, Vladislav M. Gumerov, Maxim V. Piskunov[a]
National Research Tomsk Polytechnic University, 634050 Tomsk, Russia

Abstract. Using high-speed camera, the experiments were performed to research evaporation of 10 μl water droplets containing 2 mm solid inclusions in the shape of cube, when heated (up to 850 K) in combustion products of technical ethanol. Adding solid inclusions in water droplets allowed considerably decreasing (by 70%) their evaporation times. Also, the artificial irregularities (roughness and porosity) at the surfaces of solid inclusions were manufactured to increase heat transfer area. Such approach enabled to decrease evaporation times of heterogeneous liquid droplets in high-temperature gases by 40% (when comparing inclusions with artificial irregularities and smooth surface).

1 Introduction

The experimental investigations on approaches to enhance heat and mass transfer present the broad list of possible ways discussed in scientific papers (for example, [1–3]), published over the last 150 years. In particular, existing methods to enhance heat transfer between liquid and heating surface can be divided into two types. The first one includes various changes in liquid. Among these are an addition of solid insoluble particles of different sizes and concentrations (preparation of suspensions), use of surfactants (wetting agents, plasticizers etc.), binary liquids, salt solutions and more. The second type involves mechanical change of heat-exchange surface (for example, finning well-known in thermal engineering and other artificial irregularities contributing vapor formation at the "liquid – heating surface" interfaces).

In research [4, 5], we revealed some conditions of intense vapor formation at the internal "solid inclusion – water" interfaces of heterogeneous (inhomogeneous) liquid droplet. Moreover, under such conditions, the explosive disintegration of droplets occurred. We believe that when changing geometry of heating surface, the lifetimes of heterogeneous droplets established in [4, 5] will reduce. Furthermore, the change of heat-exchange surface can be one of the key factors to reach the rapid elimination of heat. Thus, in the present research, to determine the influence of surface irregularities on heat transfer under experimental conditions as in [4, 5] is appropriate.

The purpose of the research is to establish experimentally influence of artificial irregularities of graphite inclusion inside water droplet on its lifetimes (evaporation times) in high-temperature combustion products of technical ethanol. In addition, we expect that the increase of heat flux in water layer can contribute to reach conditions of the intense vapor formation with explosive disintegration of heterogeneous liquid droplet.

[a] Corresponding author: piskunovmv@tpu.ru

DOI: 10.1051/matecconf/20168400006

2 Experimental setup and research methods

To perform the experimental research, the setup as in [4, 5] was used. Fig. 1 illustrates the scheme of the setup.

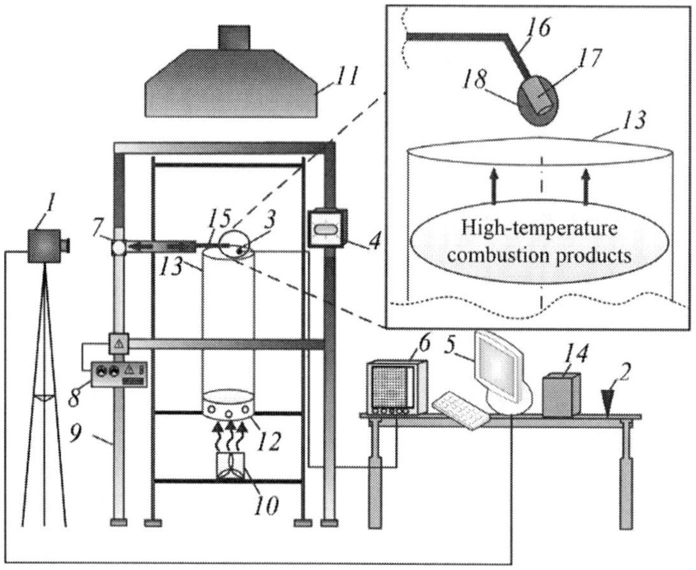

Figure 1. Scheme of the experimental setup: *1* – high-speed camera, *2* – electronic dosing devise, *3* – thermocouple, *4* – spotlight, *5* – personal computer (PC), *6* – temperature registration, *7* – moving mechanism, *8* – power supply to moving mechanism, *9* – mount, *10* – fan blower, *11* – air flow system, *12* – burner, *13* – metallic cylinder, *14* – analytical balance, *15* – guideway to fix a ceramic rod, *16* – ceramic rod, *17* – solid inclusion, *18* – water droplet

The main steps of experimental technique are similar to ones stated in [4]. Therefore, when describing the technique of the research, we will note only the significant differences and brief the necessary steps.

Recommendations and explanation of choice of such approach to fix and move heterogeneous droplets (see inset of Fig. 1) in high-temperature gas area are discussed in [4]. When experimenting, the inclusions made of pure graphite in the shape of cube (size of inclusions is l_{in}=2 mm) were used. To reach the purpose of the research, the following types of surface irregularities were manufactured:

1. Roughness (on each side of inclusion, depth – 0.15 mm, shape – cylindrical, random distribution of lines along surface with intercrossing, number of lines in 1 mm^2 – not less than 12);
2. Pores in the shape of holes (depth – from 1 mm to 1.5 mm, diameter – 1 mm, number on one side of inclusion – 1).

To perform each test, the important conditions were as follows:

1. Regulation of initial temperature of inclusion before each test;
2. Full covering of inclusion by water droplet.

These conditions were provided by dipping inclusion in a vessel with water (~298 K) during several seconds. After this procedure, water droplet of required initial volume was supplied on inclusion. Moreover, liquid mass remaining on inclusion after dipping was considered when processing the results. Experiments were carried out using the distilled water, which was taken out of a vessel by dosing device 2.

Using high-speed recording allowed solving the following problems:

1. Determining lifetimes τ_h of heterogeneous water droplets in high-temperature combustion products;
2. Finding peculiarities of phase changes of water droplets with graphite inclusions.

To provide high-temperature area in cylinder *13*, we burnt technical ethanol.

The main steps of the experimental research are as follows:

1. By using moving mechanism 7, heterogeneous liquid droplet (10 μl water droplet with graphite inclusion) was introduced into the gas area; then, video recording was carried out; after water evaporation, an inclusion was removed from the gas area.
2. Cooling and full covering of an inclusion; the formation of the next heterogeneous droplet; repeat of the first step.
3. Computing lifetimes τ_h of heterogeneous water droplets in the gas area.

Errors on lifetime τ_h are less than 10^{-2} s. Errors on measurements of solid inclusion sizes l_{in} do not exceed 0.05 mm.

3 Results and discussion

Fig. 2 presents research results that indicate the influence of addition of solid inclusions and irregularities at their surfaces on the evaporation time τ_h of heterogeneous droplets. The addition of solid inclusions in water droplets enabled to considerably decrease (by 70%) the evaporation times τ_h. Moreover, the significant decreasing (by 40%) the evaporation time τ_h of heterogeneous droplets occurred owing to manufacturing the artificial irregularities at the surfaces of solid inclusions (see Fig. 2, *a* and *b*).

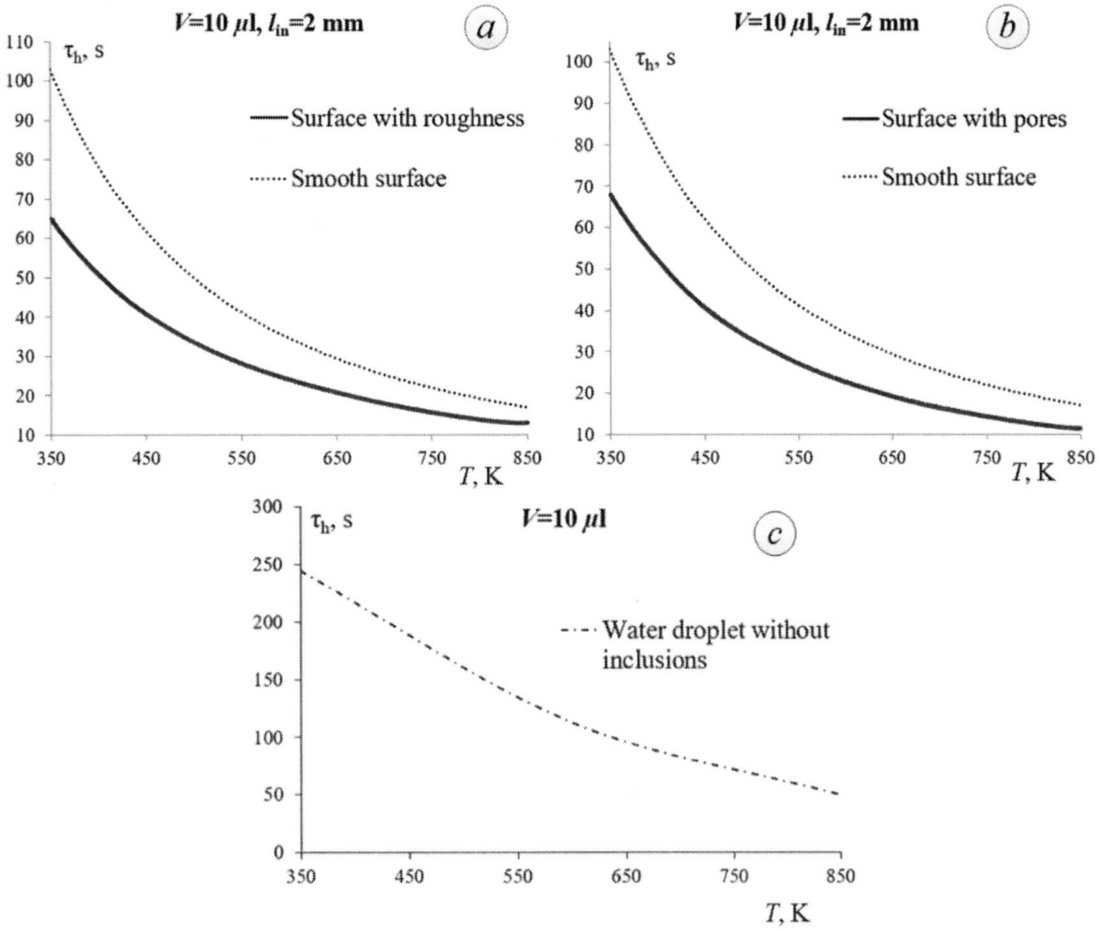

Figure 2. Comparison of lifetimes τ_h of 10 μl water droplets without (*c*) and with (*a*, *b*) solid inclusions with smooth and irregular (rough and porous) surfaces, when heated in high-temperature combustion products

The remarkable thing is that the influence of shallow and numerous roughness (Fig. 2, *a*), as well as single pores (Fig. 2, *b*) on the heat transfer enhancement is almost similar.

In the research [4], the detailed description is presented for the mechanism of phase changes in the heterogeneous droplets when heated in high-temperature gas area. Moreover, the addition of solid inclusion significantly enhanced the heat transfer in the high-temperature gases – heterogeneous liquid droplet system [4]. Such phenomenon is mainly due to the intense heat outflow from the inclusion to the water layer. Thus, all other things being equal, in the heat transfer equation $Q=kF\Delta T$ (k – heat transfer coefficient, W/m^{2}K, F – heat exchange surface area, m^2, ΔT – temperature difference, K) for the present research, the heat flux Q (from the inclusion to the water layer) will increase by the growth of heat exchange surface area F. Then, the enhancement of heat outflow to the water layer quantitatively influenced the lifetimes τ_h of heterogeneous liquid droplets (Fig. 2, *a* and *b*).

In the performed experiments, only the evaporation of liquid from the surface of heterogeneous droplet occurred. At the internal interfaces, the nucleation of vapor bubbles was not observed. From the above reasoning, adding solid inclusions and manufacturing artificial irregularities on their surfaces (increasing the heat exchange surface area inside heterogeneous droplet) contributes to the significant increasing the evaporation rate of liquid when heated. However, the used approach does not provide the necessary conditions for the intense vapor formation at the internal interfaces.

The revealed peculiarity of evaporation of heterogeneous liquid droplets can be the theoretical basis to develop recommendations for improvement of the effectiveness of waste and industrial water treatment.

4 Conclusion

Owing to manufacturing the artificial irregularities at the surfaces of 2 mm solid inclusions, evaporation time τ_h of heterogeneous droplets decreased by 40% (when comparing inclusions with artificial irregularities and smooth surface). Furthermore, adding solid inclusions in water droplets enabled to decrease evaporation times τ_h by 70% (when comparing water droplets with and without solid inclusions with artificial irregularities).

This work was supported by the grant of the President of the Russian Federation (MD-2806.2015.8).

References

1. S.-S. Hsieh, H.-H. Liu, Y.-F. Yeh, Int. J. Heat Mass Transfer **94**, 104 (2016).
2. H. Kim, Nanoscale Res. Lett. **6**, 1 (2011).
3. C.K. Huang, V.P. Carey, Int. J. Heat Mass Transfer **50**, 269 (2007).
4. G.V. Kuznetsov, M.V. Piskunov, P.A. Strizhak, Int. J. Heat Mass Transfer **92**, 360 (2016).
5. I.S. Anufriev, G.V. Kuznetsov, M.V. Piskunov, P.A. Strizhak, M.Yu. Chernetskii, Tech. Phys. Lett. **41**, 810 (2015).

The heat flux near the contact line of the droplets on heated foil

Vyacheslav Cheverda[1,4,a], Andrey Karchevsky[2,3]

[1] Kutateladze Institute of Thermophysics SB RAS, Novosibirsk 630090, Russia
[2] Novosibirsk State University, Novosibirsk 630090, Russia
[3] Sobolev Institute of Mathematics SB RAS, Novosibirsk 630090, Russia
[4] Tomsk Polytechnic University, Tomsk 634050, Russia

Abstract. The heat flux distribution near the contact lines of one or several (3) drops on a horizontal heated constantan foil is investigated. The temperature distribution of the bottom foil surface was measured by the FLIR infrared (IR) camera. To measure the heat flux distribution near the contact lines numerical solution of the Cauchy problem is used. The maximum of the heat flux density is observed near the contact line and it exceeds the average heat flux density by several times. This is due to relatively high heat conductivity coefficient of the foil material and high evaporation rate in the contact line region, so the heat flux from the foil periphery (out of the drop) is enter to the contact line region of the drop.

1 Introduction

Liquid droplets on the substrate are widely distributed in nature, techniques, and technology. The wetting angle is a fundamental macroscopic characteristic of the contact line. The wetting (contact) angle is determined by the Young equation, [1]. Analysis of papers where drop is placed on the heated substrate with different level of gravity shows that there is not enough information about the process of evaporation and wetting of a solid substrate for variable levels of gravity [2, 3]. The process in a sessile liquid drop on the heated substrates with different wettability are investigated numerically and experimentally in [4, 5]. In these papers are demonstrated that evaporation rate per drop area is inverse function of drop contact radius. The main role in the droplets evaporation is wetting of solid surface and processes in three-phase contact line gas-liquid-solid. The heat- and mass transfer processes in the contact line is important physical and technical problem which is not solved completely. The theoretical and experimental studies prove that heat transfer intensity in the contact line region can be about 1 kW/cm^2 [6]. Unfortunately, it is impossible to measure the heat flux in the microregion in the contact line, various non-contact methods and numerical procedures are applied. The results of the experiment with two heated vertical plates of metal foil with meniscus of HFE7100 liquid which is evaporated between them are demonstrated in [7]. It is shown that the local heat flux density in the contact line region was about 5 times higher than the average heat flux density. Heat transfer in a sessile liquid droplet was studied with numerical methods in [8]. The formation of the two vortexes surface-driven thermocapillary and volume-driven thermogravitational forces was studied. The heat flux distribution near the contact lines is investigated using the sessile drop configuration by numerical solution of the Cauchy problem for elliptic equations [9, 10, 11]. The heat

[a] Corresponding author: slava.cheverda@gmail.com

transfer in sessile drop on a thin metal foil is studied using thermochromic liquid crystals (TLCs) in [12]. The goal of this paper is to obtain heat flux distribution near the contact line using IR imaging for one and ensemble of several (3) droplets.

2 Experimental setup and technique of measurement

The schematic diagram of the experimental setup demonstrates in Figure 1. The basic element of the test cells is a constantan foil (CuNi) of the thickness of 25 μm, size of 80 × 35mm^2 and heat conductivity λ of 23W/mK was used as the substrate. This foil is soldered to the two brass electrodes (4). The substrate was heated by Joule heating from power supply (2). The temperature of the bottom foil surface was measured by IR camera Titanium 570M (7) with help of gold mirror (1). The video camera (3) is used to observe a drop (6) profile. Another video camera (5) is used to observe drop from top view. The average coefficient of heat transfer from the foil surface to surrounding air α_{av} was found to be within 25 W/m^2 K. These measurements were performed without droplets on the foil surface.

The heat flux on the side of IR-scanner of the foil is estimated by equation: $q_b=\alpha_{av}(T_w-T_a)$, T_W – temperature measured by IR-scanner, T_a – ambient temperature, α_{av} – average heat transfer coefficient. The heat flux on the drop side of the foil is calculated by equation $q_t=2q_{av}-q_b$ with neglecting the heat conductivity along the foil, $q_{av}=P/S$, P is electrical power, S=2 l w is the total surface area of the foil.

The shape of an axisymmetric droplet lying on a flat substrate is uniquely determined by three independent geometric parameters using Nikon D300 (3). This fact allows us to calculate the physical properties, such as capillary length and contact angle, measuring three independent values, such as height and diameter of a droplet, droplet base diameter or area of its axial cross section, or its volume [13]. The average contact angle of water with constantan foil is about 60°. The video camera (5) is used to detect drop position on the foil.

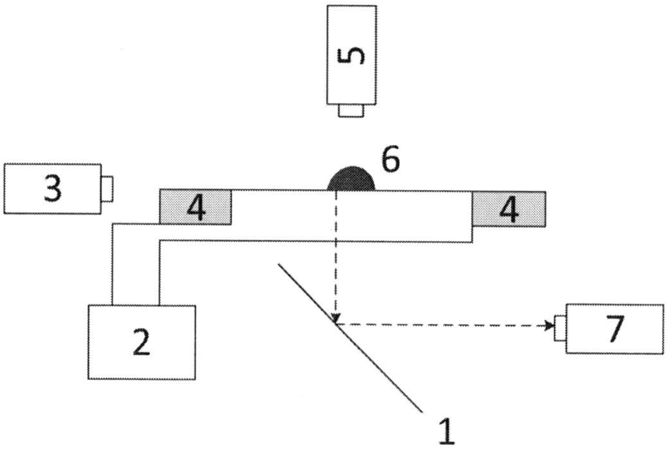

Figure 1. Schematic diagram of the experimental arrangement:

3 Results and discussion

Using the Cauchy problem solution, the distributions of heat flux density on the wetted foil surface have been obtained. The infrared data were used. The measured temperature field (Fig. 2 (a), Fig. 3(a), Fig. 4(a)), calculated distribution of the heat flux density on the substrate (Fig. 2 (b), Fig. 3(b), Fig. 4(b)), and heat flux density along the central drop cross section are shown in (Fig. 2 (c), Fig. 3(c), Fig. 4(c)). The results obtained for macro IR images are shown in Fig 3.

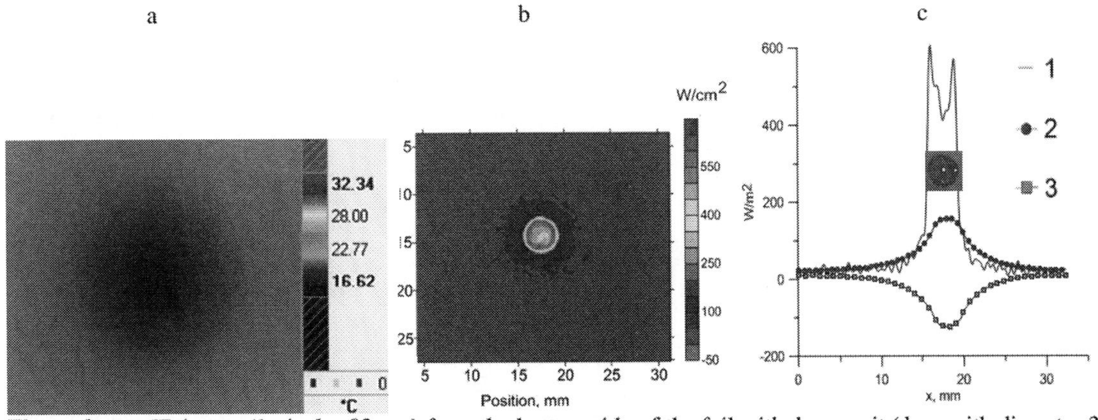

Figure 2. a – IR image (1 pixel – 83 μm) from the bottom side of the foil with drop on it (drop with diameter 3 mm, heat flux 16. 39 W/cm²), b – heat flux distribution for Cauchy problem solution, c - distributions of heat flux density for the line throw the center of the drop: 1: Cauchy problem solution. 2: calculation with neglecting of heat flux redistribution in the foil, q_t. 3: on the bottom side of foil, q_b.

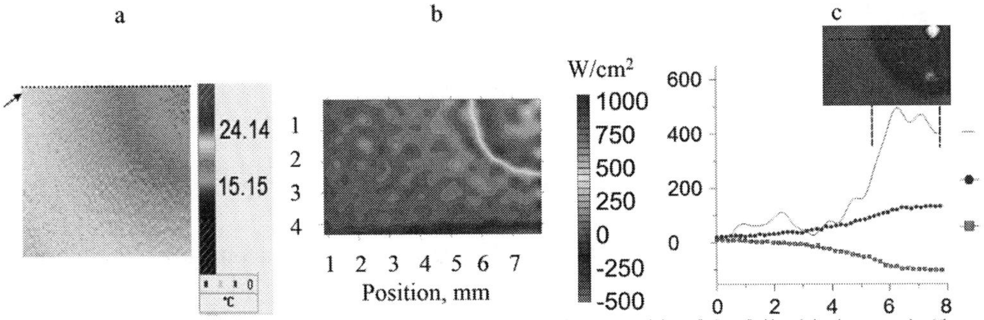

Figure 3. a – macro IR image (1 pixel – 15 μm) from the bottom side of the foil with drop on it (drop with diameter 3 mm, heat flux 16. 23 W/cm²), b – heat flux distribution for Cauchy problem solution, c - distributions of heat flux density for the line throw the center of the drop: 1: Cauchy problem solution. 2: calculation with neglecting of heat flux redistribution in the foil, q_t. 3: on the bottom side of foil, q_b.

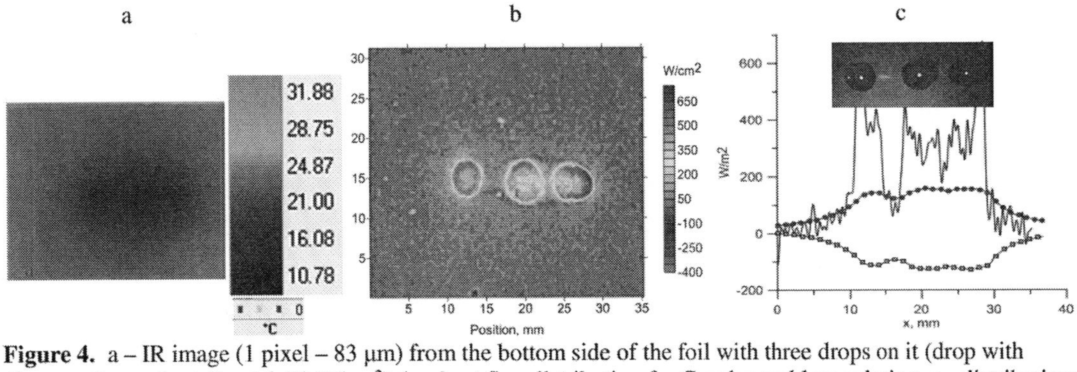

Figure 4. a – IR image (1 pixel – 83 μm) from the bottom side of the foil with three drops on it (drop with diameter 3 mm, heat flux 16. 23 W/cm²), b – heat flux distribution for Cauchy problem solution, c - distributions of heat flux density for the line throw the center of the drop: 1: Cauchy problem solution. 2: calculation with neglecting of heat flux redistribution in the foil, q_t. 3: on the bottom side of foil, q_b.

The results show that the maximum of the heat flux density is near the contact line and it exceeds the average heat flux density by several times. The results of measurements and calculations for three sessile drops on the substrate are shown in Figure 4. For these drops the symmetry condition is broken. In the contact line region between drops we do not observe the maximum of heat flux density because the heat flux from the periphery, which is now split between the two drops, decreases for each of these drops. The maximum of heat flux density is reached near the contact line at the outer edges of the three-droplet system.

Authors gratefully acknowledge support of this work by the the Russian Science Foundation (project 14-39-00003).

References

1. A. Ababneh, A. Amirfazli, and J. A. W. Elliott, Canadian Journal of Chemical Engineering, **84**, 39 (2006)
2. D. Brutin, Z.-Q. Zhu, O. Rahli, J.-C. Xie, Q.-S. Liu, and L. Tadrist, Microgravity Science and Technology, **22**, 387 (2010)
3. O. A. Kabov, D. V. Zaitsev, Doklady Physics, **58**, 292 (2013)
4. E.Ya. Gatapova, A.A. Semenov, D.V. Zaitsev, O.A. Kabov, Colloids and Surfaces A: Physicochemical and Engineering Aspects, **441**, 776 (2014)
5. A.A. Semenov, D.V. Feoktistov, D.V. Zaitsev, G.V. Kuznetsov, O.A. Kabov, Thermophysics and Aeromechanics, **22**, 771 (2015)
6. P. Stephan, C. Brandt, Heat Transfer Engineering, **25**, 78 (2004)
7. K. Ibrahem, M. F. Abd Rabbo, T. Gambaryan-Roisman, and P. Stephan, Experimental Thermal and Fluid Science, **34**, 1036 (2010)
8. M.V. Bartashevich, I.V. Marchuk, O.A. Kabov, Thermophysics and Aeromechanics, **19**, 317 (2012)
9. I. Marchuk, A. Karchevsky, A. Surtaev, O. Kabov, International Journal of Aerospace Engineering, 391036 (2015)
10. A.L. Karchevsky, Eurasian Journal of Mathematical and Computer Applications, **1**, 5 (2013).
11. A.L. Karchevsky, I.V. Marchuk, O.A. Kabov, Applied Mathematical Modelling, **40**, 1029 (2016)
12. C. Sodtke, V. S. Ajaev, and P. Stephan, Journal of Fluid Mechanics, **610**, 343 (2008)
13. I. V. Marchuk, V. V. Cheverda, P. A. Strizhak, O. A. Kabov, Thermophysics and Aeromechanics **22**, 297 (2015)

Static contact angle measurement at different volumes of a drop sitting on non-ferrous metals

Dmitriy Feoktistov[1, 2], Evgeniya Orlova[1, a], Kseniya Batischeva[1] and Andrey Semenov[2]

[1]*National Research Tomsk Polytechnic University, 30, Lenin Ave, 634050, Tomsk, Russia*
[2]*Institute of Thermophysics SB RAS, 1, Lavrentiev Ave, 63090, Novosibirsk, Russia*

Abstract. The paper presents the experimental study of the effect of the liquid volume on the static contact angle under the condition of the stationary contact line on the rough and polished surfaces of non-ferrous metals. The experiments were carried out using the shadow optical system. Comparison between two methods of drop profile processing was conducted. It was found that in addition to the friction and gravity forces, the structure of the non-ferrous metals significantly influence on the static contact angle during changing the drop volume.

1 Introduction

A contact angle is an inclination of the liquid surface to the wetted solid surface [1, 2]. There is an equilibrium contact angle, which depends only on the thermodynamic properties of the gas-liquid-solid system and non-equilibrium contact angle. This non-equilibrium contact angle is measured when the system deviates from the equilibrium. The non-equilibrium contact angles, corresponding to the constant wetted surface, are called static angles.

It is known [1] that the static contact angle depends on the drop volume if there are some physicochemical processes (evaporation, dissolution of a solid body in liquid, adsorption, chemical reactions). In their absence, the static angle is independent of volume.

In recent years, the solid surfaces with a chemical or topographic patterning attract much attention due to their potential use in the production of the electronic components, micro-channels, micro-scale cooling systems, as well as the development of self-cleaning, anti-fogging, anti-bacterial surfaces [3-6]. Studies in this field are carried out as a search of a specific geometric configuration of the microstructure, improving hydrophobic or hydrophilic properties [7-10]. At the same time, the fundamental laws of the processes, when the static angle forms on the structured surfaces have not been studied for predictive modelling.

The aim of this work was to determine experimentally the effect of a drop volume on the value of the contact angle under the condition of stationary three-phase contact line in the wetting of the non-ferrous metals.

[a] Corresponding author : lafleur@tpu.ru

2 Experimental

The experimental study was carried out on the setup [11, 12], including the shadow optical system [4, 11-13].

A drop with a set volume was placed on the surface by an electronic dosing device Single Channel Pipette (Thermo Scientific). A drop profile was obtained by the shadow optical system, which consists of the camera (NIKON D7100) equipped with a macrolens (Nikon 105mm f/2.8G IF-ED AF-S), and the source of the plane-parallel light (MI-150 Edmund). The contact angle was obtained after processing the images in the software Drop Shape Analyses (DSA) by two methods (tangential 1, Young-Laplace).

According to results of preliminary experiments the range of input parameters was defined (Table 1).

Table 1. Input parameters of the experimental research

Liquid under the study	nondeaerated distilled water
Material of substrate	aluminium, copper, brass
Volume of liquid	10, 20, 30, 40, 50, 60 μl

The metal surfaces was analysed on the profilometer "Micro Measure 3D station" and microscope Hitachi TM-3000. The parameter of roughness (arithmetic average roughness) and microstructure were defined for substrates under investigation (Table 2).

Six substrates were used in the present study made of aluminium, copper and brass. The surfaces of three substrates were rough. The roughness was formed by longitudinally arranged grooves. And the others were processed on grinding machine with diamond grit (polished surfaces).

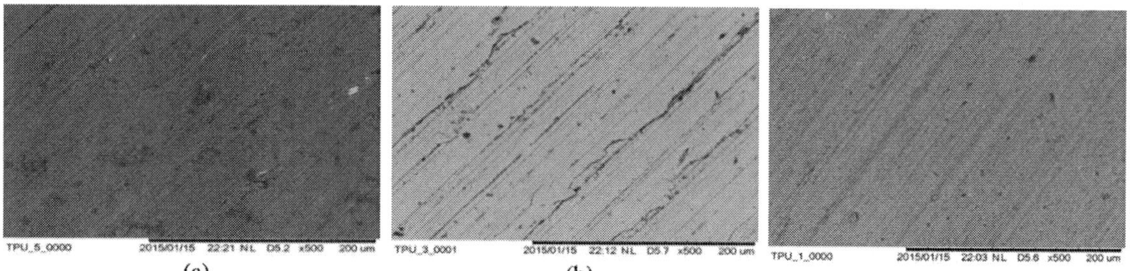

(a) (b) (c)

Figure 1. Microstructure of the rough substrates with magnification×500: (a) aluminium (Ra=0.374 μm); (b) copper (Ra=0.361 μm); (c) brass (Ra=0.939 μm).

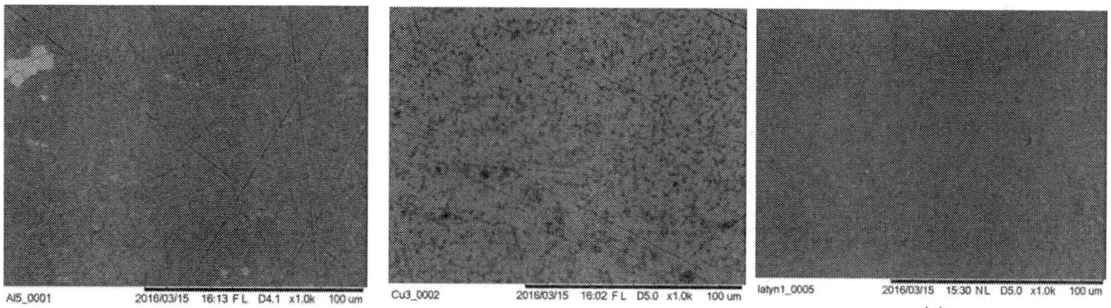

(a) (b) (c)

Figure 2. Microstructure of polished substrates with magnification×1000: (a) aluminium (Ra=0.036 μm); (b) copper (Ra=0.024 μm); (c) brass (Ra=0.032 μm).

 DOI: 10.1051/matecconf/20168400008

3 Results and discussion

According to results of the experiments the dependences of the contact angle from the drop volume were obtained (Figure 3).

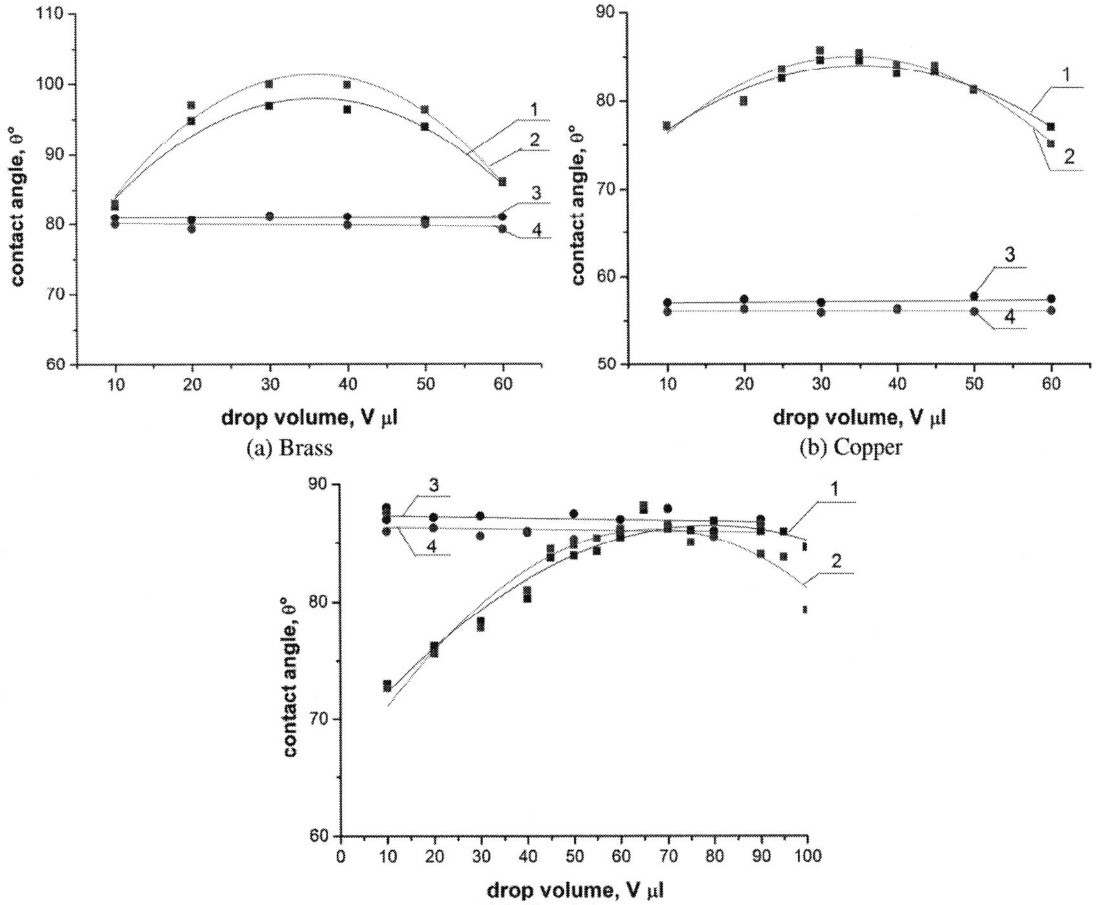

(a) Brass

(b) Copper

(c) Aluminium

Figure 3. Static contact angle versus drop volume on substrates unpolished and polished surfaces. The results were obtained by two methods of processing: 1 – tangential 1 method, unpolished substrate; 2 – Young-Laplace method, unpolished substrate; 3 – tangential 1 method, polished substrate; 4 – Young-Laplace method, polished substrate.

According to analyses of the results (Figure 3) it was found that the dependence of the contact angle from the drop volume on the surface with longitudinally arranged grooves can be fitted by parabolic curve with branches directed downwards (lines 1 and 2). The static contact angle does not depend on the drop volume on polished surfaces (lines 3 and 4).

The experimental data (Figure 1) are processed by the method of least squares. According to the first derivative of the polynomial dependence of the second order the maximum value (θ_{max}) and the corresponding drop volumes (V) were determined. Results of computations are presented in Table 2.

The maximum divergence of θ_{max} obtaining by Young-Laplace and tangential methods was 17% for the aluminium substrate. For copper and brass this value was not more than 4%. Probably, it is due to the most highly reflective surface of the aluminium substrate as compared to other non-ferrous metals used in the experiments.

Table 2. Computational results of the contact angle and the volume of the drop

Unpolished substrates	Young-Laplace method		Tangential 1 method	
	θ_{max}, °	V, μl	θ_{max}, °	V, μl
Aluminium	86.2	67.4	104.2	78.0
Copper	85.0	34.3	84.0	35.5
Brass	101.3	35.7	97.9	35.8

It should be noted, that the shape of the drop (its symmetry) was controlled by the optical system. The drop has a shape of a spherical segment in all experiments.

Difference in the nature of the obtained dependences may be explained as follows. During increasing the drop the influence of the friction and gravity forces on the static contact angle increases. In addition to these forces, the microstructure has a significant influence. Two states of a drop wetting the solid patterned substrates are known [14-16]: Cassie-Baxter state and Wenzel state. It may be assumed, that at relatively small volumes of drops liquid on the structured surface is placed on an "air cushion" [14, 15]. The weight of a drop is small; the liquid does not fill the microasperities. When the volume increases, the drop may transit to the Wenzel state due to the gravity (because of increasing the weight). It is possible that the maximum value of the contact angle on structured surfaces (Figure 1) corresponds to the transition from the Cassie–Baxter State to the Wenzel State.

The roughness (Ra) of the polished surfaces is relatively small and does not exceed 0.036 μm. Microasperities and microcavities do not affect the static contact angle.

4 Conclusion

The effect of the volume on the contact angle of a drop was studied under the conditions of stationary contact line on the rough and polished surfaces of non-ferrous metals. The comparison between two methods of drop profile processing was carried out. It was found that the maximum divergence of θ_{max} obtaining by Young-Laplace and tangential methods was 17% for the aluminium substrate. For copper and brass this value was not more than 4%. Probably, it is due to the most highly reflective surface of the aluminium substrate as compared to other non-ferrous metals, used in the experiments. It was found that in addition to the friction and gravity forces, the structure of the non-ferrous metals significantly influence on the static contact angle during changing the drop volume.

The work was realized within the research state assignment “Science” №13.1339.2014/K (Code of Federal Target Scientific and Technical Program 2.1410.2014). The shadow optical system was elaborated under the support of Russian Science Foundation (Agreement No. 14-19-01755).

References

1. B.D Summ, Yu.V. Goryunov, *The Physochemical Principles of Wetting and Spreading* [in Russian] (Khimiya, Moscow, 1976)
2. P.G. de Gennes, Rev. Mod. Phys. **57**, 827 (1985)
3. V.S. Ajaev, E.Ya. Gatapova, O.A. Kabov, Adv. Colloid Interface Sci. **228**, 92 (2016)
4. E.Ya. Gatapova, A.A. Semenov, D.V. Zaitsev, O.A. Kabov, Colloids and Surfaces A **441**, 776 (2014)
5. V.S. Ajaev, E.Ya. Gatapova, O.A. Kabov, Phys Rev E **84**, 041606 (2011)

6. R. Blossey, Nat Mater **2**, 301 (2003)
7. V.E. Nakoryakov, S.L. Elistratov, S.Ya. Misyura, Journal of Engineering Thermophysics **20**, 1 (2011)
8. E. Nakoryakov, S.Ya. Misyura, S.L. Elistratov, Int J Heat Mass Tran **55**, 6609 (2012)
9. Z.-G. Guo, W.-M. Liu, Appl Phys Lett **90**, 223111 (2007)
10. V. Hejazi, K. Sobolev, M. Nosonovsky, Sci Rep **3**, 2194 (2013)
11. E.G. Orlova, G. V. Kuznetsov, D. V. Feoktistov, EPJ Web of Conferences **82**, 01053 (2015)
12. D. Feoktistov, E. Orlova, A. Islamova, EPJ Web of Conferences **110**, 01018 (2016)
13. O.A. Kabov, D.V. Zaitsev, Doklady Physics **58**, 292 (2013)
14. A.L. Dubov, A. Mourran, M. Möller, O.I. Vinogradova, Appl Phys Lett **106**, 241601 (2015)
15. A.L. Dubov, A. Mourran, M. Möller, O.I. Vinogradova, J Phys Chem **141**, 074710 (2014)
16. O.I. Vinogradova, A.V. Belyaev, J Phys Condens Matt **23**, 184104 (2011)

MATEC Web of Conferences **84**, 00009 (2016) DOI: 10.1051/matecconf/20168400009
International Symposium IPHT 2016

Two layer dielectric-electrolyte micro-flow with pressure gradient

Georgy Ganchenko[1,a], Abhishek Navarkar[2], Michael Zhukov[3] and Sakir Amiroudine[2]

[1]*Laboratory of Micro- and Nanoscale Electro- and Hydrodynamics for the Krasnodar Branch of Financial University, 34 Luzana st., Krasnodar, 350051, Russia.*

[2]*Université Bordeaux, I2M, UMR CNRS 5295, 16 Av. Pey-Berland, 33607 Pessac, France.*

[3]*Department of computational mathematics and mathematical physics, Southern Federal University, 8a Melchakova st., Rostov-on-Don, 344090, Russia.*

Abstract. The present work considers stability of two-phase dielectric/electrolyte system, consisting of two immiscible liquids in microchannel. The system is set in motion by the constant external electric field, which causes the electroosmotic flow in the electrolyte, and by the pressure driving force. The investigation of its linear stability has shown that there are two types of instability in the system: short-wave and long-wave instability. The short-wave instability occurs for a stronger external field than the long-wave instability but the growth rate of the short-wave instability is much higher than that of the long-wave instability.

1 Introduction

Phenomena of micro- and nanoscale moving of conductive and non-conductive liquids with an interface between them requires a great deal of attention because it has numerous practical applications in bio-chemical processing: lab-on-cheap reactors, micromixers, DNA extraction, drug delivery and many others. The present work considers a two-layer micro/nanoflow of conductive (electrolyte) and non-conductive (dielectric) viscous liquids bounded by two solid walls in an external electric field and pressure gradient (see Fig. 1).

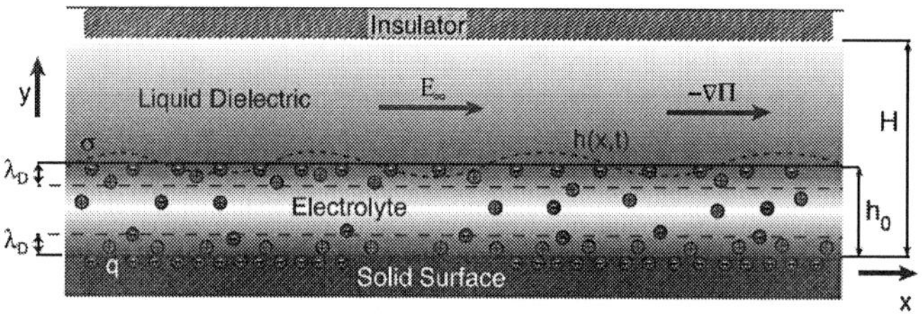

Figure 1. Electro-osmotic two-phase electrolyte–dielectric flow under an external tangential electric field and pressure gradient.

[a] Corresponding author ganchenko.ru@gmail.com

The electrolyte–dielectric and the solid–electrolyte interfaces are assumed to be charged, while the solid–dielectric interface is an electrically insulated surface. The charge near the solid is immobile, while the surface charge on the electrolyte–dielectric interface is mobile [1] and references therein. Upon a tangential electric field, these charges produce a Coulomb force and cause a movement of the liquid; experimental evidence for an electroosmotic flow at the liquid–liquid interface and in the liquid bulk is given in [2]. In spite of almost zero Reynolds numbers (negligible inertia effects), the electrohydrodynamics of liquid flows at micro- and nanoscales includes complicated instabilities [3] and bifurcations [4], and even chaotic motion at micro- and nanoscales, which is named as microturbulence.

2 Formulation of the problem

The problem is described by the Nernst–Planck–Poisson–Stokes (NPPS) system in the liquid–electrolyte phase,

$$\frac{\partial c^{\pm}}{\partial t} + \mathbf{U}\cdot\nabla c^{\pm} = \pm\nabla\cdot(c^{\pm}\nabla\Phi) + \nabla^2 c^{\pm}, \tag{1}$$

$$\nu^2\nabla^2\Phi = c^- - c^+, \tag{2}$$

$$\nabla P = \nabla^2\mathbf{U} - \frac{\kappa}{\nu^2}\nabla\Phi(c^+ - c^-), \quad \nabla\cdot\mathbf{U} = 0. \tag{3}$$

Here, $c^{\pm}$ are the molar concentrations of cations and anions; $\mathbf{U}$ is the fluid velocity vector; Φ is the electrical potential in the liquid phase; P is the pressure.
The lower solid wall has a uniform surface charge, q, and provides the following boundary condition,

$$y = 0: \frac{\partial\Phi}{\partial y} = \frac{q}{\nu} \tag{4}$$

The solid surface is assumed to be impermeable to both kinds of ions, and the velocity components obey no-slip BCs,

$$\frac{\partial c^{\pm}}{\partial y} \pm c^{\pm}\frac{\partial\Phi}{\partial y} = 0, \qquad \mathbf{U} = 0. \tag{5}$$

The BCs at the liquid–liquid interface, $y = h(x,t)$, are as follows,
(i) The surface is assumed to be impermeable to cations and anions:

$$y = h(x,t): \frac{\partial c^{\pm}}{\partial n} \pm c^{\pm}\frac{\partial\Phi}{\partial n} = 0; \tag{6}$$

(ii) The potential is continuous while its gradient has a jump connected with a mobile surface charge

$$\Phi = \phi, \frac{\partial\Phi}{\partial n} = \delta\frac{\partial\phi}{\partial n} + \frac{\sigma}{\nu}. \tag{7}$$

Here ϕ is the electrical potential in the dielectric phase.
(iii) The surface charge obeys the conservation law

$$\frac{\partial\sigma}{\partial t} + \frac{\partial(U_s\sigma)}{\partial x} = 0. \tag{8}$$

Here $U_s$is the tangential component of the liquid–liquid interface velocity; exchanges between the surface charge and the bulk charge are neglected.
(iv) The stress-balance BCs at the interface $y = h(x,t)$) are as follows,

$$(P - p) - \frac{We}{R} + [\mathbf{n}\mathbf{T}^f\mathbf{n}] + [\mathbf{n}\mathbf{T}^e\mathbf{n}] = 0, \quad [\mathbf{n}\mathbf{T}^f\mathbf{s}] + [\mathbf{n}\mathbf{T}^e\mathbf{s}] = 0. \tag{9}$$

Here p is the pressures in the dielectric, We is the Weber number, R is the radius of the curvature of the surface so that *We/R* is the capillary jump across the interface, $\mathbf{T}^f$ is the viscous , $\mathbf{T}^e$ is the Maxwell stress. Square brackets [x] are the jump of value x across the interface.
(v) The kinematic constraint at the interface, $V = \partial h/\partial t + U\,\partial h/\partial x$.

DOI: 10.1051/matecconf/20168400009

The dielectric is described by the Laplace-Stokes system,

$$\nabla^2\phi = 0, -\nabla p + \mu \nabla^2 \mathbf{u} = 0, \ \nabla \cdot \mathbf{u} = 0, \tag{10}$$

with the BCs on the upper boundary,

$$y = H: \frac{\partial \phi}{\partial y} = 0, \ \mathbf{u}=0. \tag{11}$$

The above problem is described by dimensionless control parameters: E_∞ is the external electric filed, Π is the external pressure field, ν is the Debye number δ, is the ratio of the electric permittivities of the dielectric and electrolyte, μ is the ratio of the viscosities of the dielectric and electrolyte, κ is a coefficient of coupling between the hydrodynamics and the electrostatics, We is the Weber number, q is the dimensionless surface charge, H is the dimensionless width of the channel. The surface charge at the interface is a parameter depending on x and t, $\sigma = \sigma(x, t)$, but for one-dimensional solution one need to set a mean value of surface charge σ_0, so an additional parameter occurs in the system. The characteristic quantities and value of dimension parameters can be found in [1]. We set in our calculations κ=0.2, q=−3, σ_0 =1, ν=0.1, unless otherwise indicated.

3 Basic steady-state solution

The problem has 1D steady-state equilibrium solution with a plug-like velocity profile. Since the electrostatic and the hydrodynamic parts of the problem are decoupled in one-dimensional case and can be solved independently from each other, the results obtained in electrostatic part, are the same as in [1], where the dielectric liquid was replaced by the simpler case of air. The analysis of the equations one can found in [1].

Velocity profiles for different values of μ are shown in Fig. 2(a). As illustrated, the dielectric velocity decreases with increase of μ. This is a natural behavior for the system with rather big values of viscosity of passive dielectric layer.

Electroosmotic flow has different directions, which depend on the direction of the external electric field, determined by the sign of E_∞. If $E_\infty < 0$, then the electroosmotic flow is parallel to the pressure driven flow, and vice versa, for $E_\infty > 0$, the electroosmotic flow is anti-parallel, Fig 2(b)

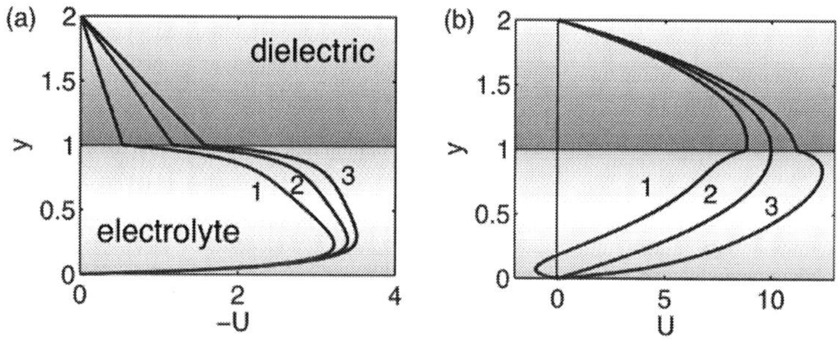

Figure 2. Velocity profile (a) for $\Pi = 0$, $E_\infty = 10$ and different values of μ 1:$\mu = 3.5$, 2:$\mu = 1$, 3:μ = 0.5 and (b) for $\Pi = 20$, μ=1 and different values of E_∞ 1: $E_\infty = 10$, 2: $E_\infty = 0$, 3: $E_\infty = -10$.

4 Numerical linear stability analysis

Linear stability of the 1D steady-state equilibria in response to sinusoidal, with wave number α, perturbations is considered. The stability investigation represents the solution to this problem in terms of a solution to an eigenvalue problem for a system of linear ordinary differential equations; the eigenvalue problem is solved numerically. The results show the coexistence of a long-wave, Fig. 3(b), and short-wave, Fig. 3(a), instabilities. The short-wave instability occurs for a stronger external field than the long-wave instability, but the growth rate of the short-wave instability is much higher than that of the long-wave instability

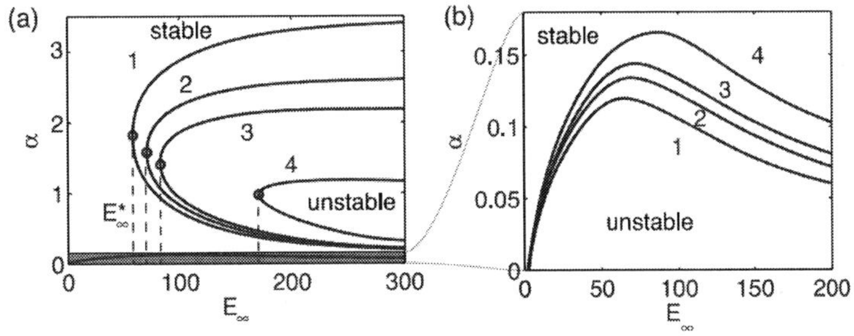

Figure 3. Marginal stability curves (a) short-wave region and (b) long-wave region for different values μ 1: $\mu = 0$ 2: $\mu = 0.5$, 3: $\mu = 1$, 4: $\mu = 3.5$

Furthermore, the application of pressure gradient results in the destabilization of the one-dimensional flow when the pressure driving force is oppositely directed to the electroosmotic flow. If the pressure driving force is codirected with the electroosmotic flow, then the one-dimensional solution is stabilized. This result is in good agreement with the experiment [2] (see Fig. 4).

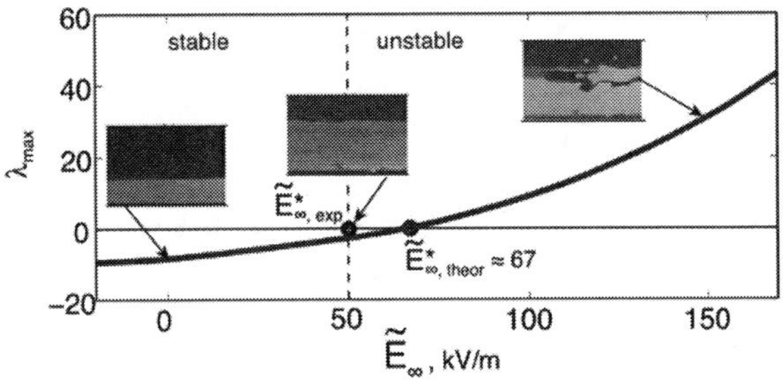

Figure 4. Comparison with experiments [2]. Maximal growth rate of the most dangerous perturbations λ_{max} vs dimensional external electric field $\tilde{E}_\infty$.

Conclusion

The stability of two-phase dielectric/electrolyte system, consisting of two immiscible liquids in microchannel, has been investigated. One-dimensional steady state has been founded. The numerical analysis provides the coexistance of long and short-wave instabilities. For the extreme cases there is a good correspondence of theoretical results in [1] and [5]. The first extreme case is the consideration of air as a dielectric liquid, and the second extreme case allows using of the Gouy–Chapman approximation. The theoretical values of critical external electric field are in good correspondence with the experimental data [2].

This work was supported, in part, by the Russian Foundation for Basic Research (Projects No. 15-08-02483-a, No. 14-08- 01171-a and 16-08-00643-a)

References

1. G.S. Ganchenko, E.A. Demekhin, M. Mayur, S. Amiroudine, Phys. Fluids **27**, 062002 (2015)
2. H. Li, T.N. Wong, N.-T. Nguyen, Micro and Nanosystems **4**, 14-24 (2012)
3. E.A. Demekhin, V.S. Shelistov, S.V. Polyanskikh, Phys. Rev. E **84**, 036318 (2011)
4. V.S. Shelistov, N.V. Nikitin, G.S. Ganchenko, E.A. Demekhin, Dokl. Phys. **56**, 538 (2011).
5. A. Navarkar, S. Amiroudine, M. Mayur, E. A. Demekhin, Microfluid Nanofluid **19**, 813 (2015)

Two-dimensional numerical simulation of microscale thermoelectrokinetic instability

Nataly Ganchenko[1,a], Georgy Ganchenko[2], Evgeny Kalaydin[3] and Evgeny Demekhin[3]

[1] *Department of Mathematical Modelling,Kuban State University,Krasnodar, 350040, Russia*

[2] *Laboratory of Micro- and Nanoscale Electro- and Hydrodynamics for the Krasnodar Branch of Financial University, 34 Luzana st., Krasnodar, 350051, Russia.*

[3]*Department of Mathematics and Computer Science for the Krasnodar Branch of Financial University, 34 Luzana st., Krasnodar, 350051, Russia.*

Abstract. Two-dimensional direct numerical simulation of thermoelectrokinetic instability in electrolyte bounded by semiselective surfaces was conducted for the first time. The results of numerical experiments show that the convective motion in the electrolyte takes place due to the stability loss of one-dimensional steady state solution. This motion manifested itself in two-dimensional structures similar to electroconvective rolls. What fundamentally differs is formation of thin lines connecting the membranes, where the salt concentration reaches its maximum value. Electric current in the system, obviously, will firstly flow through these lines.

1 Introduction

Problems of electrokinetics have recently captured attention of scientists from different fields of knowledge because of rapid developments in micro-, nano- and biotechnology. It tends to be very promising for practical issues to use ion-exchange surfaces, permeable only for one type of ions, cation or anion. Among such surfaces are semiselective electric membranes, electrodes and systems of micro- and nanochannels. It is well known that electric current in electrolyte in microgap between semiselective membranes is proportional to voltage ΔV, if the values of ΔV are rather small. This is a manifestation of Ohm's law, so such a regime is usually called ohmic. With increase of ΔV transition to limited mode takes place (electric current weakly depends on voltage) [1]. Further magnification of external electric field leads to reappear linear dependence of the electric current from the voltage. Transition from limited to overlimiting mode is caused by appearance of liquid movement owing to instability development, theoretically predicted in [2] and numerically confirmed in [3]. However, it must be said, that numerical and experimental data used to match only qualitatively, but not quantitatively, e.g. the experimental critical value of potential drop was less than that obtained numerically. As it has been shown in [4], neglecting of thermic effects caused by Joule heating may be the reason of such an unpleasant mismatch. A new type of microscale instability, occurring then electric current at the layer is oppositely directed to gravity vector, was found out. The critical potential drop can be smaller than that one for electokinetic instability threshold, if special combination of new introduced parameters is considered. In this work we add to Nernst-Planck-

[a] Corresponding author nataly.ganchenko@gmail.com

Poisson-Stocks system of equation an energy equation with source term caused by Joule heating of the electrolyte owing to electric current in it. Also, an additional term, respected for bouncy force, occurs in the Stocks equation. Membranes are assumed to be partly insulted (Fig. 1). For two-dimensional direct numerical simulation of the problem, the special algorithm based on one suggested in [5] and developed in [3] was designed.

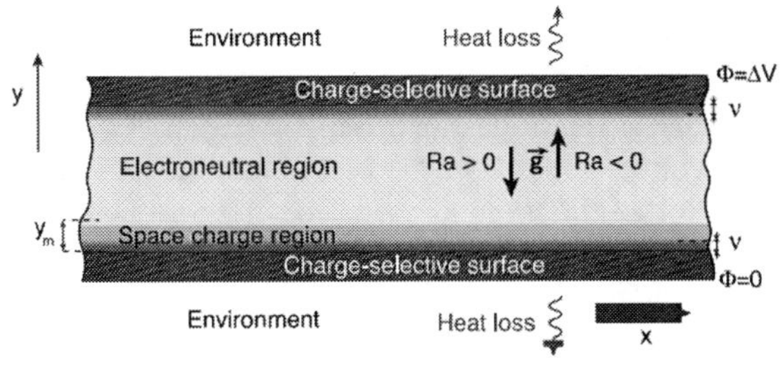

Figure 1. Schematics of the system.

2 Formulation of the problem

In dimensionless form the problem is described by the Nernst–Planck–Poisson–Stokes (NPPS) system

$$\frac{\partial c^{\pm}}{\partial t} + \mathbf{u} \cdot \nabla c^{\pm} = \pm \nabla \cdot \left(c^{\pm} \nabla \Phi\right) + \nabla^2 c^{\pm}, \tag{1}$$

$$\nu^2 \nabla^2 \Phi = c^- - c^+, \tag{2}$$

$$\nabla P = \nabla^2 \mathbf{u} - \frac{\kappa}{\nu^2} \nabla \Phi (c^+ - c^-) + Ra\, T\, \boldsymbol{e}_y, \quad \nabla \cdot \mathbf{u} = 0. \tag{3}$$

Here, $c^{\pm}$ are the molar concentrations of cations and anions; **U** is the fluid velocity vector; Φ is the electrical potential in the liquid phase; P is the pressure.

The Joule heating effect is described by energy equation

$$Le\left(\frac{\partial T}{\partial t} + \mathbf{u} \cdot \nabla T\right) = \nabla^2 T - Bi\, \boldsymbol{I} \cdot \nabla \Phi \tag{4}$$

with $I = -K\Delta\Phi - \Delta\rho,\ K = c^+ + c^-, \rho = c^+ - c^-$

The boundary conditions are as follows:

$$y = 0: \quad c^+ = p,\ \frac{\partial c^-}{\partial y} + c^- \frac{\partial \Phi}{\partial y} = 0, \quad \Phi = \Delta V,\ \mathbf{u} = 0, \quad -\frac{\partial T}{\partial y} + Bi\, T = 0; \tag{5}$$

$$y = 1: \quad c^+ = p,\ \frac{\partial c^-}{\partial y} + c^- \frac{\partial \Phi}{\partial y} = 0, \quad \Phi = 0, \quad \mathbf{u} = 0, \quad \frac{\partial T}{\partial y} + Bi\, T = 0. \tag{6}$$

The above problem is described by seven dimensionless control parameters: ν is the Debye number, κ is a coupling coefficient between the hydrodynamics and the electrostatic (it is essential that the coupling coefficient depends only upon the physical properties of the electrolyte), ΔV is the potential drop between the membranes, Ra is the Rayleigh number, Bi is the Biot number (characterizing the system's thermal insulation with respect to environment), Le is Lewis number, p is the membrane interface concentration in the boundary conditions. The characteristic quantities and value of dimension parameters can be found in [4]. We set in our calculations κ=0.2, Bi=0.01, p =5, ν=0.1,Le=0.013 unless otherwise indicated.

3 Theoretical flashback

It is worthwhile to remind here some basic results from [4]: the system is stabilized by the thermal effects at $Ra > 0$ and destabilized at $Ra < 0$. Changing of the sign of Ra is nothing more than the rotation of the system at 180° relative to the direction of gravity. In the problem, there are two competing mechanisms of instability, which can be clearly elucidated by the analytical solution (7). The first one is the Zaltzman–Rubinstein electrokinetic instability, characterized by the first term in the right-hand side of (7). The second is connected with the Joule heating, and is characterized by the second term in the right-hand side of (7)

$$.\ 1 = \frac{\kappa}{8}\Delta V^2 \frac{k^3 \cosh k - \sinh^3 k}{4 \sinh k\,(k^2 - \sinh^2 k)} - Ra\frac{\Delta V}{8}\,Bi\,\frac{k^2(\sinh k - k\cosh k)^2 - (k^2 - \sinh^2 k)^2}{k^4 \sinh^2 k (k^2 - \sinh^2 k)}. \qquad (7)$$

there k is a wave number of superimposed small perturbations.

For the case when the heating is absent ($Ra = 0$) the instability is caused by the Coulomb force in the space charge region and the corresponding formation of a slip velocity at its border. The nonuniformity of the slip velocity leads to the electrokinetic instability [2] The competing mechanism is associated with the Joule heating, mainly in the electroneutral region. The second mechanism is connected with heating and the thermic expanding of the liquid, but it is totally different from the Rayleigh–Benard convection and the instability is caused by an induced nonuniformity of the conductivity in the electrolyte.

4 Direct numerical simulation

In this paper we present the results of two-dimensional direct numerical simulation of the problem. The designed algorithm is based on one suggested by Nikitin in [5] and adopted for electrokinetic problems in [3]. A finite-difference method with second-order accuracy is applied for the spatial discretization. A uniform grid is used in the homogeneous tangential x direction; the grid is stretched in the normal y direction via a $tanh$ stretching function in order to properly resolve the thin double layers attached to the membrane surfaces. For solving stiff problems implicit methods require the inversion of rather large matrices and thus are extremely costly, while explicit methods of time advancement require a very small time step and, hence, are prohibitively ineffective. A semi-implicit method is found to be a reasonable compromise: only a part of the right-hand side of the system is treated implicitly. A special semi-implicit $3\,\frac{1}{3}$-step Runge-Kutta scheme is used for the eventual integration in time. The details of the numerical scheme will be presented elsewhere.

For the natural "room disturbances", the infinite spatial domain is changed to a large enough finite domain that has lengths $lx = l$ in both spatial dimensions, with the corresponding wave number $k = 2\pi/l$. The condition that the solution be bounded as $x \to \infty$ is changed to periodic boundary conditions. The length of the domain l has to be taken large enough to make the solution independent of the domain size.

5 Simulation results

Two-dimensional modeling confirms the results of investigation conducted in [4]: two competing mechanisms of instability are determined by the parameters κ and Ra. The relation between these parameters determines which of the instability mechanisms will be decisive for the destabilization of the system. The case Ra = 0 separates the destabilizing and stabilizing effects of the Joule heating. In contrast to the Rayleigh–Benard convection, the second mechanism stabilizes the system with heating at the bottom ($Ra > 0$) and destabilizes it with heating at the top ($Ra < 0$).

For $Ra < 0$, with decreasing or increasing $|Ra|$, the heat effects prevail over the electrokinetic effects and a drastic change of instability modes occurs: the critical voltage V^* decreases dramatically. Moreover, the short-wave instability changes to a long-wave instability.

The convective motion in the electrolyte takes place due to the stability loss of one-dimensional steady state solution. Electroconvective vortices observed in [3] differ from those obtained during our calculations: they are short-wave and their centres are situated on the border of electric double layer, and our vortices are long-wave and situated in the centre of the channel. (Fig. 2). If the heat-insulation of the system is good enough, overlimiting mode might occur at essentially smaller values of potential drop than in isothermal model. Two-dimensional modelling shows that for supercritical values of potential drop two-dimensional structures take place. These structures (Fig. 3 (b)), are similar to electrokinetic "spikes" (Fig. 3 (a)), but have a long-wave behaviour. The interesting feature of such structures is formulation of thin lines with high ion concentration between two membranes. Because of higher ion concentration these lines have a higher conductivity, and, hence, the electric current in the system goes mostly through these lines.

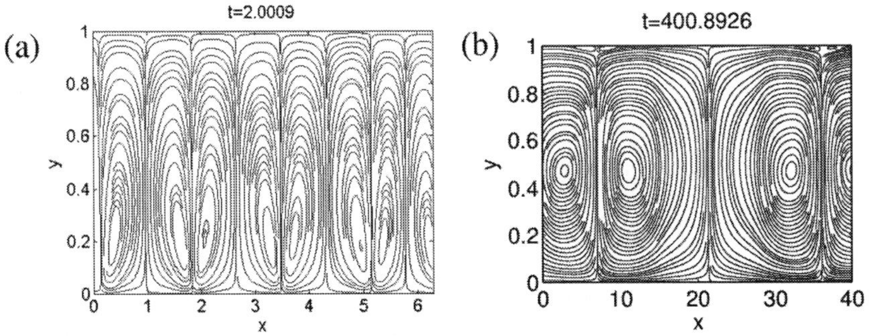

Figure 2. Stream lines, (a) Ra=0 (isothermal), $\Delta V = 30, (b) Ra = -50, \Delta V = 20$.

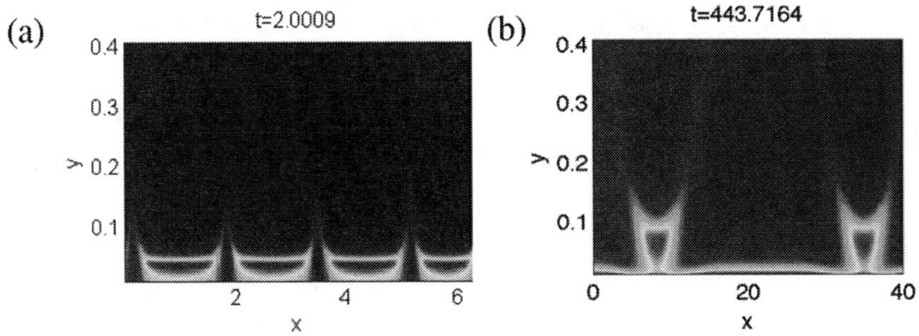

Figure 3. Charge density, (a) $Ra = 0$ (isothermal), $\Delta V = 30, (b)\ Ra = -50, \Delta V = 20$.

Conclusion

A new kind of instability caused by Joule heating near charge-selective surfaces and its influence on the electrokinetic instability are investigated theoretically by two-dimensional direct numerical simulation. The numerical experiments confirm the results of one-dimensional analysis [4] and allow clarifying the physical mechanism of thermoelectrokinetic instability.

This work was supported, in part, by the Russian Foundation for Basic Research (Projects No. 14-08-00789-a, No. 15-58-45123-IND_a and 16-08-00643-a)

References

1. I. Rubinstein and L. Shtilman, J. Chem. Soc. Faraday Trans. II **75**, 231 (1979)
2. B. Zaltzman and I. Rubinstein, J. Fluid Mech. **579**, 173 (2007)
3. E. A. Demekhin, N. V. Nikitin, and V. S. Shelistov, Phys. Fluids **25**, 122001 (2013)
4. E. A. Demekhin, S. Amiroudine, G. S. Ganchenko, N. Yu. Khasmatulina, Phys. Rev. E **91**, 063006 (2015)
5. N. V. Nikitin, Int. J. Numer. Methods Fluids **51**, 221 (2006)

MATEC Web of Conferences **84**, 00011 (2016) DOI: 10.1051/matecconf/20168400011
International Symposium IPHT 2016

Technique for the measurements of the temperature profile near liquid-gas interface

Elizaveta Gatapova[1,a], Maxim Filipenko[1,2]

[1]*Kutateladze Institute of Thermophysics SB RAS, 630090 Novosibirsk, Russia*
[2]*Novosibirsk State University, 630090 Novosibirsk, Russia*

Abstract. The technique for the experiments with moving microthermocouple across the liquid-gas systems with interfaces is presented. The experimental setup, apparatus and automation technique for temperature profile measurements are described. The special software for automation, processing and synchronization of the experimental procedure and data is developed. The software allows one to control the progress of the experiment, such as evaluation of various equipment delays. The data can be collected and stored in an accessible and comprehensive form. The typical experimental values are automatically obtained. The use of program minimizes the human factor.

1 Introduction

Nowadays, data centers and mobile devices have many problems with thermal regulation [1]. The micro-sized two-phase cooling systems with phase transition are of a particular interest because of the high heat transfer coefficients [2]. The study of the heat and mass transfer in system with evaporation, boiling and condensation is required. Study of heat transfer in a two-layer system with liquid-gas interfaces is one of the topical scientific problems, for instance, in developing the technologies for heat pipe production and in investigation of dry spot formation [3]. The temperature jumps at in the Knudsen layer at phase transition are known for a long time in the kinetic theory, since the phase transitions occur under the non-equilibrium conditions [4,5]. An approach to the description of heat and mass transfer in the two-phase system, based on the Navier–Stokes equations with the temperature and pressure jumps boundary conditions, was suggested in [6]. However, the experimental data on the pressures and temperatures jumps at the interfaces is limited [7,8,9].

The experimental setup, apparatus and automation technique for temperature profile measurements in two-layer liquid-gas system are presented in this paper. The purpose of work is creation of the software for the automation of experimental process of the temperature field measurements across two-layer systems with a moving interface and heating elements. The data synchronizing with sufficient accuracy should be an important procedure of the software.

[a] Corresponding author : gatapova@itp.nsc.ru

 DOI: 10.1051/matecconf/20168400011

2 Experimental setup and measurements technique

A photo of the experimental setup is presented in Figure 1. The main part is a round test cell with a local heater in the centre. Ultrapure water (MilliQ) is used as a working fluid, which evaporates into the atmosphere under normal conditions. The main measuring device is a self-made microthermocouples of type-E with a bead thickness of less than 4 microns. Also there are standard microthermocouple of type-K located in the heater and the air, which provide additional temperature control. Cold junctions of a microthermocouples are soldered to the copper wires and lowered into an electronic zero point with error of 0.001 °C (Fluke). Accuracy of all thermocouples is 0.02 °C. Calibration is performed using two references RTD (ETS 100). Microthermocouples receives data from the data collection terminal (NI TB-9214). Precise micropositioner (Zaber) with a minimum step size of about 50 nanometers moves the main microthermocouples fixed in a special holder. Special software determines the position of microthermocouple bead. The shadow technique is used to control the movements of the microthermocouple. The experiments are carried out for the different heating modes. The experimental setup is insulated by protective box. The data is processed in real time using a computer.

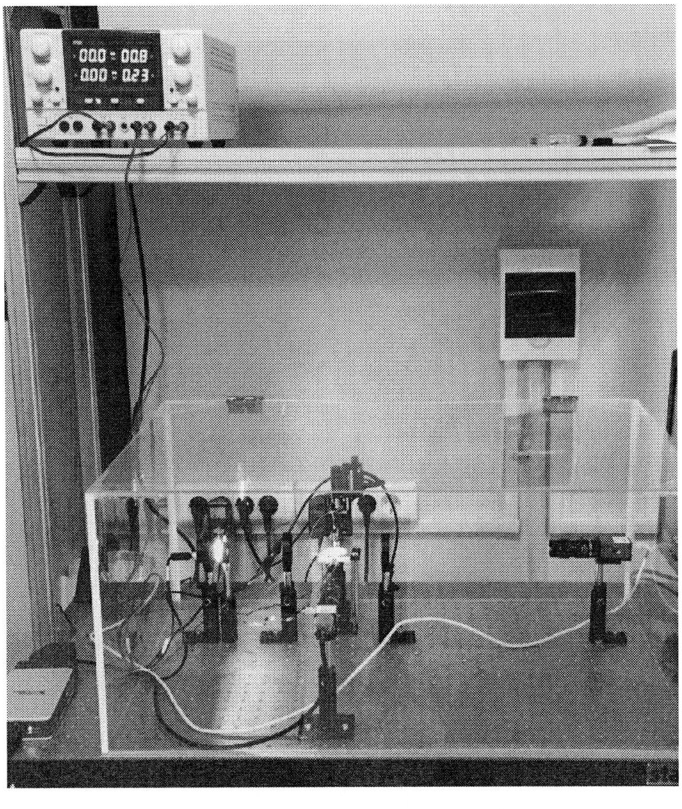

Figure 1. Photo of experimental setup.

3 Automation and synchronization

The experiment is as follows: microthermocouples moves to a position above the water and then a passes through the liquid-gas interface with constant speed. The experiment is controlled by specially developed on C++ software, which allows carrying out all experiments in the same operating condition. The program records the time and position of micropositioner, as well as the time and temperature values from the thermocouples data collection terminal. The comparison and correlation of collected data with respect to time are made by software. A coordinate of the liquid-gas interface is defined in two ways: first method is by shadow technique visualization with CCD-camera in the moment of contact of the microthermocouple and liquid; second is by the definition by the temperature jump when microthermocouple passes through the liquid-gas interface.

Finally, as a result of the measurements and data processing by software we obtain a graph. The graph (Figure 2) presents the temperature profile (yellow line) across the liquid and gas layers, including interface, via microthermocouple position, where zero point is a position of liquid-gas interface. Green line in Figure 2 is a delay in data synchronization. The software collects, stores and processes the data from other thermocouples located in different parts of the experimental setup. Timing error is calculated as the difference between the times of signal acquisition module from microthermocouple and of micropositioner signal. Figure 2 shows the delay time not exceeding 35 ms, which is sufficient for the present experiment because the size of microthermocouple bead is an order of magnitude greater than the distance travelled by microthermocouples during this time.

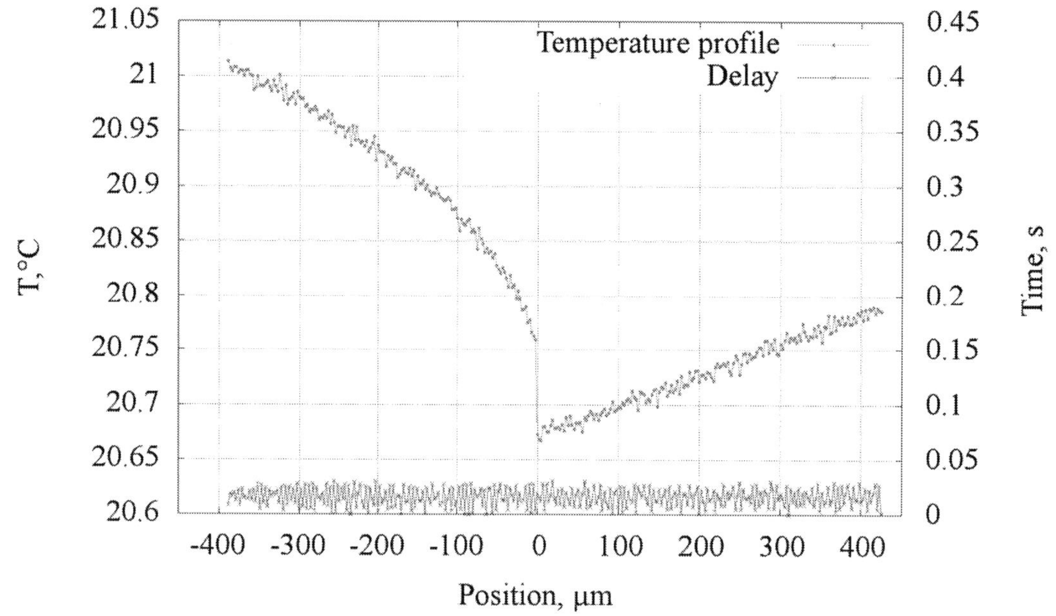

Figure 2. Temperature profile and delay of measurements. Position 0 corresponds to the liquid-gas interface. Ambient temperature 25.2 °C, humidity 20.2 %, pressure 99 kPa, heating power is 18 mW. Speed of microthermocouple displacement is 5 μm/s. The microthermocouple signal reads every 0.5 seconds.

The position and coordinate of the heater surface, situated in bottom of cuvette, are determined once for all experiments. The determination of the heater coordinate allows automatically calculate the liquid layer thickness and hence mass flow rate of evaporated liquid. The temperature difference between readings of two steps of microthermocouple is calculated automatically by software and obtained, see Figure 3. The value of the maximum temperature jump of about 0.09 °C is obtained for the operating condition. The time of the temperature jump occurring from the experiment beginning is 176.8 s. The position of the maximum jump in the micropositioner coordinates is 18422.7 μm, The bottom position of micropositioner coordinates is 20621.5 μm. Hence, the liquid thickness of 2198.8 μm is obtained.

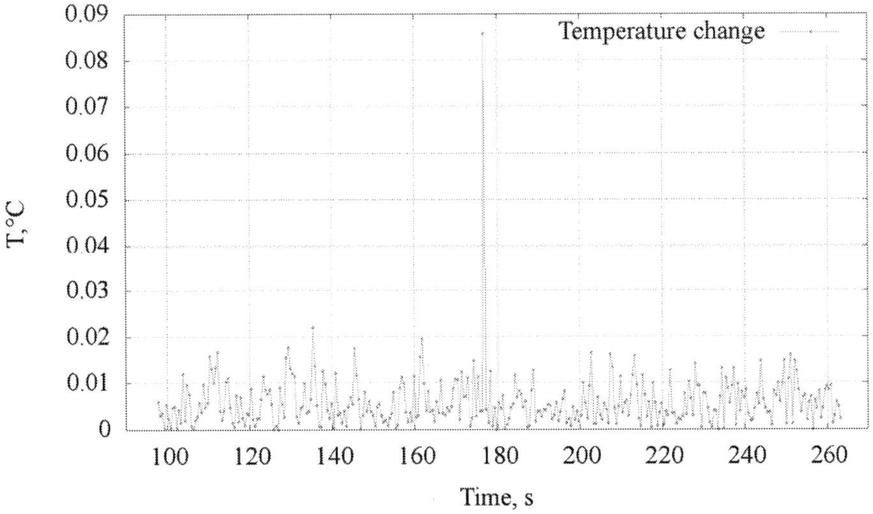

Figure 3. The absolute change in temperature after each read. Ambient temperature 25.2 °C, humidity 20.2 %, pressure 99 kPa, heating power is 18 mW. Speed of microthermocouple displacement is 5 μm/s. The microthermocouple signal reads every 0.5 seconds.

4 Conclusions

The technique for the experiments with moving microthermocouple across the liquid-gas systems with interfaces is described. The experimental setup, apparatus and automation technique for temperature profile measurements are developed.

The specially developed software allows one to control the progress of the experiment, such as evaluation of various equipment delays. It reduces the number of different options for some experiments. The software collects and stores comprehensive experimental data in an accessible and visual form. The typical experimental values are automatically obtained. The use of program minimizes the human factor.

Acknowledgements

Authors gratefully acknowledge support of this work by the Ministry of Education and Science of the Russian Federation (project identifier RFMEFI61614X0016).

References

1. A. Bar-Cohen, P. Wang, Journal of Heat Transfer, **134 (5)**, 051017 (2012)
2. S.G. Kandlikar, S. Colin, Y. Peles, S. Garimella, R.F. Pease, J.J. Brandner, D.B. Tuckerman, Journal of Heat Transfer, **135 (9)**, 091001 (2013)
3. D.V. Zaitsev, O.A. Kabov, Microgravity Science and Technology, **19**, 174-177 (2007)
4. Y.-P. Pao, Physics of Fluids, **14 (7)**, 1340–1346 (1971)
5. K. Aoki, C. Cercignani, Physics of Fluids, **26 (5)**, 1163–1164 (1983)
6. E.Ya. Gatapova, I.A. Graur, F. Sharipov, O.A. Kabov, International Journal of Heat and Mass Transfer, **83**, 235 – 243 (2015)
7. P.N. Shankar, M.D. Deshpande, Physics of Fluids A, **2 (6)**, 1030–1038 (1990)
8. C.A. Ward, D. Stanga, Physical Review E, **64**, 051509 (2001)
9. E.Ya. Gatapova, R.A. Filipenko, Yu.V. Lyulin, I.A. Graur, I.V. Marchuk, O.A. Kabov, Thermophysics and Aeromechanics, **22 (6)**, 701-706 (2015)

Increasing of the liquid flow rate due to contrast wettability in the microchannel

Dmitry Gluzdov[1,2,a]

[1]*Kutateladze Institute of Thernophysics SB RAS, 630090 Novosibirsk, Russia*
[2]*Novosibirsk State University, 630090 Novosibirsk, Russia*

Abstract. The global miniaturization leads to microchannels flows be perspective for novel cooling technologies. The problem of the drag reduction occurs in such systems. The two-dimensional Poiseuille flow in microchannel with Navier slip condition is considered. The presence of bubbles with 1 μm size on hydrophobic parts on the bottom wall of the channel is assumed for the problem. The calculations of the liquid flow rate and velocity depending on the slip length at the bottom wall have been done.

1 Introduction

Increasing the power of modern microprocessors is linked with the solution of problems with thermal regulators and the creation of a new type of cooling systems. The microchannels one or two-phase flow is seems to be perspective for novel cooling technologies. However, the problem associated with the drag reduction arises in such microchannels flow [1,2]. One of the solutions to this problem is use of micro-/nano-structured surfaces. It was showed that the grooves and cavities in the structure filled by a gas phase can reduce the viscous resistance of more than 40% [3,4]. Note, that there is a problem with film rupture and instability in such systems [5,6,7,8].

Considering textured surfaces the term of effective slip length should be investigated [9]. As well as some possible effect inside a grooves, which can affect on the main flow should be carefully considered [10].

An alternative solution for drag reduction has been proposed in [11]. To use contrast wettability and the bubble accumulation on hydrophobic parts to reduce viscous resistance in microchannel flow has been investigated. Nanocoating may be deposited discretely on the surface of a microchannel that may allow to vary performance and stability of the flow resistance, since it allows fluid to slip in areas with a layer of bubbles, and to stick to portions without bubbles. The coating such as described in [12,13] can be considered as a stable one.

In this paper calculations of the liquid flow rate and velocity depending on the slip length at the bottom wall have been done using the model proposed in [11], where the bubble size, channel height and slip length were the parameters of flow.

[a] Corresponding author : gluzdov1@yandex.ru

DOI: 10.1051/matecconf/20168400012

2 Problem statement and results

It has been proposed to use the effect of concentration of gas bubbles on hydrophobic segments of the surface of the channel with contrast wettability for ensuring drag reduction [11]. Two regimes of flow with the concentration of bubbles in the hydrophobic region were analyzed. The first one is a regime without heating, where small gas bubbles in the flow precipitate in the hydrophobic region. The nanocoating ensured the contrast of wettability of about 20°–25° between the surface of the silicon plate and coating. The second regime was produced specially for the enhancement of the effect: a heater mounted beneath the hydrophobic region was switched on and the process of nucleation of bubbles began.

The scheme of the flow in microchannel is presented in Figure 1. D - the length of the nanocoated section, L - the length of the uncoated section, P_0 - inlet pressure, P_1 - outlet pressure, H – channel height.

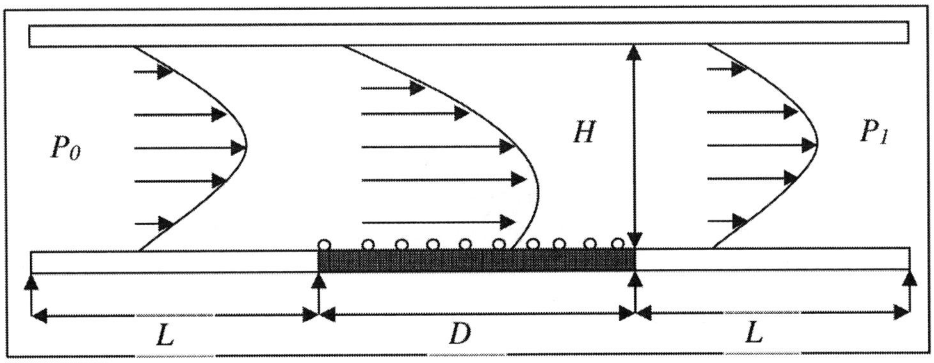

Figure 1. Sketch of the flow: D - the length of the nanocoated section, L - the length of the uncoated section, P_0 - inlet pressure, P_1 - outlet pressure, H – channel height.

Two-dimensional model of Poiseuille flow in microchannel with Navier slip condition is considered. The equation for the flow velocity is written as follows [11]:

$$u = \frac{1}{2\mu}\frac{\partial P}{\partial x_\beta}\left[y^2 - y\frac{H^2 - \delta^2 + 2\delta\beta}{H - \delta + \beta} - H\frac{\beta(H - 2\delta) - \delta(H - \delta)}{H - \delta + \beta} \right], \quad (1)$$

where

$\partial P / \partial x_\beta$ - pressure drop on section with slip condition,
β – effective slip length,
μ – viscosity,
δ – the bubble size.

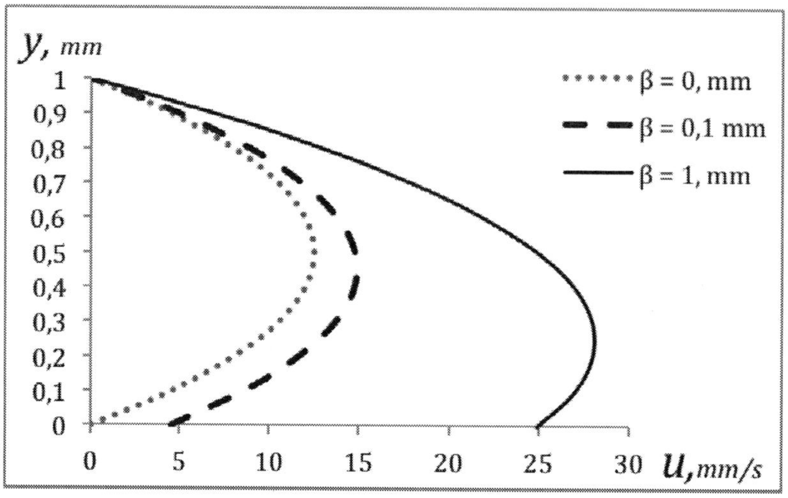

Figure 2. Velocity distribution for different effective slip length.

The dependence of the velocity distribution on the effective slip length is shown in Figure 2, where liquid is water, the channel height $H = 1$ mm, bubbles size $\delta = 1$ µm, viscosity $\mu = 0.001$ kg/ms, pressure drop $\partial P / \partial x_\beta = 100$ Pa/m. Figure 2 shows how the flow front varies with the different slip length, β.

Figure 3 presents the flow rate depending on slip length. The dependence is not linear. When the slip length is 1 mm, the flow rate is equal to 22 mm²/s, which is significantly higher than the flow rate in the absence of slippage equal to 8.25 mm²/s (Figure 3). The surfaces with contrast wettability can increase the fluid flow rate at a fixed pressure gradient.

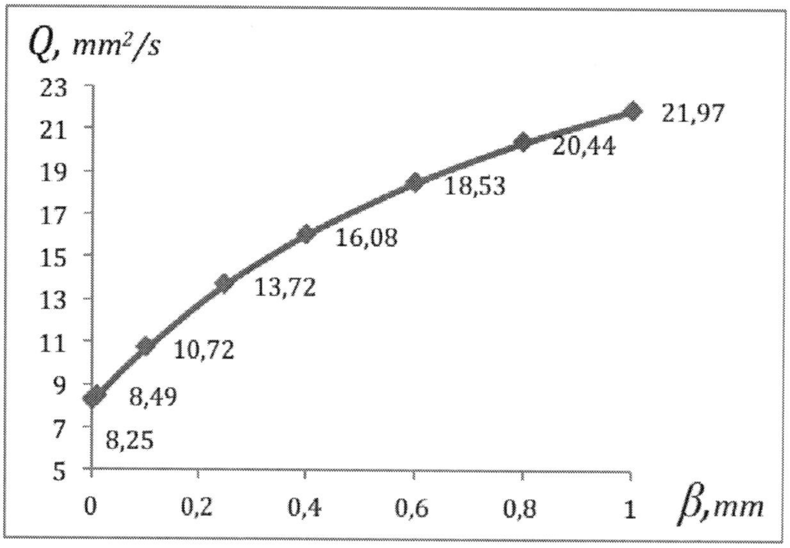

Figure 3. Flow rate for different effective slip length.

Acknowledgements

This work was supported by a grant from the Russian Science Foundation (project 14-19-01755).

References

1. T. M. Squires and S. R. Quake, Reviews of Modern Physics, **77**, 977 (2005)
2. H. A. Stone, A. D. Stroock, and A. Ajdari, Annual Reviews of Fluid Mechanics, **36**, 381 (2004)
3. J. Ou, B. Perot, and J. P. Rothstein, Physics of Fluids, **16**, 4635 (2004)
4. J.P. Rothstein, Annual Reviews of Fluid Mechanics, **42**, 89 (2010)
5. O.A. Kabov, D.V. Zaitsev, Multiphase Science and Technology, **21**, 249 (2009)
6. V.S. Ajaev, E.Ya. Gatapova, O.A. Kabov, Physics of Fluids, **25**, 122102 (2013)
7. V.S. Ajaev, E.Ya. Gatapova, O.A. Kabov, Microgravity Science and Technology, **24** (1), 33 (2012)
8. V.S. Ajaev, E.Ya. Gatapova, O.A. Kabov, Physical Review E, **84** (4), 041606, (2011)
9. A.V. Belyaev, O.I. Vinogradova, Journal of Fluid Mechanics, **652**, 489 (2010)
10. E. Karatay, A. S. Haase, C. W. Visser, C. Sun, D. Lohse, P. A. Tsai, and R. G. H. Lammertnik, Proceedings of the National Academy of Sciences of the United States of America, **110**, 8422 (2013)
11. E.Ya. Gatapova, V.S. Ajaev, O.A. Kabov, JETP Letters, **101** (3), 160 (2015)
12. European Patent No. 2028432A1 (2009)
13. R. Rioboo, M. Marengo, S. Dall'Olio, M. Voue, and J. de Coninck, Langmuir **25** (11), 6005 (2009)

MATEC Web of Conferences **84**, 00013 (2016) DOI: 10.1051/matecconf/20168400013
International Symposium IPHT 2016

Mathematical Modelling of the Evaporating Liquid Films on the Basis of the Generalized Interface Conditions

Olga Goncharova[1, 2], Ekaterina Rezanova [1, 2, a]

[1]*Altai State University, Faculty of Mathematics and Information Technologies , pr. Lenina, 61, 656049 Barnaul, Russia*
[2]*Kutateladze Institute of Thermophysics SB RAS, pr. Lavrentyeva, 1, 630090 Novosibirsk, Russia*

Abstract. The two-dimensional films, flowing down an inclined, non-uniformly heated substrate are studied. The results contain the new mathematical models developed with the help of the long-wave approximation of the Navier-Stokes and heat transfer equations or Oberbeck-Boussinesq equations in the case, when the generalized conditions are formulated at thermocapillary interface. The evolution equations for the film thickness include the effects of gravity, viscosity, capillarity, thermocapillarity, additional stress effects and evaporation.

1 Introduction

Modelling of the convective processes caused by impact of various forces on the fluid media is important nowadays [1-4]. The tangential stresses due to flow in the gas phase on a thermocapillary gas-liquid interface and the evaporation effects should be taken into consideration in the case when the fluid flows are accompanied by the adjacent gas flows. One of the most critical points in the mathematical modelling of the above mentioned processes is the statement of the boundary conditions on the interface. The conditions at the interface between two interacting gas-liquid phases have been presented in [6-8]. The conditions at the interface are the result of the conservation laws of mass, momentum and energy. The additional relations concerning a continuity of some flow characteristics, laws of the heat and mass fluxes, of the dynamic, energy and phenomenological properties etc. are usually assumed to be fulfilled in order to obtain the closed problem statements. The generalized interface conditions allow to model the evaporative processes in the liquid and gas-vapor phases with interface in the full problem statement and in the long-wave approximation of the governing equations.

The effects of evaporation on the nonlinear stability of uniformly heated films have been investigated experimentally, numerically and analytically in many papers [1-5, 9-11]. In this paper we present new mathematical models of flows of the thin evaporating liquid layers on the basis of the long-wave approach (see also [9-11]). These models are constructed on the basis of the Navier-Stokes equations or their Boussinesq approximation and generalized kinematic, dynamic and energetic interface conditions. One-sided two-dimensional models of the evaporating falling film are considered here in the case of moderate Reynolds numbers. The evolution equations of the film thickness take into account the effects of gravity, viscosity, capillarity, thermocapillarity, evaporation and of the

[a] Corresponding author : katerezanova@mail.ru

additional tangential stresses at interface. Numerical results demonstrate the differences in the flow characteristics obtained with the help of alternative mathematical models.

2 Mathematical models of the problem

Let the impermeable substrate be inclined at an angle α to the horizon. The coordinate system is chosen so that the direction of the x-axis coincides with the direction of fluid flow (see Fig. 1a), $\mathbf{g}=(g_1, g_2)=(g \sin \alpha, -g \cos \alpha)$ is the gravitation vector, $g = |\mathbf{g}|$. The position of the interface is determined by the equation $z=h(x,t)$, $z=0$ is the equation of solid plate. If a longitudinal characteristic length (length of the surface waves) l is much larger than a transverse characteristic length (the mean film thickness) d then $\varepsilon = d/l$ is a small parameter. The one-sided mathematical model based on the Navier-Stokes and heat transfer equations or the Oberbeck-Boussinesq equations in dimensionless form can be written as follows [9-11]:

$$Re\varepsilon^2\left(u_t + uu_x + wu_z\right) - \varepsilon^2 u_{xx} = u_{zz} - p_x + f_u \sin\alpha, \tag{1}$$

$$Re\varepsilon^4\left(w_t + uw_x + ww_z\right) - \varepsilon^4 w_{xx} - \varepsilon^2 w_{zz} = -p_z + f_w \cos\alpha, \tag{2}$$

$$u_x + w_z = 0,$$

$$RePr\varepsilon^2\left(T_t + uT_x + Tw_z\right) - \varepsilon^2 T_{xx} = T_{zz}.$$

Here $\mathbf{v}=(u, w)$ is the velocity vector; p is the pressure or the modified pressure in the case of the Oberbeck-Boussinesq equations; T is the temperature; ρ is some relative value of liquid density; $Re=u_*l/\nu$ is the Reynolds number; $Pr=\nu/\chi$ is the Prandtl number; $f_u = \gamma_1$, $f_w = -\gamma_2$ in the case of the Navier-Stokes and heat transfer equations (here $\gamma_1 = Ga/(Re\varepsilon)$, $\gamma_2 = Ga/Re$) and $f_u = -\gamma_1 T$, $f_w = \gamma_2 T$ in the case of the Oberbeck-Boussinesq equations (here $\gamma_1 = Gr/(Re\varepsilon)$, $\gamma_2 = Gr/Re$); $Ga=gd^3/\nu^2$ is the Galileo number; $Gr=gT_*\beta d^3/\nu^2$ is the Grashof number; ν, χ are the kinematic viscosity and thermal diffusivity coefficients; β is the thermal expansion coefficient; u_* is the characteristic velocity, T_* is the characteristic temperature drop.

We restrict ourselves to the case of moderate Reynolds numbers ($Re=O(1)$). The generalized kinematic, dynamic and energetic conditions are satisfied at the thermocapillary interface [6-8] and take into account evaporation:

$$-\varepsilon\left(h_t + h_x - w\right) = J_{ev}E,$$

$$\varepsilon^2 h_x\left(w_z - u_x\right) + 0.5\left(u_z + \varepsilon^2 w_x\right) = \alpha_\tau\left[\varepsilon h_x\left(w_z^g - u_x^g\right) + 0.5\left(w_x^g + u_z^g\right)\right] - 0.5\alpha_{Ma}(T_x + h_x T_z),$$

$$-p + 2\varepsilon^2\left(w_z - h_x u_z\right) = -p^g + \alpha_E J_{ev}^2 + \alpha_{Ca} h_{xx}\left(1 - \alpha_\sigma T\right), \tag{3}$$

$$-T_z + \varepsilon^2 h_x T_x + \beta_2\left\{T div_\Gamma \mathbf{v}\right\} = \beta_3 J_{ev} + \beta_4 J_{ev}\left[-p + 2\varepsilon^2\left(w_z - h_x u_z\right)\right] + 0.5\beta_5 J_{ev}^3 + \beta_6 \varepsilon h_{xx} J_{ev}\left(1 - \alpha_\sigma T\right).$$

The conditions are presented in the form, where the terms of order higher that ε^2, are neglected. The local mass flux is determined with the help of the Hertz-Knudsen equation [4]: $J_{ev} = \alpha_J T\,|_{z=h(x,t)}$.

Here J_{ev} is the local mass flux at the thermocapillary interface, $E=\kappa T_*/(\lambda_U \rho\nu)$ is the evaporation coefficient, κ is the heat conductivity coefficient, λ_U is the latent heat of vaporization, $\alpha_\sigma=MaCa/(RePr)$, $Ma=\sigma_T T_* l/(\rho\nu\chi)$ is the Marangoni number, $Ca=u_*\rho\nu/\sigma_0$ is the capillary number, σ_T is the temperature coefficient of surface tension, σ_0 is the value of the coefficient of surface tension σ at the same temperature ($\sigma=\sigma_0-\sigma_T(T-T_0)$), $\alpha_{Ma}=Ma\varepsilon/Pr$, $\alpha_{Ca}=\varepsilon/Ca$, $\alpha_\tau = \bar{\rho}\bar{\nu}\bar{\upsilon}\varepsilon / \bar{h}$, $\bar{\rho}, \bar{\nu}$ are the ratio of the densities and coefficients of kinematic viscosity of the gas and liquid, $\bar{\upsilon}$ is the ratio of the characteristic gas velocity to u_*, $\bar{h}$ is the ratio of the characteristic scale of gas layer to l, $\alpha_J = \alpha\rho_s\lambda_U T_* / J_*\left(M/\left(2\pi R_g T_s^3\right)\right)^{1/2}$, α is the accommodation coefficient, ρ_s is the vapor density, J_* is the characteristic value of the vapor mass flux, M is the molecular weight, R_g is the universal gas constant, T_s is the saturated vapor temperature, $\beta_2 = Ma\varepsilon / (RePrE\bar{U})$, $\bar{U} = \lambda_U / u_*^2$, $\beta_3 = 1$, $\beta_4 = (1/\bar{\rho} - 1)/(Re\bar{U}\varepsilon^2)$, $\beta_5 = (1 - 1/\bar{\rho})^2 E^2 / (Re^2\bar{U}\varepsilon^2)$, $\beta_6 = (1 - 1/\bar{\rho})(Re\bar{U}Ca)$, $div_\Gamma \mathbf{v}$ is the surface divergence of velocity.

The no-slip conditions on the non-uniformly heated solid substrate are fulfilled. Statement of the problem must be added by the initial interface position and the initial values of the functions. The results of the parametric analysis of the problem are demonstrated in Table 1, where we present the values of the dimensionless parameters in (3).

3 The evolution equation for the thin layer

In the case when the mathematical model is based on the equations of Navier-Stokes, we have the evolution equation of the film thickness (lubrication theory approach):

$$h_t + h_x \left\{ \frac{h^2}{2}((C_0)_x - \gamma_1 \sin\alpha) + hC_1 \right\} + \left\{ \frac{h_3}{6}(C_0)_{xx} + \frac{h^2}{2}(C_1)_x \right\} + \frac{E\alpha_J}{\varepsilon}\{Ah + \Theta_0\} = 0. \tag{4}$$

Here the functions A and Θ_0 determine the heat flow regime. The function Θ_0 is given, the function A satisfied a certain differential equation. Functions C_0, C_1 can be found as follows:

$$C_0(x,t) = p^g - \alpha_{Ca}h_{xx}\left(1 - a_\sigma\Theta^0\right) + \gamma_2 h\cos\alpha, \quad C_1(x,t) = a_\tau \tau(x,t) - \alpha_{Ma}\tilde{\Theta} - (C_0)_x h + \gamma_1 h \sin\alpha.$$

In the case when mathematical modelling is based on the Oberbeck-Boussinesq equations, the evolution equation of the film thickness has the form:

$$\begin{aligned} h_t + h_x &\left\{ \frac{h^2}{2}((C_0)_x + \gamma_1 \sin\alpha\Theta_0) + hC_1 + \frac{h^4}{24}\gamma_2 \cos\alpha A_x + \frac{h^3}{6}(\gamma_2\cos\alpha(\Theta_0)_x + \gamma_1 \sin\alpha A) \right\} + \\ &+ \left\{ \frac{h^3}{6}(C_0)_{xx} + \frac{h^2}{2}(C_1)_x + \frac{h^5}{120}\gamma_2\cos\alpha A_{xx} + \frac{h^4}{24}(\gamma_2\cos\alpha(\Theta_0)_{xx} + \gamma_1 \sin\alpha A_x) + \right. \\ &\left. + \frac{h^3}{6}(\Theta_0)_x \gamma_1 \sin\alpha \right\} + \frac{E\alpha_J}{\varepsilon}\{Ah + \Theta_0\} = 0. \end{aligned} \tag{5}$$

Here the functions C_0, C_1 are found with the help of the relations

$$C_0(x,t) = p^g - \alpha_{Ca}h_{xx}\left(1 - a_\sigma\Theta^0\right) - \gamma_2\cos\alpha\left(A\frac{h^2}{2} + \Theta_0 h\right) - \frac{\gamma_1}{Bu}x\sin\alpha + \gamma_2 h\cos\alpha,$$

$$C_1(x,t) = a_\tau\tau(x,t) - \alpha_{Ma}\tilde{\Theta} - (C_0)_x h - \gamma_1 h\sin\alpha\left(A\frac{h^2}{2} + \Theta_0 h\right) - \gamma_2\cos\alpha\left(A_x\frac{h^3}{6} + (\Theta_0)_x\frac{h}{2}\right).$$

Table 1. The values of the parameters α and β

"α"	Ethanol-nitrogen	HFE-7100-nitrogen	FC-72-nitrogen	"β"	Ethanol-nitrogen	HFE-7100-nitrogen	FC-72-nitrogen
α_σ	10^{-2}	10^{-1}	10^{-1}	β_2	ε	ε	ε
α_{Ca}	$\varepsilon^3 10^6$	$\varepsilon^3 10^6$	$\varepsilon^3 10^5$	β_3	1	1	1
α_E	-10^{-3}	-10^{-2}	-10^{-2}	β_4	$\varepsilon^{-2}10^{-3}$	$\varepsilon^{-2}10^{-3}$	$\varepsilon^{-2}10^{-3}$
α_τ	$\varepsilon 10^{-2}$	$\varepsilon 10^{-2}$	$\varepsilon 10^{-2}$	β_5	$\varepsilon^{-2}10^{-6}$	$\varepsilon^{-2}10^{-5}$	$\varepsilon^{-2}10^{-5}$
α_{Ma}	$\varepsilon 10^3$	$\varepsilon 10^4$	$\varepsilon 10^3$	β_6	-10^{-2}	-10^{-2}	-10^{-2}

4 The numerical results

The evolution of the evaporating film has been investigated with use of two mathematical models (see Section 3, eqs. (4) and (5)). The periodic problems of moving of the liquid layers (ethanol and

HFE7100) accompanied by the gas flow (nitrogen) have been solved numerically on the interval [$-L$, L] with the initial disturbance $h_0=1-\delta_1 \cos(kx)$ (δ_1=0.01, $k=\pi/2$) and periodic conditions at $x=-L$, $x=L$. The non-uniform heating is realized with the help of tempearture distribution $\Theta_0 = 1 + \delta_0 \cos k_1 x \cdot \cos k_2 t$ (δ_0=0.25, $k_1=\pi/2$, k_2=2). Numerical investigations have been carried out on the basis of the finite-difference scheme of the second order of approximation.

Figure 1b presents the results of numerical study of the liquid layer dynamics in the case of low gravity. We have observed the qualitatively similar results using the alternative mathematial models and found the quantitative differences. The evaporation process is significantly more intensive for "HFE7100-nitrogen" system (see lines 3 and 4 on the Fig. 1.b). A small increase in the intensity of evaporation is observed in the case of mathematical model based on the Oberbeck-Boussinesq equations (see (5) and lines 2 and 4 on the Fig. 1.b).

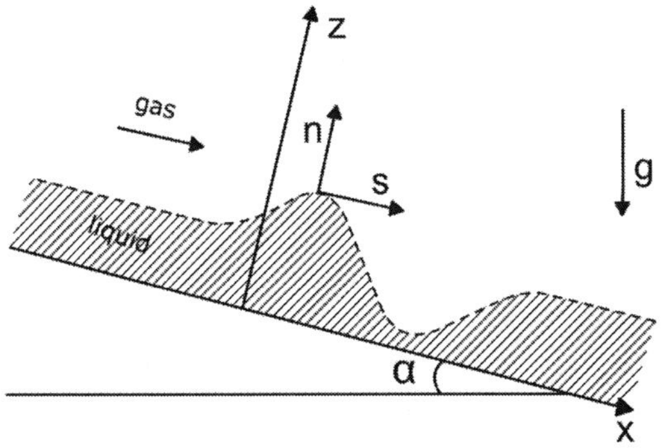

a

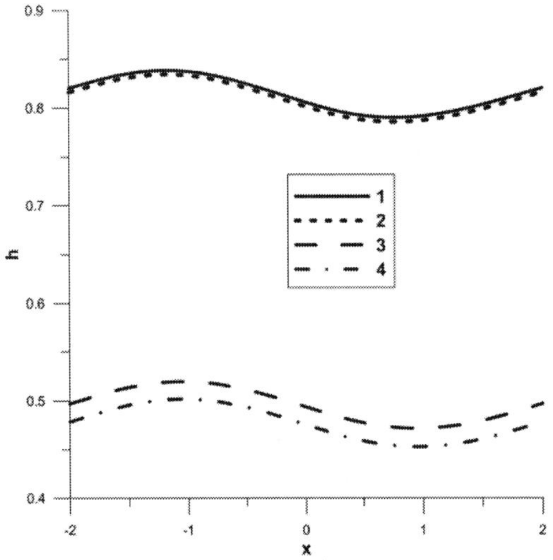

b

Figure 1. a) Geometry of flow domain; b) The behaviour of the liquid layer thickness, the case of low gravity (g=9.81·10^{-2} m/s^2), $\alpha = \pi/6$, t = 0.1: 1 – the model (4), "ethanol-nitrogen" system; 2 – the model (5), "ethanol-nitrogen" system; 3 – the model (4), "HFE7100-nitrogen" system; 4 – the model (5), "HFE7100-nitrogen" system.

Acknowledgments. The research has been supported by the Russian Foundation for Basic Research (Project No. 14-08-00163).

References

1. Yu.V. Lyulin, O.A. Kabov, Int. J. of Heat and Mass Transfer, **70**, 599-609 (2014)
2. A. Oron, S.H. Davis, S.G. Bankoff, Reviews of Modern Physics, **69** (**3**), 931-980 (1997)
3. J. Klentzman, V.S. Ajaev, Physics of Fluids, **21**, 122101-1 - 122101-9 (2009)
4. S. Miladinova, G. Lebon, Acta Mechanica, **174**, 33-49 (2005)
5. Yu.O. Kabova, V.V. Kuznetsov, O.A. Kabov, T. Gambaryan-Roisman, P. Stephan, International Journal of Heat and Mass Transfer, **68**, 527-541 (2014)
6. C.S. Iorio, O.N. Goncharova, O.A. Kabov, Microgravity Sci. Technol, **21**(**1**), 313-320 (2009)
7. C.S. Iorio, O.N. Goncharova, O.A. Kabov, Computational Thermal Sci., **3**(**4**), 333-342 (2011)
8. O.N. Goncharova, Izvestia AltGU, **73** (**1/2**), 12-18 (2012)
9. O.N. Goncharova, E.V. Rezanova, Ya.A. Tarasov, Izvestia AltGU, **81** (**1/1**), 47-52 (2014)
10. O.N. Goncharova, E.V. Rezanova, Izvestia AltGU, **81** (**1/2**), 21-25 (2014)
11. O.N. Goncharova, E.V. Rezanova, Izvestia AltGU, **85** (**1/1**), 70-74 (2015)

Investigation of behavior of the dynamic contact angle on the basis of the Oberbeck-Boussinesq approximation of the Navier-Stokes equations

Olga Goncharova[1,2,a], Alla Zakurdaeva[1,2] and Jean-Claude Legros[2]

[1] *Altai State University, Faculty of Mathematics and Information Technologies , pr. Lenina, 61, 656049 Barnaul, Russia*
[2] *Kutateladze Institute of Thermophysics SB RAS, pr. Lavrentyeva, 1, 630090, Novosibirsk, Russia*

Abstract. Flows of a viscous incompressible liquid with a thermocapillary boundary are investigated numerically on the basis of the mathematical model that consists of the Oberbeck-Boussinesq approximation of the Navier-Stokes equations, kinematic and dynamic conditions at the free boundary and of the slip boundary conditions at solid walls. We assume that the constant temperature is kept on the solid walls. On the thermocapillary gas-liquid interface the condition of the third order for temperature is imposed. The numerical algorithm based on a finite-difference scheme of the second order approximation on space and time has been constructed. The numerical experiments are performed for water under conditions of normal and low gravity for different friction coefficients and different values of the interphase heat transfer coefficient.

1 Introduction

The problems of the fluid flows with a dynamic contact angle are very important at present. Such problems occur in many applications by study of the convective multi-phase flows, flows through porous medium, of evaporative liquid drops, wetting hydrodynamics etc. (see, for example, [1]). The contact angle is defined as an angle between the two surfaces, where the free boundary touches the solid boundary. The non-stationary problems remain to be rather difficult for investigations in connection with an open problem of the dynamic contact angle, due to an appearing problem of motion of the contact line of three phases, which does not have unique closure (see [2-7]). The main question of such types of the problems relates to a non-consistency of the condition on the free boundary and the classical no-slip boundary condition on the solid wall in a neighborhood of the contact point. From the mathematical point of view this non-consistency has been considered in [8].

If the boundary walls of a liquid containing domain are moving with a constant speed, the contact angle is changing with the velocity [4, 5, 9]. Changing of this angle will also depend on a character of thermal regimes on the free and solid boundaries. Interaction of the gravitational and thermocapillary convection and their influence on behavior of the contact angle and shape of the free boundary should be investigated.

[a] Corresponding author: gon@math.asu.ru

2 Statement of the problem

Two-dimensional motion of a viscous incompressible liquid with a thermocapillary boundary is investigated numerically on the basis of the mathematical model that consists of the Oberbeck-Boussinesq approximation of the Navier-Stokes equations, kinematic and dynamic conditions at free boundary [5, 9]. The slip conditions (conditions of proportionality of the tangential stress to the difference of the tangential velocities of liquid and wall) are prescribed on the solid boundaries of the channel supporting by constant temperature. On the thermocapillary gas-liquid interface the condition of the third order for temperature is imposed. The problem is studied in the quasi-stationary formulation when the coordinate system is moving with the liquid. The flow is assumed to be symmetric that has to be taken into account in determination of the boundary conditions.

Let Ω be the flow region with the boundary $\partial\Omega$ which consists of three solid Γ_0, Γ_s and one open part Γ_f (free thermocapillary boundary). Herewith the Cartesian coordinate system is chosen so that the gravity acceleration vector **g** is directed along the longitudinal axis (here Ox, see Fig. 1a). Let the linear size of the flow domain in the y-direction l be the characteristic length. We introduce the characteristic values for the problem of the fluid flow as follows: u_* is the characteristic velocity, T_* is the characteristic characteristic temperature drop, $p_* = \rho u_* \nu / l$ is the characteristic pressure, ρ is the density, ν is the coefficient of kinematic viscosity.

After reformulation the boundary value problem in the terms of new unknown functions ψ (stream function) and ω (vorticity) we will have the following governing equations:

$$\begin{gathered}\Delta\psi = -\,\omega,\\ Re(u\omega_x + \upsilon\omega_y) = \Delta\omega + T_y\, Gr/Re,\\ RePr(uT_x + \upsilon T_y) = \Delta T.\end{gathered} \tag{1}$$

Here $u = \psi_y$, $\upsilon = -\psi_x$, $\omega = \upsilon_x - u_y$, $\mathbf{v} = (u, \upsilon)$ is the velocity vector, T is the temperature, $Re = \upsilon_* l/\nu$ is the Reynolds number, $Gr = \beta T_* g l^3/\nu^2$ is the Grasgof number, $Pr = \nu/\chi$ is the Rrandtl number, χ and β are the coefficients of thermal diffusivity and thermal expansion, respectivelly, $g=|\mathbf{g}|$.

Direct calculations lead to the boundary conditions (see [4-6]) written in the terms of ψ and ω :

$$\psi = 0 \text{ on } \partial\Omega, \tag{2}$$

$$\psi_{nn} - H\psi_n = -MaT_\tau\,,\quad \frac{1}{\sqrt{1+f'^2}}\left(-\bar{\sigma}\sigma H' + \mathrm{Ma}\,T_\tau H + \bar{g} f'\right) = Q(\psi) \ \text{ on } \Gamma_f, \tag{3}$$

where $Q(\psi) = Re(\psi_y(\psi_x)_n - \psi_x(\psi_y)_n) - (\Delta\psi)_n - 2(\psi_n)_{\tau\tau}$.

$$\bar{\alpha}\psi_{nn} + \psi_n = -S \text{ on } \Gamma_s \text{ (at } y = 0),\ \bar{\alpha}\psi_{nn} + \psi_n = S \text{ on } \Gamma_s \text{ (at } y = 1),\ \bar{\alpha}\psi_{nn} + \bar{\gamma}\psi_n = 0 \text{ on } \Gamma_0\ (x = x_0), \tag{4}$$

$$T_n = -Nu(T - T_{ex}) \text{ on } \Gamma_f, \tag{5}$$

$$T = T_s \text{ on } \Gamma_s \text{ at } y = 0 \text{ and at } y = 1,\ T = T_0 \text{ on } \Gamma_0\ (x = x_0), \tag{6}$$

$$f(0) = f(1) = 0.$$

The contact angle Φ at $y=0$ is computed with the help of formula $\Phi = \arccos\left(f'(0)/\sqrt{1+f'^2(0)}\right)$.

Here $x=f(y)$ is the parametrization of the free boundary, $H = \left(f'/\sqrt{1+f'^2}\right)'$ is the curvature of the boundary Γ_f, $Ma = \sigma_T T_*/(\rho\nu\upsilon_*)$ is the Marangoni number, $Nu = \delta l/\kappa$ is the Nusselt number, $\bar{\alpha} = \rho\nu/(\gamma l)$, $\bar{\gamma} = \gamma_0/\gamma$, $\bar{\sigma} = \sigma_0/(\rho\nu\upsilon_*) = 1/Ca$, $\bar{g} = gl^2/(\nu\upsilon_*) = Gr/(Re\beta T_*)$, $Ca = \rho\nu\upsilon_*/\sigma_0$ is the capillary number, σ is the non-dimensional surface tension ($\sigma = 1 - MaCa(T - T_0)$), S is the contact point velocity, κ is the thermal conductivity, σ_0 is the surface tension value (dimensional) at some reference temperature, σ_T is the temperature coefficient of surface tension, γ, γ_0 are the friction coefficients of the liquid and solid walls, δ is the interphase heat transfer coefficient. T_s, T_0 and T_{ex} are the given constant values of the temperature at solid walls and on the outside. The indexes n and τ denote the derivatives with respect to the normal and tangential directions. The conditions (2)-(6) written in the terms of ψ and ω are: (2) is the nonpermeability condition including the kinematic condition at free boundary, (3) are the dynamic conditions at free boundary, (4) are the slip conditions on solid walls, which follow from the assumption of momentum conservation and denote the proportionality of shear stresses to the tangential velocity on the walls). The thermal boundary regime is determined by (5) and (6). It is assumed that the constant temperature (5) is kept on the solid walls, on the free boundary the heat exchange with the atmosphere is prescribed by the Newton's law (6).

3 Numerical results

The numerical algorithm for the problem (1)-(6) based on a finite-difference scheme of second-order approximation in time and space has been constructed (see [5, 7]). This algorithm includes computation of the free boundary position f and the initial position f_0 of Γ_f defined by a rest state (at $S=0$, $\mathbf{v}=0$, $T=const$) at a given value of the static contact angle Φ_0. This value Φ_0 is determined by the type of a liquid, material of the solid wall and external gas medium. Starting with an approximation of f_0 we fix a sufficiently small $\Delta S>0$ and organize "step-by-step" computations to reach a necessary value of S. In each step we organize the iteration processes to compute the unknown functions ψ, ω, T.

The numerical experiments are performed for water under conditions of normal and low gravity. The velocity field and the size of the contact angle, as a function of the velocity of the contact point, have been computed (see in figures 1, 2 the examples of the velocity field, the free boundary positions (f) and the dependence of the contact angle (Φ) on the contact point velocity (S); the angle values are given in degrees; here the static contact angle has the value $\Phi_0=87$). Numerical experiments have been carried out for such liquid as water under moderate heating (the characteristic temperature difference is equal to $T_*=10^0\ C$). The value of the Nusselt number Nu ($Nu=0.17$) is dictated by new experiments with a drop of water (see [1]). The other value ($Nu=17$) corresponds to the interphase heat transfer coefficient (δ) that exceeds the values in [1], but allows us to compare the effects of intensity of the thermal regime at a free boundary.

The vortex structure of the flow (see Fig. 1a) and the behaviour of the free boundary (Fig. 1b) are investigated. We observe two-vortex symmetric flow structure. For higher values of the contact point velocity S the flow topology is characterized by displacement of the vortex centres to the free boundary (as shown in Fig. 1a). The flow picture (Fig. 1a) and the examples of the free boundary shape (Fig. 1b) are presented in the case of thermal boundary regime characterized by the non-dimensional values $T_s = 8$, $T_0 = 1$ (or of type "8x1") and $T_{ex} = 8$, $Nu=0.17$ (see (5), (6)) under conditions of low ($g=9.81$(cm/s^2); $Gr=230$) and normal ($g=981$(cm/s^2); $Gr=23000$) gravity.

The influence of changing of the values of friction coefficients γ_0, γ coefficients ($\gamma_0=\gamma=100$ and $\gamma_0=\gamma=1$ (g/(cm^2 s))), gravity acceleration g ($g=9.81$ and $g=981$ (cm/s^2)) and Nusselt number Nu ($Nu=0.17$ and $Nu=17$) on the dynamics of the contact angle has been studied numerically (Fig. 2).

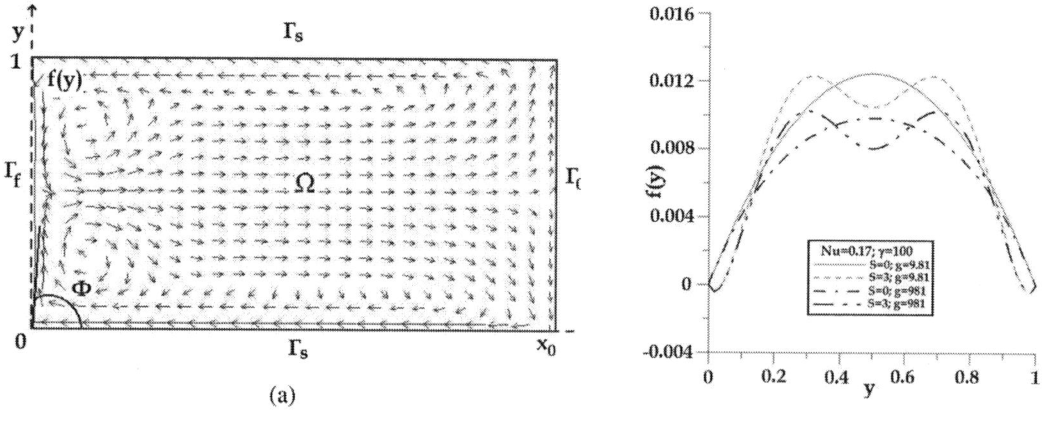

(a)

(b)

Figure 1. (a) Velocity field: Nu = 0.17, $g=981$ (cm/s^2), $T_{ex}=T_s=8$, $T_0=1$, $\gamma_0=\gamma=100$ (g/(cm^2 s)). (b) Free boundary: $Nu=0.17$; $\gamma_0=\gamma=100$ (g/(cm^2 s)); $T_{ex}=T_s=8$, $T_0=1$; $g=9.81$ (cm/s^2): $S=0$ – solid line, $S=3$ – dashed line; $g=981$: $S=0$ – dash-dot line, $S=3$– dash-two-dots line.

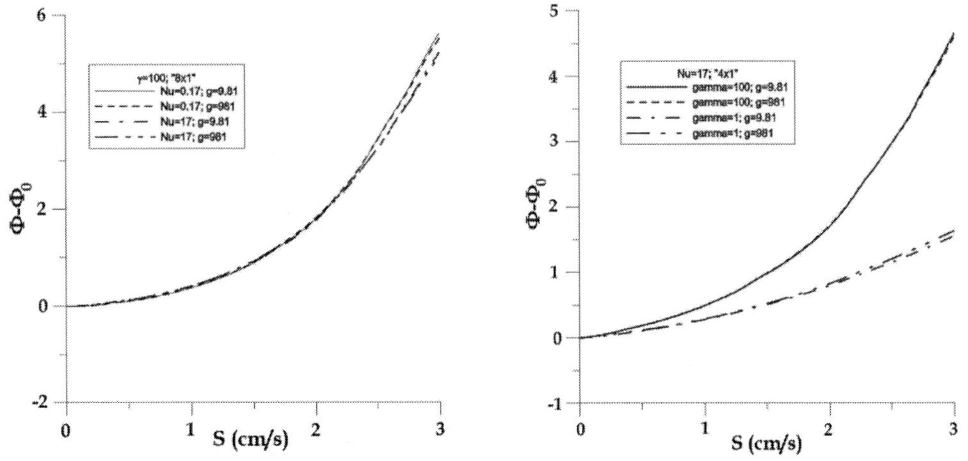

(a) (b)

Figure 2. Dynamic contact angle. Comparison of the effects of gravity acceleration. (a) "8x1" thermal regime on solid walls, $T_{ex}=8$, $\gamma_0=\gamma=100$ (g/(cm^2 s)), $Nu=0.17$; $Gr=230$ – solid line, $Gr=23000$ – dashed line. (b) "4x1" thermal regime on solid walls, $T_{ex}=1$, $Nu=17$: $\gamma_0=\gamma=100$ (g/(cm^2 s)), $g=9.81$ (cm/s^2) – solid line; $\gamma_0=\gamma=100$ (g/(cm^2 s)), $g=981$ (cm/s^2) – dashed line; $\gamma_0=\gamma=1$ (g/(cm^2 s)), $g=9.81$ (cm/s^2) – dash-dot line; $\gamma_0=\gamma=1$ (g/(cm^2 s)), $g=981$ (cm/s^2) – dash-two-dots line.

The presented results demonstrate the differences in the contact angle behavior with respect to the different values of the contact point velocity, friction coefficients $\gamma_0=\gamma=100$ and $\gamma_0=\gamma=1$ (g/(cm^2 s)), gravity acceleration $g=981$ and $g=9.81$ (cm/s^2) and to an intensity of the thermal boundary regimes (Fig. 2a, 2b). We observe the increasing character of the contact angle dependence on the contact point velocity. In the case of investigating boundary regimes with rather high values of the external temperature $T_{ex}=8$ and high temperature on the lateral walls Γ_s (Fig. 2a) we have found slight discrepancies in the contact angle dynamics under conditions of normal (Fig. 2a, dashed and dash-three-dots lines) and low (Fig. 2a, solid and dash-dot lines) gravity. The similar behavior is observed both for $Nu=0.17$ and $Nu=17$ in Fig. 2a and in the case of more moderate thermal regime presented in Fig. 2b for $Nu=17$ (here $T_{ex}=1$ and the thermal regime with $T_s=4$, $T_0=1$ or of type "4x1" on solid walls has been modeled). At the same time we can speak about more intensive increase of the dynamic contact angle with respect to the contact point velocity in the case of larger values of the friction coefficient (here for $\gamma_0=\gamma=100$ in comparison with $\gamma_0=\gamma=1$ (g/(cm^2 s)), see Fig. 2b).

Acknowledgments. The study was supported by the Russian Science Foundation (grant RSF 15-19-20049).

References

1. E. Ya. Gatapova, A.A. Semenov, D.V. Zaitsev, O.A. Kabov, Colloids and surfaces A, Psysicochem. Eng. Aspects **441**, 776 (2014)
2. V.S. Ajaev, G.M. Homsy, Annu. Rev. Fluid Mech. **38**, 277-307 (2006)
3. C. Baiocci, V.V. Pukhnachov, J. Appl. Mech. Tech. Phys. **31**, 185-196 (1990)
4. D. Kroener, Asymptotische Entwicklungen fuer Stroemungen von Fluessigkeiten mit freiem Rand und dynamischem Kontaktwinkel, Bonn, Preprint/Univ. Bonn **N. 809** (1986)
5. W. Doerfler, O. Goncharova, D. Kroener, ZAMM **82**, 1-10 (2002)
6. P. G. de Gennes, Rev. Mod. Phys. **57**, p. 827 (1985)
7. E. B. Dussan V, Annu. Rev. Fluid Mech. **11**, p. 371 (1979)
8. V.V. Pukhnachov, V.A. Solonnikov, J. Appl. Mech. Tech. Phys. **46(6)**, 771-779 (1982)
9. O.N. Goncharova, A.V. Zakurdaeva, Journal of Siberian Federal University. Mathematics & Physics (accepted for publication) (2016)

The Effect of Film Temperature on crater morphology during the impact of a single droplet

D. B. Hann, E.R. Underhill and A. Mitchell

Faculty of Engineering, university of Nottingham, UK

Abstract. The effect of varying the film temperature from 15C to 60 C is investigated using high speed video imaging. It is demonstrated that increasing the temperature difference between the droplet and film liquid changes the size and frequency of the secondary droplets. The impact can be split into three regimes. In the crater expansion regime, the crater follows a self-similar behaviour. In the second stage, the crater becomes deeper and wider at higher temperatures possibly due to decrease in the viscosity or surface tension. It is seen that the crater collapse is less dependent on the temperature and occurs at fixed time for a particular Webber number.

1 Introduction

The phenomenon of droplets impacting onto both liquid and solid surfaces has been of great interest since the early 1900s, one of the first to investigate the characteristics of splashing being [1]. The interest in droplet impact has stemmed from its appearance in many engineering applications, a few examples listed by [2] include; ink-jet printing, spray cooling of turbine blades and internal combustion engines. Within internal combustion engines the lubricating oil is sprayed from the rotating parts onto the interior walls, quickly creating an oil film. The following oil droplets will then inevitably splash when impacting onto the film, creating secondary drops. Minimising these secondary drops is of great importance to engine manufacturers as they are believed to reduce efficiency. This study is therefore relevant to industry by developing an understanding to how temperature may affect the outcome of droplet impact onto liquid films.

There are several reviews of the droplet impact outcomes, such as [3] , and there are a number of theories about the crater evolution [4] for droplets of same temperature and liquid impacting on a film [5]. However there are few studies on droplets of different temperature to the film [6, 7]. This work is part of an experimental study to provide data for comparison with models to confirm or deny if they are still valid in the case where the film liquid has different properties.

2 Experimental methodology

The apparatus for the experiment is shown in figure 1. The control variables that were kept constant throughout the experiment were; the droplet diameter and the film thickness. The independent variables were; the impact velocity and the film temperature, and the dependent variables were the crater dimensions and the splash regimes.

Five different film temperatures were investigated; 15°C, 30°C, 40°C, 50°C and 60°C. At each temperature five different initial heights of the primary droplet were used in increments of 50 mm from 100 mm to 300 mm. Each droplet was repeated five times in order to demonstrate the repeatability and the reliability of the results. In total there were 125 droplets analysed.

The videos from the high-speed camera were analysed using MATLAB software similar to that of the work from [5]. Any measurements from the images were originally taken in pixels and then calibrated to mm. The precision was 1 pixel which translated on average to 0.065 mm. Each droplet size and velocity was individually measured to eliminate errors from movement of the camera or film between tests. The dimensions as well as time (t) were non-dimensionalised for comparison with the work from [8] and [4]using the following relationships.

$$z^* = \frac{Z}{D}, \Omega = \frac{w}{D}, \tau = \frac{tU}{D}$$

Where; z is the crater depth, w is the crater width, t is the time after impact, D is the droplet diameter

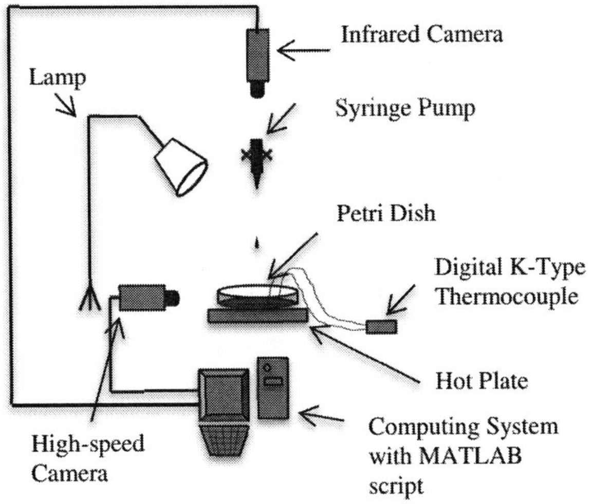

Figure 1: Diagram of experimental set up used

and U is the droplet velocity measured just before the impact.

3 Results

The effect of the film temperature on the impact can be easily determined in the sequences of images shown in figure (2). At 50 C it is clear that the secondary droplets generated from the jet formed after collapse of the impact crater are smaller and more frequent. This might be expected since the surface tension and viscosity of the film liquid are decreased by an increase in temperature.

For each of the five sequences of droplet impact parameters, the location of the base of the impact crater and the width of the crater can be determined from the Matlab code. These were averaged and it was seen that the impact was highly repeatable with little variation between each of the repeats. The average of the crater depth and width are shown in figure 3. In both cases it can be seen that in the inertial self-similar region ($\tau = \frac{tU}{D} < 5$) is not affected by the variation in the film temperature. The effect of the film temperature becomes evident during the collapse of the impact crater which is dominated by the Weber number. When the film temperature increases, the crater becomes deeper and wider before it starts to collapse in response to the lower surface tension and viscosity. There is an indication that the time for crater collapse is the same for all temperatures.

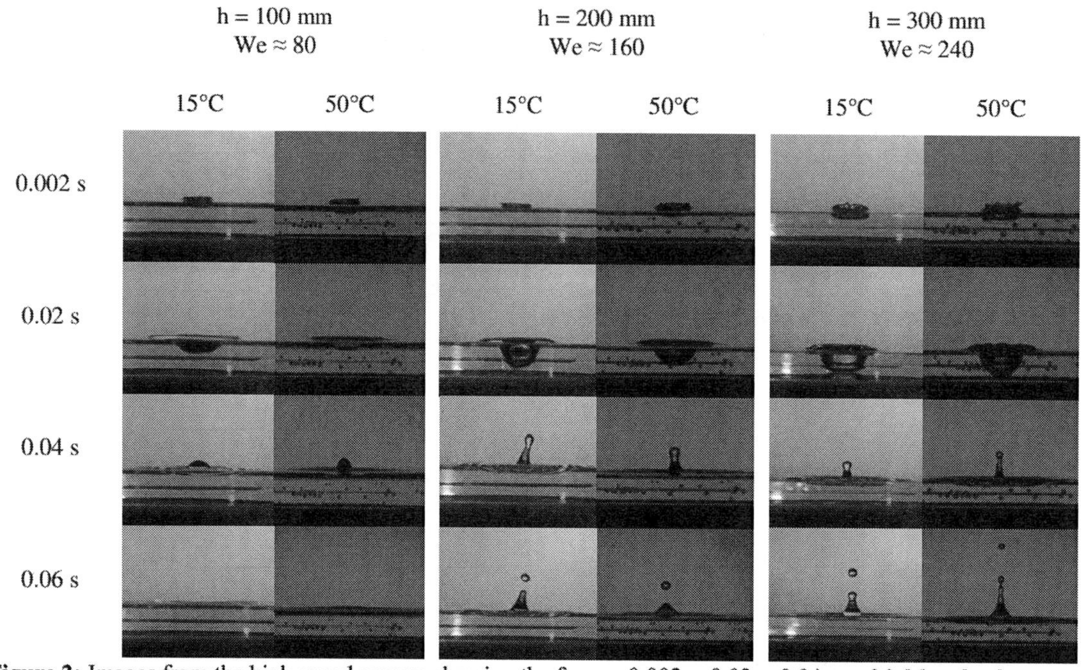

Figure 2: Images from the high-speed camera showing the frames 0.002 s, 0.02 s, 0.04 s and 0.06 s after impact for the 15°C and 50°C films using droplets with Weber numbers 80, 160 and 240

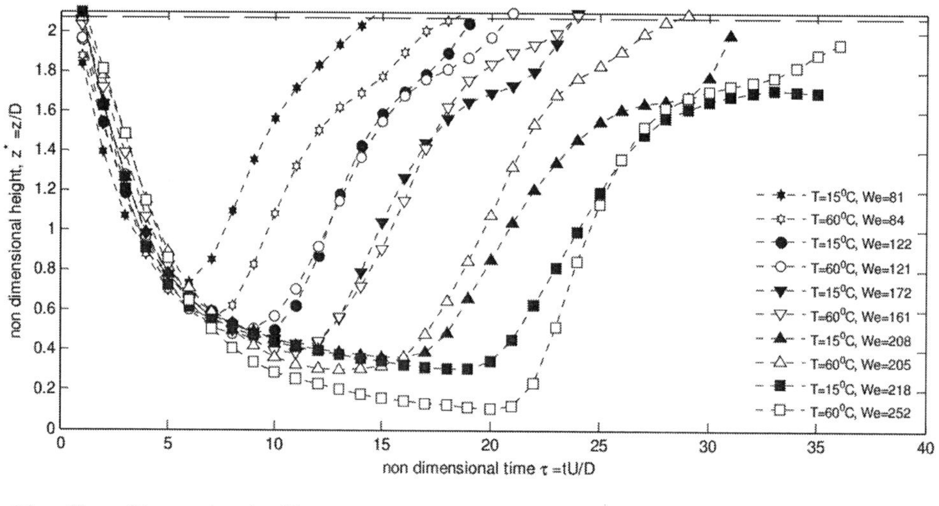

Figure 3. The effect of increasing the film temperature on the depth of the crater and its evolution with time.

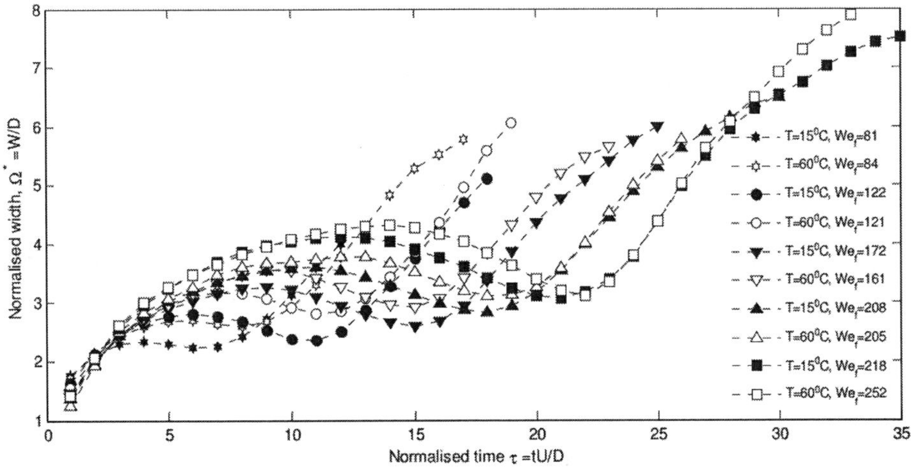

Figure 4. The effect of film temperature on the crater width and evolution of the crater with time.

4 Conclusions and discussion

The experiment outlined in this paper has demonstrated that changing the film temperature has an effect on the droplet impact outcome. Video evidence shows that secondary droplets are more frequent, but smaller if the temperature is increased. Three regimes can be identified in the impact; the crater expansion, crater slowing and crater collapse. Film temperature is shown to have little effect in the initial stages of the droplet impact. In this region, more commonly called the "inertial self-similar" region, the impact is dominated by momentum exchange [4], and the track of the crater evolution follows a similar pattern for all impact velocities and film temperatures. In the second region, the crater becomes deeper and wider as the temperature increase. In the collapsing regime, the crater becomes self-similar again for constant Webber number. These results will be used to understand in better detail the effect of cold droplets impacting on a hotter film.

References

1. Worthington, A.M., Nature,. **16**, 165-166 (1877).
2. Rein, M., Fluid Dynamics Research, **12**(2), 61 (1993).
3. Yarin, A.L., Annual Review of Fluid Mechanics. 159 (2006).
4. Bisighini, A., et al., Physical Review E - Statistical, Nonlinear, and Soft Matter Physics, **82** 2010.
5. Hann, D.B., et al., Experiments in Fluids, **57**, 1-12 (2016).
6. Manzello, S.L. and J.C. Yang, Experiments in Fluids, **32**, 580-589 (2002).
7. Manzello, S.L. and J.C. Yang, Physics of Fluids, **15**, 257-260 (2003).
8. Berberović, E., et al., Physical Review E - Statistical, Nonlinear, and Soft Matter Physics, **79**, 2009.

Electrostatic effects in the apparent contact line region under a vapor bubble

Christiaan Ketelaar[1] and Vladimir Ajaev[2,3,a]

[1]*Department of Mathematical Sciences, University of Delaware, Newark, DE 19716, USA*
[2]*Department of Mathematics, Southern Methodist University, Dallas, TX 75275, USA*
[3]*Kutateladze Institute of Thermophysics SB RAS, Lavrentyev Ave., 1, Novosibirsk, 630090, Russia*

Abstract. Evaporation and local interface shapes are investigated in the apparent contact line region formed when an isolated vapor bubble is attached to a heated substrate. The model takes into account the electrostatic effects due to presence of ions in the liquid, as appropriate for water and aqueous solutions. For the apparent contact line region under a stationary bubble, local interface shapes are obtained for different parameter values. The model is then applied to the case of an expanding vapor bubble.

1 Introduction

Studies of heat transfer and viscous flow dynamics near apparent contact lines are important for a number of applications such as pool boiling, flow boiling, micro heat pipes, microfluidic actuation, and many others. Present work is motivated by a configuration of a vapor bubble on a heated substrate, as sketched in Fig. 1. Since pool boiling process starts with nucleation and growth of individual bubbles at the solid substrate, the configuration considered in the present study can provide important insights into the physics of boiling phenomena. Previous studies indicate that the part of the substrate under the bubble that appears dry on macroscale is often covered by a micro- or nanoscale adsorbed film. Thus, what appears to be a line of contact between the bubble surface and the dry substrate is in fact a transition region between the curved bubble surface and the flat ultra-thin film, as schematically shown in the figure. Suppression of evaporation in the ultra-thin film covering the heated substrate is explained by the action of the London-van der Waals dispersion forces. The transition region in this context is referred to as an apparent contact line. Numerous theoretical and experimental studies of evaporation near apparent contact lines are reviewed e.g. in Wayner [1] and Ajaev and Homsy [2]. While most of these studies address the situation when the apparent contact line is stationary, the approach can be extended to the case of moving contact lines as well, as was first shown by Ajaev et al. [3]. An important advantage of the apparent contact line approach is that it allows one to avoid dealing with a non-integrable shear-stress singularity encountered in other mathematical formulations. Such singularity leads to the unphysical conclusion about the resistance

[a] Corresponding author : ajaev@smu.edu

force near the moving contact line being infinite; the approach of Ajaev et al. [3] allows one to relax the singularity and thus leads to finite value of resistance force at the contact line.

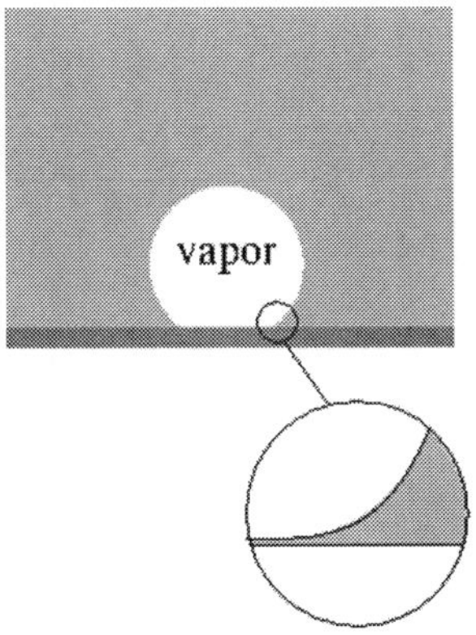

Figure 1. A sketch of a small isolated vapor bubble of radius R_0 formed on a heated substrate; the enlarged view of the apparent contact line region is shown.

The assumption of ultra-thin films being dominated by the London-van der Waals forces [1,2] is appropriate for describing experimental data with liquids such as pentane and hexane, but not in water and aqueous solution. The latter situation was investigated in the pioneering experimental study of Mazzoco and Wayner [4]. They considered evaporation of water and aqueous solutions of different concentrations. The general framework for mathematical modeling for this situation was developed in our previous work [5]. The objective of the present study is to use this framework to conduct additional parametric studies of the case of stationary contact line and then extend the approach to the case when the bubble is slowly expanding.

2 Formulation

Since the key novel aspect of our modeling approach is the addition of electrostatic effects, let us first discuss the model of electric field in the liquid. Liquid is assumed to be an aqueous solution, so electrical double layers of characteristic thickness equal to the Debye length are formed at interfaces. We use the standard definition of the Debye length, appropriate for 1:1 symmetric electrolyte solution at a temperature T,

$$\lambda_D = \left(\frac{\varepsilon k_B T}{2 n_0 e^2} \right)^{1/2} \tag{1}$$

Here e is the elementary charge, k_B is the Boltzmann constant, n_0 is the equilibrium concentration of ions in the bulk, ε is the dielectric permittivity. The electric field in the liquid is determined from the solution of the classical Poisson-Boltzmann equation,

$$\nabla^2 \psi = \kappa^2 \sinh \psi \tag{2}$$

We use $k_B T / e$ as the scale for the electric potential, so that the variable ψ is nondimensional, with its value at the liquid-vapor interface denoted by $\tilde{\psi}$; κ is the ratio of the characteristic vertical length scale d and the Debye length. Equation (2) is solved numerically with the conditions of constant scaled substrate potential (equal to -3 in the present study) and liquid-vapor interface charge density q; we use $\kappa=3$, $q=-0.04$ in all simulation results reported below.

Standard lubrication approximation [6, 7] is used for the viscous flow, with the assumptions of small capillary number and small ratio d/R_0. The film thickness (scaled by d) is then determined by the evolution equation,

$$h_t + J + \frac{1}{3}\left[h^3\left(h_{xx} + \kappa^2 Q\cosh\tilde{\psi} + \frac{\alpha}{h^3}\right)_x\right]_x - \frac{1}{2}\kappa Q q\left[h^2\tilde{\psi}_h h_x\right]_x = 0 \quad (3)$$

Here the scaled evaporative flux is given by

$$J = \frac{\delta\left(\frac{1}{2}\kappa^2 Q q^2 - h_{xx} - \kappa^2 Q\cosh\tilde{\psi} - \frac{\alpha}{h^3}\right) + T_0}{K + h} \quad (4)$$

The parameters K, δ arise from the nondimensional form of the so-called non-equilibrium condition at the interface, discussed e.g. in Ajaev and Homsy [2], α is the scaled Hamaker constant, T_0 is the nondimensional superheat. Detailed derivation of equations (3)-(4) can be found in our previous study [5]. Note that we account for the electrostatic contribution to stresses in the fluid at the interface not only in the normal stress, but also in the shear stress. The effects of London-van der Waals forces and the electrostatic interactions are added together in the formula for disjoining pressure, an approach that has an advantage due to its simplicity but also some limitations since in reality different components of disjoining pressure can be coupled together in a complicated manner. An alternative approach, free from these limitations, would be to consider potentials of intermolecular interactions which incorporate all relevant physics, as was done e.g. in Gatapova et al. [8].

Equation (3) with the expression for flux given by (4) in our approach has been solved numerically using a finite-difference approach for both steady and expanding/contracting bubbles.

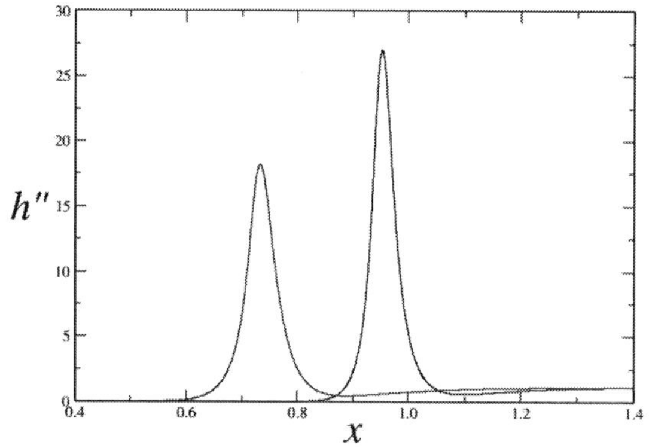

Figure 2. Interfacial curvature in the lubrication-type approximation as a function of local coordinate for $Q = 0.1$, $T_0=0.1$, $\delta=10^{-4}$, $\alpha=10^{-3}$, $K=0.1$ (black line) and $K=0.2$ (blue line).

3 Results and discussion

Changes in the interface shapes in the apparent contact line region are best visualized by plotting the curvature as a function of coordinate. Let us first discuss such plots for the case of steady bubbles. Figure 2 illustrates how the curvature profile changes with the non-equilibrium parameter, K. In our previous work, the value of this parameter was fixed at 0.1, based on the assumption that the accomodation coefficient for evaporation at the interface is equal to unity [5]. However, in reality this coefficient can vary over a wide range, possibly several orders of magnitude, as discussed e.g. in the well-known paper of Marek & Straub [9]. Thus, it is important to study the effect of variation of the accomodation coefficient on the interface shape; in our formulation such variation is equivalent to changes in K. Fig. 2 clearly shows that increasing this parameter leads to significantly smaller value of curvature and to widening of the profile, i.e increase in the width of the transition region between the macroscopic meniscus and the adsorbed film. Also, the apparent contact line shifts to the left. The decrease in the curvature can be explained by the fact the local evaporative flux, defined above by equation (4), is not as large as at the smaller K and therefore the local curvature increase needed to maintain the steady-state configuration is less. Overall, the qualitative shape of the curvature profile remains the same, with a well-defined maximum in the transition region and a point of local minimum to the right from the location of the maximum.

For heat transfer applications, especially pool boiling, it is important to study the dynamics of contact line evaporation for bubbles which are expanding rather than steady. Such expansion results in a reduction of curvature away from the transition region with time. In our formulation, such curvature is the boundary condition at the right end point of the computational domain. To account for possibility of bubble expansion, we made this boundary condition time-dependent, with the constant rate of reduction of curvature equal to 0.01 in our nondimensional units. The initial curvature is set to unity. The resulting curvature profiles are shown in Fig. 3 at three different moments in time. Clearly, the location of the contact line region shifts to the left, but the shape of the profile does not change significantly, with just a small increase in the local curvature. This suggests that the local interface profile on that time scale quickly adjusts to the shape which is close to the steady one. In fact, the transient type behaviour can be seen during the initial stages of evolution, when the shape changes are slightly faster than in the quasi-steady regime corresponding to the blue and green lines in Fig. 3.

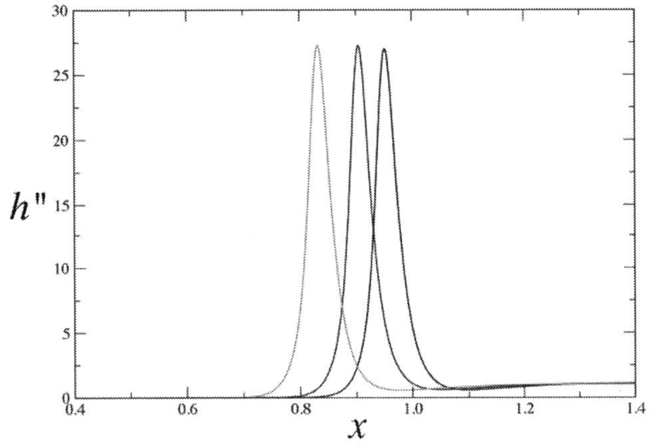

Figure 3. Local interfacial curvature profiles for the case of expanding bubble at time moments $t = 0.1$ (black line), $t = 5$ (blue line), and $t = 10$ (green line) for $K = 0.1$ and all other parameter values being the same as in Fig. 2.

The work was supported by the Russian Ministry of Education and Science (project identifier RFMEFI61314X0011).

References

1. P.C. Wayner, AIChE J. **45**, 2055 (1999)
2. V.S. Ajaev and G.M. Homsy, Annu. Rev. Fluid Mech., **38**, 277 (2006)
3. V.S. Ajaev, G.M.Homsy, S. J. S. Morris, J. Colloid Interface Sci., **254,** 346 (2002)
4. R.R.Mazzoco and P. C. Wayner, J. Colloid Interface Sci., **214**, 156 (1999)
5. C. Ketelaar and V.S. Ajaev, Phys. Fluids , **27,** 112110 (2015)
6. A. Oron, S.G.Bankoff and S.H. Davis, Rev. Mod. Phys., **69**, 931 (1997)
7. R.V. Craster, O.K. Matar, Rev. Mod. Phys., **81**, 1131 (2009)
8. E.Y.Gatapova, I.A. Graur, F. Sharipov, O.A. Kabov, Int. J. Heat and Mass Transfer, **83**, 235 (2015)
9. R. Marek, J. Straub, Int. J. Heat Mass Transfer, **44**, 39 (2001)

Convection flow study within a horizontal fluid layer under the action of gas flow

Aleksei Kreta[1,2,a]

[1] Institute of Thermophysics SB RAS, 1, Lavrentiev Ave, 630090, Novosibirsk, Russia
[2] Novosibirsk State University, 2, Pirogov Ave, 630090, Novosibirsk, Russia

Abstract. Experimental investigation of convective processes within horizontal evaporating liquid layer under shear–stress of gas flow is presented. It is found the structures of the convection, which move in opposite direction relative to each other. First convective structure moves in reverse direction with the flow of gas, and the second convective structure moves towards the gas flow. Convection flow within the liquid layer is registered with help of PIV technique. Average evaporation flow rate of Ethanol liquid layer under Air gas flow is measured. Influence of the gas velocity, at a constant temperature of 20 ^{0}C, on the evaporation flow rate has been studied.

1 Introduction

The processes of convection, accompanied by evaporation at the interface, are actively studied experimentally [1-4], numerically [4-6] and theoretically in the present time [6-8]. Scientific activity in this direction is determined by the experiments in the frame of the CIMEX project of the European Space Agency, [9]. Most often the liquid evaporates into air or more generally an inert gas, which implies a limitation of the evaporation rate by diffusion of vapor into the inert gas, [10-12]. However, as studied in [10], the presence of an inert gas also strongly favors surface-tension-driven (thermal Marangoni) instabilities in the liquid, which can enhance the heat transfer through the liquid phase, hence the evaporation rate. Cross-influence of the evaporation process on convection within the liquid layer under the action of the inert gas flow can have a quite complicated character. The evaporation from the liquid surface induces a cooling of the gas-liquid interface. It causes a temperature gradient between the heated bottom and the gas-liquid interface. The temperature gradient leads to the appearance of the buoyancy convection within the liquid layer. The inert gas flow induces the shear-stress at the gas-liquid interface. The liquid surface can be moved in the co-current direction of the gas flow. A strong evaporation at the initial section of the gas-liquid interface leads to an interfacial temperature gradient. Thermocapillary effect due to the interfacial temperature gradient induces motion of the liquid at the interface countercurrent to the gas flow. It is evident that the structure of the convective flow within the liquid layer should significantly depend on the various parameters of experiment like the inert gas flow rate, the gas and liquid temperature and the size of the liquid layer.

The main aim of this work is to study of the features of the convective fluid flows within a horizontal layer under the action of gas flow.

[a] Corresponding author: alexy.slav@yandex.ru

DOI: 10.1051/matecconf/20168400017

2 Experimental setup

A schematic of the experimental rig is shown in Fig. 1. The experimental rig consists of the fluid cell, gas/liquid supply circuits, the data acquisition system, thermal stabilization system and optical systems. The pure gas arrives into the inlet of the gas channel of the test cell from the compressor. The flow controller maintains the flow rate at the inlet of the fluid cell. To supply the working liquid into the fluid cell the syringe pump is used. The liquid is evaporated into the gas phase. The flow rate of the gas-vapor mixture is measured by the flow meter at the outlet of the test cell. Temperature in the experimental rig is measured by thermistors and thermocouples. The temperature difference between the liquid and the gas flows at the entrance to the test cell is maintained less than 0.1°C. An optical Shlieren technique is used to control the flatness of the liquid surface [1]. Optical PIV method is used for the visualization of convection in the liquid layer.

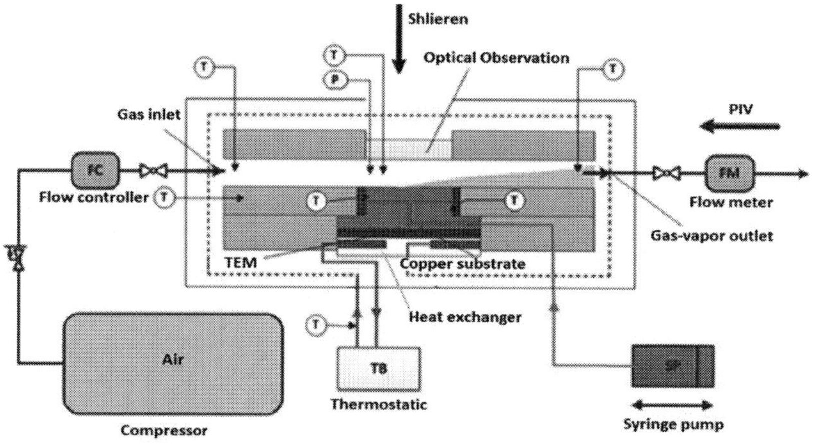

Figure 1. Scheme of the experimental rig.

The test cell consists of several interconnected units forming a rectangular gas channel with height of 3 mm and width of 40 mm, as well as a liquid pool with size of 40x40 mm. A metallic plate of stainless steel is installed between the gas channel and the liquid cell. The metallic plate has a square cut out in the center. The working fluid is filled to liquid pool through special channels in the test cell. The temperature of the liquid seems to be equal to the temperature of the copper substrate. The Pelletier elements and water-cooling heat exchanger thermostabilize the copper substrate. The walls of the liquid cell are made of transparent Plexiglas. This design gives the possibility to use the PIV technique. The height of the liquid in the cell varies from 1 to 10 mm.

3 Results

In Fig. 2 the visualization of the convection flow is presented. The images were obtained using PIV technique and demonstrated circulation of the liquid in the liquid cell. Size of the particles (white dotes on the images) was 5 micrometre. The experiment has been performed for the gas flow rate of 1000 ml/min (gas velocity - 0.13 m/s) and the gas/liquid temperature of 20 °C. Ethanol and Air were used as working fluids. The area of evaporation was 100 mm^2. As one can see in Fig. 4 there are two convective vortexes within the liquid layer. The first vortex at the initial area of the gas-liquid contact is moving in counter-current direction of the gas flow. Governing factor of this convective vortex is the thermocapillary effect due to the intensive evaporation, i.e. temperature decreasing near the attack border of the liquid. Such convective flow has been theoretically predicted in [5]. In [5] the exact solutions for different types of thermal boundary conditions have been obtained. An evaporation effect through the gas–liquid interface is modelled qualitatively with the help of a heat transfer condition.

Thus, the fact of the motion of gas-liquid interface in counter-current direction to the gas flow has been proved experimentally. The shear stress on the liquid surface τ_{sur} caused by the thermocapillary effect can be estimated as:

$$\tau_{sur} = \frac{\partial \sigma}{\partial T} \cdot \frac{\partial T}{\partial x}$$

The value of this shear stress has to be proportional to the evaporation rate. This thermocapillary stress acts opposite to the gas flow and moves liquid at the gas-liquid interface in the direction of the attack border of the liquid opening. The motion of the second vortex is coincide with the gas flow direction. It is induced by the gas shear stresses and by reverse flow of the first vortex.

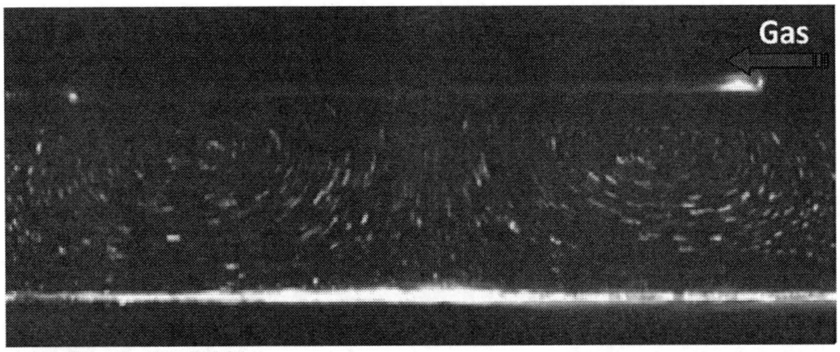

Figure 2. Convective flow in the fluid layer

Dependency of the evaporation mass flow rate of liquid layer on the average gas velocity is plotted in Fig. 3. The evaporation mass flow rate is calculated per unit area. As shown in Fig. 3 the evaporation flow rate significantly depends on the gas flow velocity and growth of the evaporation flow rate is observed with increasing of the gas flow velocity. First of all it is because the gas flow in the channel removes the vapor generating near the gas-liquid interface at the evaporation process. The vapor concentration on the free interface corresponds to the pressure of the saturated vapor at the temperature of the interface. The vapor is transported by forced convection and diffusion from the gas-liquid interface. The concentration boundary layer is formed. Evaporation rate is enhanced with increasing of the gas flow velocity due to increase of the vapor concentration gradient in the gas phase.

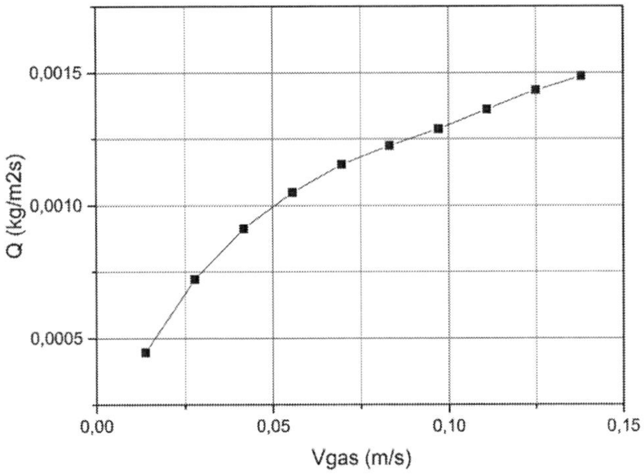

Figure 3. Evaporation flow rate versus the average gas velocity. Liquid depth is equal to 3mm.

4 Conclusion

Experimental study of the convection within the liquid layer caused by intensive evaporation at the interface and action of the gas flow has been performed. Visualization of the convection in the liquid layer has been carried out with the help of PIV method. It is found formation of two symmetrical and inverse convective vortexes within the liquid layer. The first vortex at the initial area of the gas-liquid contact is moving in counter-current direction of the gas flow. Governing factor of this convective vortex is the thermocapillary effect due to the intensive evaporation, i.e. temperature decreasing near the attack border of the liquid. The fact of the motion of gas-liquid interface in counter-current direction to the gas flow has been proved experimentally. The motion of the second vortex is coincide with the gas flow direction. It is induced by the gas shear stresses and by reverse flow of the first vortex.

The work was financially supported by the Russian Ministry of Education and Science (Project identifier RFMEFI61314X0011).

References

1. Yu.V. Lyulin, O.A. Kabov, Int. J. Heat and Mass Transfer, **70** 599 (2014)
2. Yu.V. Lyulin, D. V. Feoktistov, I. A. Afanas'ev, E. S. Chachilo, O. A. Kabov, G. V. Kuznetsov, Technical Physics Letters **41** (7) 665 (2015)
3. Yu.V. Lyulin, O. A. Kabov, Technical Physics Letters, **39**, (9) 795-797 (2013)
4. B. Scheid, J. Margerit, C.S. Iorio, L. Joannes, M. Heraud, P. Queeckers, P.C. Dauby, P. Colinet, Experiments in Fluids, **52** 1107 (2012)
5. H. Machrafi, C.S. Iorio, P.C. Dauby, Interfacial Phenomena and Heat Transfer, **2**, 199 (2014)
6. O.N. Goncharova, O.A. Kabov, Int. Journal of Heat and Mass Transfer, **53**, 2795 (2010)
7. O.N. Goncharova, M. Hennenberg, E.V. Rezanova, O.A. Kabov, Interfacial phenomena and heat transfer, **1**, 317 (2013)
8. O.N. Goncharova, E.V. Rezanova, Yu.V. Lyulin, and O.A. Kabov, Thermophysics and Aeromechanics **22** (5), 631-637 (2015)
9. B. Toth, Microgravity and Technology **24** 189 (2012)
10. B. Haut, P. Colinet, Journal of Colloid and Interface Science, **285**, 296 (2005)
11. E. Sultan, A. Boudaoud, M. Ben Amar, Journal of Fluid Mechanics **543** 183 (2005)
12. P. Colinet, J.C. Legros, M.G. Velarde, *Wiley–VCH Berlin* (2001)

Dynamics of droplets moving on a solid surface: Lattice Boltzmann simulations

Alexander Kupershtokh[1,2,a]

[1]*Lavrentyev Institute of Hydrodynamics SB RAS, Acad. Lavrentyev ave., 15, Novosibirsk, 630090, Russia*
[2]*National Research Novosibirsk State University, Pirogova Str., 2, Novosibirsk, 630090, Russia*

Abstract. The three-dimensional dynamics of droplets on inclined and vertical walls in a gravity field is simulated. The dependence of a contact angle on a velocity of contact line is obtained for slow regime and smooth surface. The simulations of a rupture of the liquid films on the nonwettable solid substrate due to the thermocapillary effect (Marangoni effect) are also carried out. The lattice Boltzmann method is successfully implemented for computer simulation of these three-dimensional problems.

1 Introduction

The understanding of the behaviour of thin liquid films and droplets placed on the surface of solid substrates is very important for modern technologies.

A static contact angle is closely related with forces of interaction of molecules of fluid and a solid substrate. Even for sessile and pendant drops on inclined surfaces, a contact angle depends on an azimuthal angle and changes from the advancing point to the receding point. In general case, the contact angle depends on a local curvature radius of the contact line. For the stable shape of 3D static drop on an inclined surface, the Young–Laplace equation is valid

$$2\sigma\kappa_{\rm m} = \Delta p_* + (\rho_{\text{ж}} - \rho_{\text{г}})(\mathbf{g}\cdot\mathbf{z} + \mathbf{g}\cdot\mathbf{x}), \qquad (1)$$

where σ is the surface tension of liquid, $\kappa_{\rm m} = (\kappa_1 + \kappa_2)/2$ is the mean curvature of the liquid-air interface, $\mathbf{z} = \mathbf{n}z$, $\mathbf{x} = \boldsymbol{\tau}x$. Here, $\mathbf{n}$ and $\boldsymbol{\tau}$ are the unit normal vector and unit tangent vector to the solid surface.

A large number of experiments showed that for motion of contact line along the horizontal solid substrate, the advancing contact angle is greater than the receding angle (hysteresis). In general case, the contact angles depend also on the velocity of a contact line. Moreover, in several works, the phenomenon of dynamic contact angles and a singularity at a contact line were investigated.

A non-stationary motion of a drop can only be simulated numerically. The viscous flow of fluid with surface tension of liquid and the interaction of a liquid with a solid wall should be described in computer simulations. The most important problem is the interaction forces between fluid and solid at the interface. This phenomenon can be described better using the mesoscopic methods that are closer to the nature of such interaction than macroscopic approaches.

[a] Corresponding author : skn@hydro.nsc.ru

2 The lattice Boltzmann method

The lattice Boltzmann method (LBM) is successfully implemented for computer simulation of these three-dimensional problems. The lattice Boltzmann method describes the viscous flows of fluids with an arbitrary equation of state and simulates the interfaces between vapor and liquid phases with a surface tension.

The three-dimensional version of lattice Boltzmann method D3Q19 [1] with nineteen vectors $\mathbf{c}_k$ of pseudo-particle velocities on a cubic lattice is realized. The evolution equation for the distribution functions N_k can be written in the form

$$N_k(\mathbf{x}+\mathbf{c}_k\Delta t, t+\Delta t) = N_k(\mathbf{x},t) + \Omega_k(N) + \Delta N_k , \tag{2}$$

where Δt is the time step, Ω_k is the collision operator, and ΔN_k is the change of the distribution functions due to the action of the internal and external body forces.

The hydrodynamic variables (the density ρ and the velocity $\mathbf{u}$ of fluid) in a node are calculated as

$$\rho = \sum\nolimits_{k=0}^{b} N_k \quad \text{and} \quad \rho\mathbf{u} = \sum\nolimits_{k=1}^{b} \mathbf{c}_k N_k . \tag{3}$$

The collision operator is usually used in Bhatnagar–Gross–Krook (BGK) form

$$\Omega_k = (N_k^{eq}(\rho,\mathbf{u}) - N_k(\mathbf{x},t))/\tau , \tag{4}$$

where τ is the dimensionless relaxation time.

The Exact Difference Method (EDM) [2] is used for the implementation of the body forces (internal forces and gravity) in the LBM:

$$\Delta N_k(\mathbf{x},t) = N_k^{eq}(\rho,\mathbf{u}+\Delta\mathbf{u}) - N_k^{eq}(\rho,\mathbf{u}) , \tag{5}$$

where the value of the velocity after the action of the total force $\mathbf{F}$ on a node is equal to $\mathbf{u}+\Delta\mathbf{u} = \mathbf{u}+\mathbf{F}\Delta t/\rho$. The corresponding equilibrium distribution functions N_k^{eq} are calculated as

$$N_k^{eq}(\rho,\mathbf{u}) = \rho w_k \left(1 + \frac{\mathbf{c}_k\mathbf{u}}{\theta} + \frac{(\mathbf{c}_k\mathbf{u})^2}{2\theta^2} - \frac{\mathbf{u}^2}{2\theta}\right). \tag{6}$$

To simulate boundaries between liquid and gas, the internal forces acting between nodes of fluid were introduced by Shan and Chen [3]. For equation of state $P(\rho,T)$, the total force acting on a node was introduced by Qian [4] as a gradient of the pseudopotential $\mathbf{F} = -\nabla U$, where $U(\rho,T) = P(\rho,T) - \rho\theta$. The isotropic finite-difference approximation of the gradient operator was proposed by Kupershtokh et al. [5]. We use the quite simple well-known "bounce-back" rule to implement the no-slip boundary conditions at the solid walls in the LBM simulations.

The effect of solid walls on fluids is simulated by interaction forces acting on a node $\mathbf{x}$ belonging to the fluid from the nearest nodes $\mathbf{x}+\mathbf{e}_k$ representing the solid boundaries [6]

$$\mathbf{F}_k = w_k \psi(\rho(\mathbf{x}))\, B(\mathbf{x}+\mathbf{e}_k)\, \mathbf{e}_k . \tag{7}$$

Thus, contact angles are not prescribed in lattice Boltzmann method but are simulated in natural way.

The Graphics Processing Units (GPUs) are exploited for all computer simulations.

3 Droplets moving on solid surface

The three-dimensional dynamics of droplets on inclined and vertical walls in a gravity field was simulated. The results of computer simulations for the droplet on inclined substrate ($\alpha = 40^\circ$) are shown in Fig. 1. The important parameter for the problem is the Bond number $\mathrm{Bo} = \rho g H^2/\sigma$. It represents a dimensionless parameter used to characterize the ratio of gravity force to surface tension force. Here ρ is the density of liquid droplet. The Bond number in dimensionless variables has the form $\mathrm{Bo} = \tilde{\rho}\tilde{H}^2\tilde{g}\cos\alpha/(k\tilde{\sigma})$. Here, the dimensionless parameter is $k = P_{\mathrm{cr}}\Delta t^2/(\rho_{\mathrm{cr}}h^2)$. The reduced values $\tilde{\rho} = 1.93$, $\tilde{g}\cos\alpha = 0.000007$, $\tilde{\sigma} = 1.8$ are chosen. The initial reduced height of droplet is $\tilde{H} = \tilde{R}_0 \approx 115$. Hence, the initial value of the Bond number is $\mathrm{Bo} \approx 10$. At the final stage of simulations (Fig. 1,b), the Bond number is $\mathrm{Bo} \approx 1$.

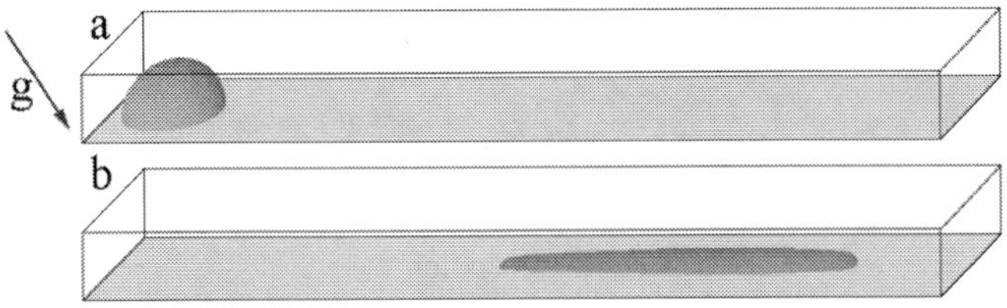

Figure 1. Droplet flowing down the inclined surface ($\alpha = 40°$). $\tilde{R}_0 = 115$. Grid 2048x368x160.

The second parameter W is the ratio of gravity force to viscous forces $W = \mathrm{Ga}/\mathrm{Re} = \tilde{g}\sin\alpha\,\tilde{H}^2/(\tilde{\nu}\tilde{U})$ (the ratio of the Galilei number to the Reynolds number). Here, $\tilde{\nu}$ is the dimensionless kinematic viscosity, $\tilde{U}$ is the dimensionless characteristic velocity of the liquid along the surface.

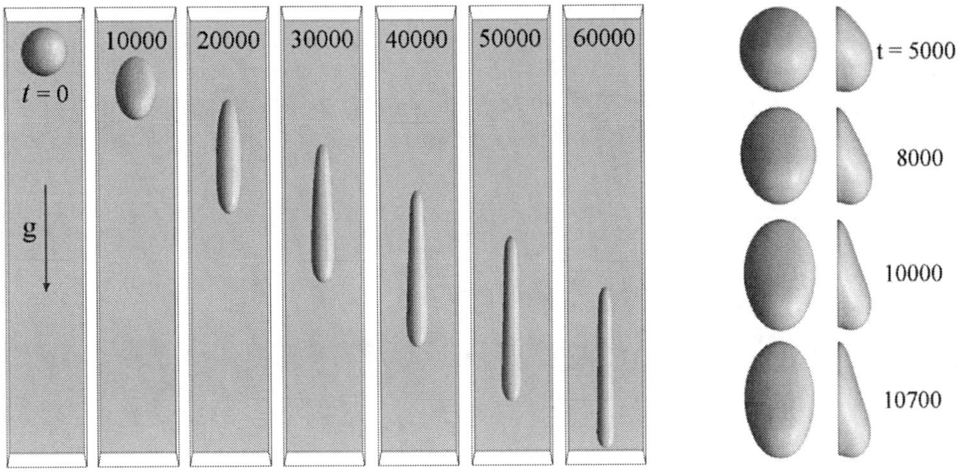

Figure 2. Droplet flowing down the vertical wall in gravity. $\tilde{R}_0 = 115$. Grid 368x2048x160.

The results of three-dimensional computer simulations for the droplet flowing down the vertical wall in gravity are shown in Fig. 2. The initial spreading of the droplet on the surface is determined by the Bond number. The steady velocity of the droplet is determined by the parameter $W = \mathrm{Ga}/\mathrm{Re}$. The initial value is $W \approx 15$. The final value is $W \approx 3$ at $t = 60000$. Hence, the flow tends to the quasi-stationary velocity of the droplet.

4 Dynamic contact angles

The dependence of a contact angle on a velocity of contact line was studied for a smooth surface. A liquid droplet is placed on the horizontal solid surface ($B = 0.4$). Then liquid is removed slowly ($dm/dt \approx -180\rho_{\mathrm{cr}}h^3/\Delta t$) from the spherical central region of the droplet as shown in Fig. 3a. Hence, the volume of droplet is reduced (Fig. 3a). We obtain the values of receding contact angle α vs. velocity of contact line $\tilde{V}$ (Fig. 3b).

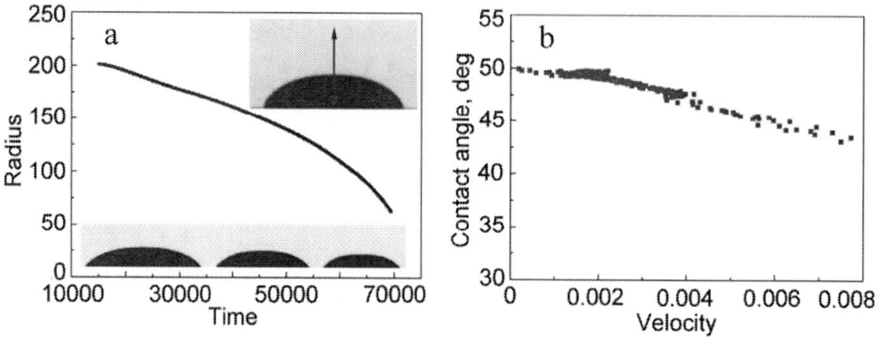

Figure 3. Reduced radius of contact line vs. time (a). Receding contact angle vs. velocity of contact line (b). Slow regime.

5 A rupture of the thin liquid films

The simulations of a rupture of thin liquid films on the nonwettable solid substrate due to the thermocapillary effect (Marangoni effect) were also carried out. For the prescribed axisymmetric temperature distribution $T = T(r)$ with a flattened vertex, a diverging flow in the film in the vicinity of the hot spot is generated because of the gradients of the surface tension along the film (Fig. 4a-f). The rupture of the liquid film occurs not in the center of symmetry but along a certain circle where the gradient of the temperature is more pronounced [6]. As a result, a central liquid disk is formed (Fig. 4a,b). The disk transforms initially into the toroidal figure due to surface tension (Fig. 4c,d) and then into an oscillating droplet (Fig. 4e,f). The rupture of free-hanging thin film for the non-axisymmetric temperature distribution $T = T(x, y)$ is shown in Fig. 4g.

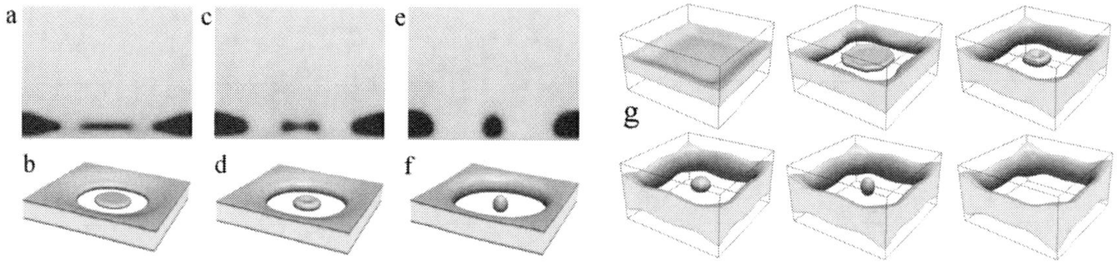

Figure 4. Thin film on nonwettable solid substrate (a-f). Free-hanging thin film at prescribed non-axisymmetric temperature distribution (g). Periodic boundary conditions are used along the x and y directions.

References

1. Y.H. Qian, D. d'Humières, P. Lallemand, Europhys. Lett., **17**, 479–484 (1992)
2. A.L. Kupershtokh, Computers and Mathematics with Applications, **59** 2236–2245 (2010)
3. X. Shan, H. Chen, Phys. Rev. E, **47** 1815–1819 (1993)
4. Y.H. Qian, S. Chen, Int. J. Mod. Phys. C, **8** 763–771 (1997)
5. A.L. Kupershtokh, D.A. Medvedev, D.I. Karpov, Computers and Mathematics with Applications, **58** 965–974 (2009)
6. A.L. Kupershtokh, E.V. Ermanyuk, N.V. Gavrilov, Comm. Comp. Phys., **17** 1301–1319 (2015)

Contact angle hysteresis of a drop spreading over metal surfaces

Geniy Kuznetsov[1], Dmitriy Feoktistov[1, 2] and Evgeniya Orlova[1, a]

[1]*National Research Tomsk Polytechnic University, 30, Lenin Ave, 634050, Tomsk, Russia*
[2]*Institute of Thermophysics SB RAS, 1, Lavrentiev Ave, 63090, Novosibirsk, Russia*

Abstract. The paper presents experimental data on the contact angle hysteresis of the distilled water drop spreading over the surfaces of non-ferrous metals. The measurements of the advancing and receding contact angles were carried out by method of sitting drop on the horizontal surface during increasing and decreasing drop volume with a syringe pump. It was found that the contact line speed has a great influence on the hysteresis of the polished non-elastic substrates. The mechanism of spreading was described using the balance of the forces from the physical point of view.

1 Introduction

The fundamental characteristic of a drop sitting on a solid surface is the contact angle (CA). According to Young-Laplace equation CA depends only on the surface tensions at the interfaces $\sigma_{SG,}\ \sigma_{SL,}\ \sigma_{LG}$. In fact, the angle depends on temperature, shape and arrangements of microasperities, roughness of a surface [1]. The last factor substantially affects the spreading of the droplet with a characteristic size of 1 mm and less [2, 3, 4]. Because during spreading over such substrates, a drop pins on microasperities, and it leads to non-uniformly movement of the contact line in all directions. Physical mechanisms of these phenomena have not been studied yet for carrying out the the prognostic assessment. To study these mechanisms, analyses of existing research [5-9] and new results needs to be done.

A contact angle hysteresis is the dependence of the contact angles from the conditions of their formation. Hysteresis can be measured directly by measuring the advancing and receding contact angles (Fig. 1). The advancing contact angle θ_A is the largest angle observed before the contact line begins to move towards the gas phase. The receding contact angle θ_R is characterized by the smallest contact angle observed before the contact line begins to move towards the liquid phase. If the contact line is stationary, the static conact angle θ can be equal any value between the advancing and receding angles $\theta_R \leq \theta \leq \theta_A$.

Systematic study of the effect of the surface roughness on the contact angle hysteresis is carried out in [10]. The water drop spreading over surfaces with fluorocarbon and paraffin coatings was under investigation. It was found, that in the initial period of time the dynamic CA increased and the receding CA decreased with increasing the surface roughness. With further increase of the roughness

[a] Corresponding author : lafleur@tpu.ru

the nature of the curve changed: the receding CA increased, the advancing CA raised slowly. At the final stage θ_A is constant with a slight increase of θ_R. It leads to small values of the contact angle hysteresis as compared to smooth surfaces.

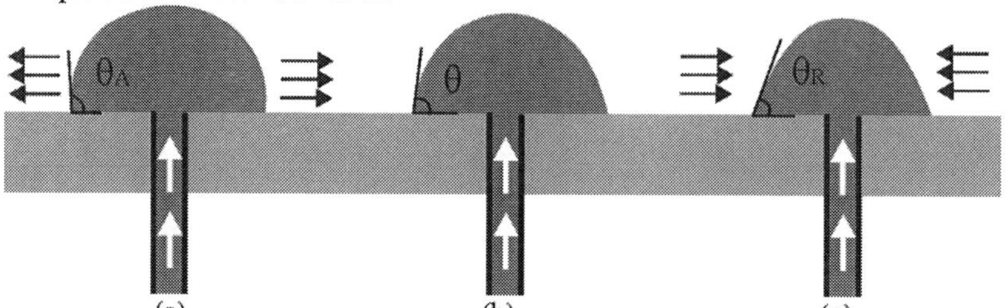

Figure 1. Interpretation of the contact angle hysteresis: (a) advancing contact angle; (b) static contact angle; (c) receding contact angle.

Study of the contact angle hysteresis in different gravity conditions were carried out at spreading a drop over surface with Teflon and polyvinyl-acetate coatings [5]. These surfaces have small and large contact angle hysteresis, respectively. The gravity in the experiment was varied from 0.03 m/s^2 to 1.8g (g is the terrestrial gravity). The influence of the gravity conditions on drop shape and spreading was observed. The shape of the drop changed from the circular segment under microgravity to flattened profile under hypergravity.

Spreading of a drop on a surface may be realized under the action of external forces, the forced spreading [11, 12], and by placing the droplet on a substrate without application of any forces (gravitational or spontaneous spreading) [13].

Dependences of the advancing contact angle (θ_A) from the contact line speed (U) were obtained as a result of the forced spreading study [11, 12]. The nature of the curves is found to change significantly, if one factor changes (for instance, the liquid flow rate). It follows that the contact angle is substantially independent of the contact line speed.

In the second case (gravitational spreading) a drop spontaneously spreads on a surface only by capillary action, the contact angle will decrease, reaching static values in equilibrium state [13]. The speed of the contact line tends to zero.

The measurement of the dynamic CA and the receding CA during forced spreading may be carried out with different contact line speed. For very small speeds (near one μm/s) the angles will be close or equal to the static values. However, the nature of the influence of the contact line speed on the hysteresis value has not been established yet. The obtained experimental dependences may subsequently be used as a basis of the drop spreading models verification.

The aim of this study is to determine the effect of the liquid flow rate (the contact line speed) and the material of the substrate to the value of the contact angle hysteresis.

2 Experimental

To investigate the dynamics of a drop spreading on substrates, the experimental set-up presented in Figure 2 was used [14, 15]. The study is based on the use of the shadow optical method for obtaining the drop profiles. The main equipment for the implementation of this method are the source of plane-parallel light and high-speed video camera (Fastvideo 500M) with a maximum resolution of 1280×1024 pixels and with frequency of 10 frames per second, equipped with a macro lens (Sigma AF 105 mm f / 2.8 EX DG).

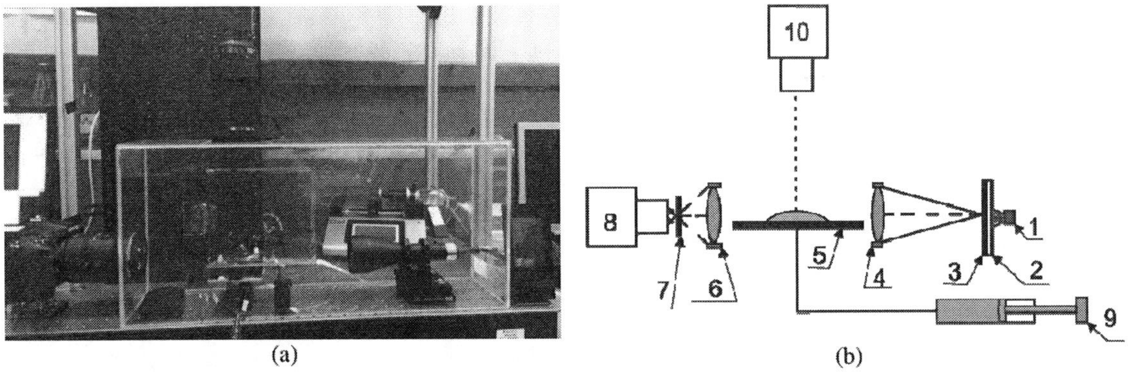

Figure 2. General view (a) and schematic diagram (b) of an experimental setup: 1 – light source; 2 – ground glass; 3, 7 – transparent shields with an opening; 4 – collimating lens; 5 – substrate; 6 – condensing lens; 8, 10 – high-speed video cameras; 9 – syringe pump.

The second high-speed video camera Fastvideo 500M 10 with lens (Figure 2) was used for visualizing drop formation and controlling drop symmetry. If the drop became asymmetrical due to random influence, the experiment was repeated.

The contact angle hysteresis measurement was carried out by method of sitting drop on the horizontal surface during increasing and decreasing drop volume. To obtain the advancing and the receding contact line, liquid was supplied through the opening in the substrate by an electronic syringe pump. Such nontraditional method of the drop formation (Figure 3) allows implementing the controlled spreading, in particular, changing the liquid flow rate, thereby the contact line speed changes.

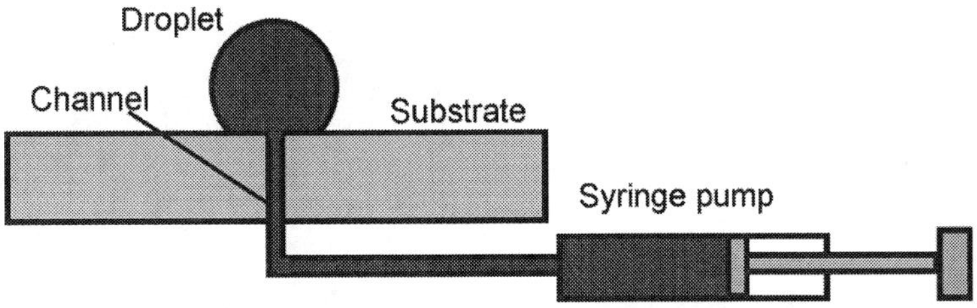

Figure 3. Method of a drop formation by an electronic syringe pump.

A test cell (Figure 4) of the experimental setup consists of the goniometer for lining the surface in two planes and substrate. The disk-shaped substrates (54 mm in diameter and 4 mm thick) made of aluminium and brass with a centered through opening were used in the experiment. The size of the opening was chosen so that it does not affect the wetting of the surface under study. As the wetting liquid nondeaerated distilled water was chosen.

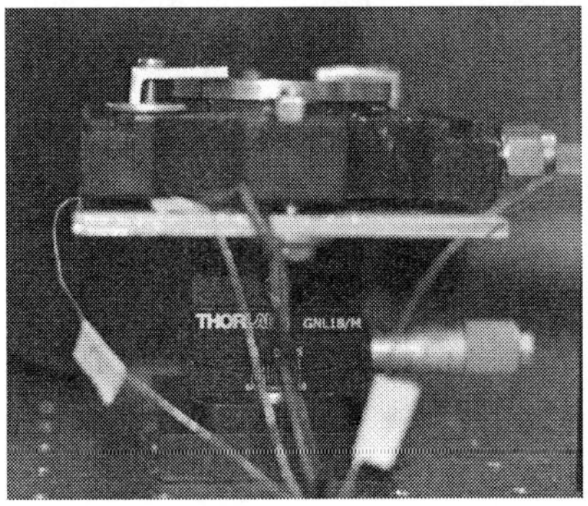

Figure 4. Photo of the test cell.

The surfaces of the substrates before experimental study were processed on grinding machine with diamond grit. The surface profile and roughness parameter (arithmetic average roughness Ra) are obtained on the profilometer "Micro Measure 3D station". Photos of the surfaces are made on scanning electronic microscope Hitachi TM-3000 with multiple zooming (Figures 5, 6).

The flow rate of the liquid supplied through the plastic tube connected with the syringe pump and the opening in the substrate was varied: 0.04 ml/sec, 0.05 ml/sec, 0.07 ml/sec, 0.11 ml/sec.

The main characteristics of spreading drop were obtained after processing shadow images by using software of KRUSS Company. Drop profile was calculated from a shadow photo by using Tangential 1 and Tangential 2 methods.

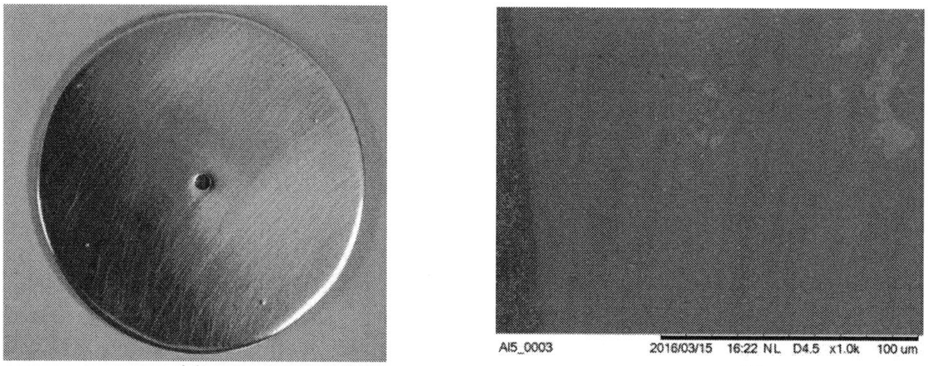

(a) (b)

Figure 5. Aluminium (Ra 0.0287 μm): (a) - substrate view; (b) - surface microstructure (Magnification: x1000).

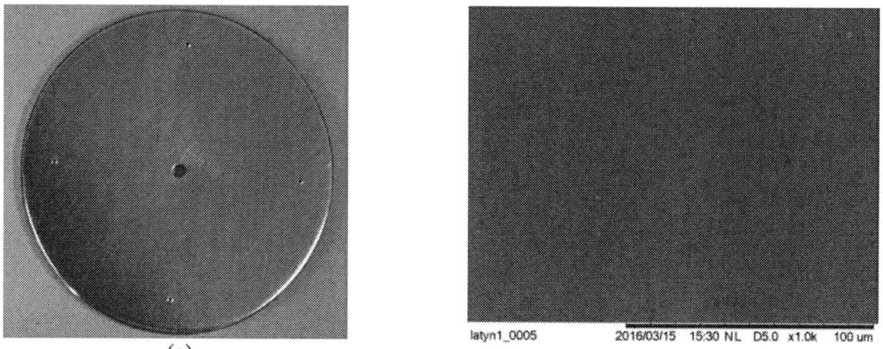

(a) (b)

Figure 6. Brass (Ra 0.033 μm): (a) - substrate view; (b) - surface microstructure (Magnification: x1000).

The main characteristics of spreading drop were obtained after processing the shadow images by using software of KRUSS Company. Drop profile was calculated from a shadow photo by using Tangential 1 and Tangential 2 methods.

Random error was calculated from the results of repeated experiments. To do this, from four to six experiments were carried out at constant values of the studied factors. Thereafter, the mean values of θ_A, θ_R, U and root-mean-square deviations. The fractional error of the contact angle is 2.5%, the error of the contact line speed is 1.9%.

3 Results and discussion

As a result of experimental studies the values of the advancing CA and receding CA of the water drop on the surfaces of aluminium and brass substrates.

Figures 7 and 8 present the dependence of the dynamic contact angle of a drop from the three-phase contact line speed on aluminium and brass at different values of the liquid flow rate (0.04 ml/sec, 0.05 ml/sec, 0.07 ml/sec, 0.09 ml/sec, 0.11 ml/sec). Dependences were plotted according to the values obtained by processing with two different methods: tangential method 1 and tangential method 2. These methods differ in the equations describing the drop profile.

Thus, when the speed $U > 0$ contact line moves toward the gas phase, i.e. it is advancing, the advancing contact angle was measured. When the contact line speed $U < 0$ the contact line moves toward the liquid phase, i.e. the contact angle is receding. The difference between advancing and receding angles at $U = 0$ is the contact angle hysteresis.

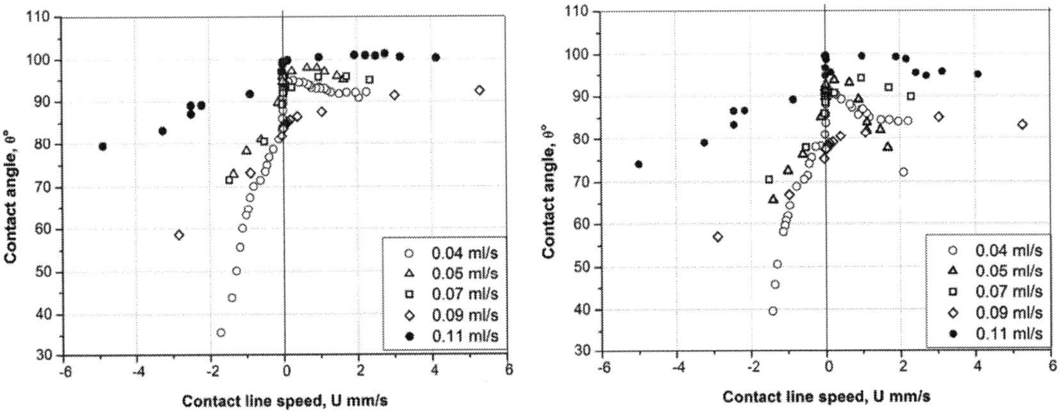

(a) (б)

Figure 7. Dynamic contact angle versus three-phase contact line speed over the substrate made of aluminium with varying liquid flow rate. The results were obtained by two methods of processing: (a) tangential method 1 and (b) tangential method 2.

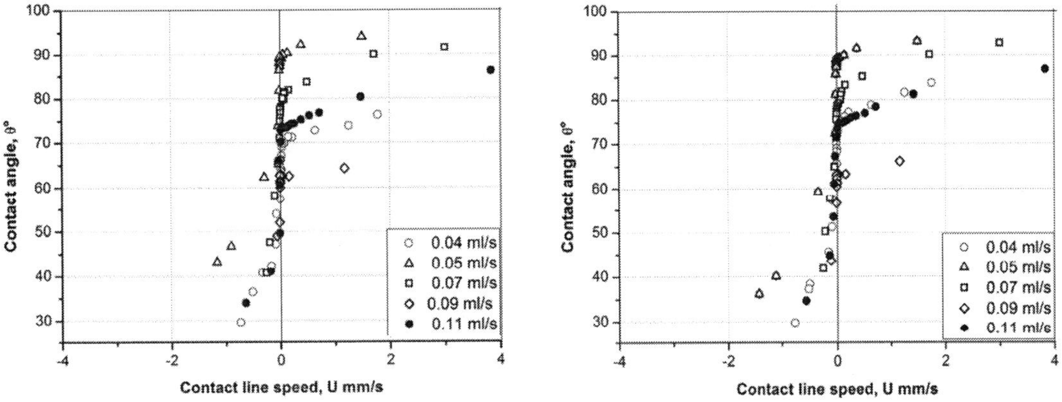

(a) (б)

Figure 8. Dynamic contact angle versus three-phase contact line speed over the substrate made of brass with varying liquid flow rate. The results were obtained by two methods of processing: (a) tangential method 1 and (b) tangential method 2.

It was found that the contact angle hysteresis changes with varying the liquid flow rate. The obtained values of the hysteresis on aluminium and brass are presented in Tables 1 and 2.

The contact angle hysteresis also varies depending on the flow rate on the substrate made of brass.

According to the obtained results it was found that the differences in the contact angle hysteresis determined by two methods fall within the limits of the measurement error.

Table 1. The contact angle hysteresis on aluminium obtained by two methods.

Liquid flow rate, ml/sec	Methods of processing	
	Tangential 1	Tangential 2
0.04	8.2 °	7.4 °
0.05	1.9 °	2.8 °
0.07	3.7 °	4 °
0.09	4 °	3.1 °
0.11	3.7 °	4,9 °

Table 2. The contact angle hysteresis on brass obtained by two methods.

Liquid flow rate, ml/sec	Methods of processing	
	Tangential 1	Tangential 2
0.04	9.9 °	8.8 °
0.05	8 °	7.7 °
0.07	7.2 °	6.6 °
0.09	6.5 °	5.8 °
0.11	6.3 °	7.5 °

It is known [16] that the surface roughness has a great impact on the hysteresis, particularly when microasperities have height not less than 0.5 μm. The value of hysteresis obtained in the experiment (Tables 1, 2) is relatively small (less than 10°) for substrates with roughness parameter 0.033 μm and 0.0287 μm for aluminium and brass, respectively. It should be noted that at low contact line speeds the hysteresis is greater than at high speeds (Tables 1, 2). Therefore, it may be affirmed that besides the roughness, the contact line speed affects the hysteresis of polished non-elastic solids. As a rule, the value of the speed is caused by the external forces. In this series of experiments, the force depended on the set values of the flow rates on a high-precision syringe pump. Carrying out the analyses of advancing and receding of contact lines, it should be noted that at low contact line speeds the inertial force, which depends on a set flow rate, is smaller in the force balance (the inertia, viscosity, friction, gravity, surface tension), acting on the drop. The inertial forces are small to overcome the energy barriers (occurring due to microasperities and microcavities on the surface of metals). As a result the hysteresis increases. When the flow rate is relatively high, the inertial forces are sufficient to overcome microroughness, the hysteresis in this case will be smaller.

For a visual representation of the physics of advancing and receding contact line on a solid substrate the typical dependences of the dynamic CA from the contact line speed are presented in Figure 9. The advancing contact angle and the contact line speed are found to decrease monotonically during infuse of liquid to a certain value, which is different for brass and aluminium surfaces. Further, after turning off the syringe pump, the contact line does not stop, i.e. the value of the contact angle and the speed continued to decrease. Probably, this is due to the fact that the drop has continued to spread under the influence of the residual pulse due to the hydrodynamic flows in the drop volume. Spreading in this period may be called inertial or spontaneous, which lasts until the establishment of equilibrium state. Then, a syringe pump is switched on and the liquid is withdrawn for the receding contact line formation, accompanied by a decrease in the receding contact angle and the contact line speed.

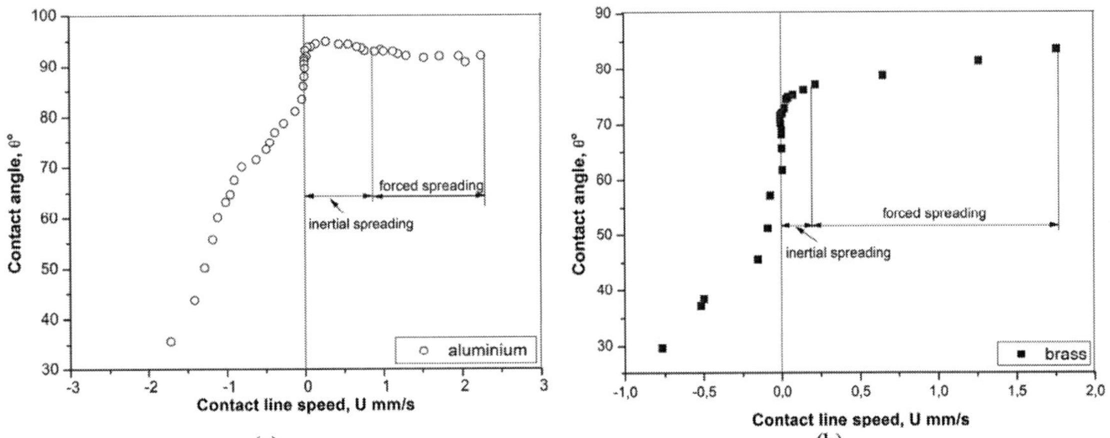

(a) (b)

Figure 9. Typical experimental results for the dependence of the dynamic contact angle on the contact line speed for the liquid flow rate of 0.04 ml/s on the substrates made of: (a) aluminium and (b) brass.

4 Conclusion

According to results of the experimental study it was found the effect of the liquid flow rate on the contact angle hysteresis of polished metal substrates. In addition to the surface roughness, the contact line speed significantly affects the hysteresis. The influence of the inertia forces on the hysteresis of non-elastic solids was explained. The mechanism of spreading was described from the physical point of view according to obtained dependences of the dynamic CA and receding CA from the contact line speed.

It was found that the differences in the contact angle hysteresis determined by two methods (tangential 1 method and tangential 2 method) fall within the limits of the measurement error.

The reported study was supported by the grant of the President of the Russian Federation for the government support of young Russian scientists (MK-6810.2016.8).

References

1. P.G. de Gennes, Rev. Mod. Phys. **57**, 827 (1985)
2. D. Sivakumar, K. Katagiri, T. Sato, H. Nishiyama, Phys. Fluids **17**, 100608 (2005)
3. H. Song, Y. Lee, S. Jin, H.-Y. Kim, J.Y. Yoo, Microelectron Eng **88**, 3249 (2011)
4. V.S. Ajaev, E.Ya. Gatapova, O.A. Kabov, Adv. Colloid Interface Sci. **228**, 92 (2016)
5. Zaitsev D.V., Kabov O.A., Microgravity science and technology, **19**, 174 (2007).
6. D. Bonn, J. Eggers, J. Indekeu, J. Meunier, E. Rolley, Rev. Mod. Phys. **81**, 739 (2009)
7. C.L. Moraila-Martinez, F.J. Montes Ruiz-Cabello, M.A. Cabrerizo-Vilchez, M.A. Rodriguez-Valverde, Colloids Surf., A **404**, 63 (2012)
8. G. McHale, N.J. Shirtcliffe, M.I. Newton, Langmuir **20**, 10146 (2004)
9. J.W. Krumpfer, T.J. McCarthy, Faraday Discuss. **146**, 103 (2010)
10. R.H. Dettre, R.R. Johnson jr. *Contact Angle, Wettability and Adhesion* (American Chemical Society, Washington, DC, 1963)
11. T.D. Blake, A. Clarke, K.J. Ruschak, AIChE J. **40**, 229 (1994)
12. T.D. Blake, M. Bracke, Y.D. Shikhmurzaev. Phys. Fluids **11**, 1995 (1999)
13. M.J. de Ruijter, J. De Coninck, T.D. Blake, A. Clarke, A. Rankin, Langmuir **13**, 7293 (1997)
14. E.G. Orlova, G. V. Kuznetsov, D. V. Feoktistov, EPJ Web of Conferences **82**, 01053 (2015)
15. D. Feoktistov, E. Orlova, A. Islamova, EPJ Web of Conferences **110**, 01018 (2016)
16. B.D Summ, Yu.V. Goryunov, *The Physochemical Principles of Wetting and Spreading* [in Russian] (Khimiya, Moscow, 1976)

MATEC Web of Conferences 84, 00020 (2016) DOI: 10.1051/matecconf/20168400020
International Symposium IPHT 2016

Modelling of the joint motion of nonisothermal liquid film and gas flow in a microchannel: numerical simulation of full Navier-Stokes equations

Vladimir V. Kuznetsov[1] and Yulia O. Kabova[2,a]

[1]*Lavrentyev Institute of Hydrodynamics, SB RAS, 630090, 15 Lavrentyev prosp., Novosibirsk, Russia*
[2]*Institute of Thermophysics, SB RAS, 630090, 1 Lavrentyev prosp., Novosibirsk, Russia*

Abstract. New full-statement 3D mathematical model of joint motion of thin liquid film and gas in a microchannel at local heating developed by taking into account the heat transfer by flows, evaporation and condensation, as well as the heat transfer at the gas-liquid interface is derived. The model is based on the full system of the Navier-Stokes equations, taking into account the convective terms of motion equations in the phases. Comparison of the numerical results obtained using the model based on the full Navier-Stokes equations and using the simplified model developed in the framework of the thin layer approximation has been performed. The comparison shows that at low Reynolds numbers, simplified model well describes all the main characteristics of the gas and liquid motion. With the gas Reynolds numbers significant increase difference between numerical results starts to grow.

1 Introduction

Flows of thin liquid films driven by various forces such as gas flow, gravity, capillarity, thermocapillarity and intermolecular forces are encountered in cooling devices, heat-exchangers, microfluidic devices, in wetting and spreading, in condensers, as well as in biomedical and geophysical applications. Over the past decades, numerous theoretical and experimental research works have been focused on the dynamics and heat transfer in thin liquid films, evaporation and condensation processes were taken into account in many studies [1, 2, 3, 4, 5]. Particularly, big variety of mathematical models taking into account different features of the liquid motion and heating has been developed using the lubrication theory [6, 7, 8], and also numerical simulation of the full Navier-Stokes equations under certain conditions [9] has been performed. Method of the integral proportions proposed by Prof. V.Ya. Shkadov also gives good results [10], but the question about impact of the gas phase on the free surface phenomena remains poorly studied. Generally, investigation of joint influence of thermocapillary and mass forces is conducted within the framework of generally accepted models, while full understanding of mechanisms of the structures formation and of influence of heat and mass transfer effects requires the use of new and more complex mathematical models taking into account the convective terms of equations. The first model of joint liquid and gas motion taking into account convective heat transfer in liquid and gas phases, as well as evaporation, which has been

[a] Corresponding author: kabova@itp.nsc.ru

described by the convective diffusion equation in the gas phase has been proposed in [11] for laminar stratify flow with nondeformable free interface. In [12, 13] a 3D simulation of evaporating liquid and gas joint motion in a microchannel at local heating is presented. The hydrodynamics has been described in the framework of the lubrication approximation, whereas convective heat transport in both phases and convective diffusive vapor transport in the gas phase have been taken into account. It should be noted that full statement three-dimensional models, describing the liquid films motion taking into account the gas phase and the phase changes at the free boundaries are still very rare in literature. In the present work new full statement three-dimensional mathematical model of joint motion of a viscous liquid film and gas in a microchannel at local heating developed by taking into the inertial terms in motion equations as well as convective heat transport in both phases and convective diffusive vapor transport in the gas phase. Gas compressibility is neglected.

2 Problem statement

Consider a thin film of incompressible viscous liquid sheared by gas inside a microchannel of given height H_C. The microchannel is assumed to be unbounded in x and y directions. A part of wall at the side of liquid film is heated (Fig. 1). The transport processes in the liquid and in the gas are described by the Navier-Stokes, continuity and energy equations, as well as by diffusion equation in the gas phase.

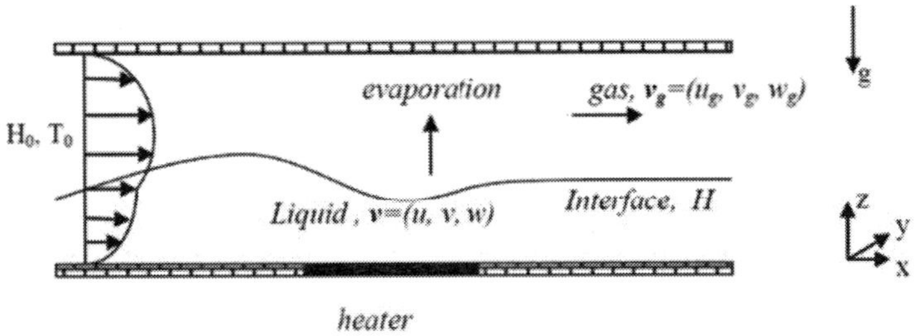

Figure 1. A sketch of the physical model.

The initial film thickness, tangential stresses, the pressure drop, and velocity fields in the liquid and gas phases are determined from the exact solution of the problem of isothermal laminar co-current flow in the channel with straight streamlines. The upper wall is assumed to be adiabatic and impermeable. At the bottom wall no slip condition and thermal boundary conditions, prescribing the constant heat flux, q, or constant temperature distribution, T_1, at the heater, are posed. At the gas-liquid interface the condition of continuity of the tangential component of the velocity vector (no-slip condition of two viscous media), mass conservation for the fluid comprising evaporating film and mass conservation conditions, dynamic condition with the term expressing the mechanical effect of the evaporated matter on the liquid, thermal boundary condition, taking into account the energy spending on the phase transition, as well as condition of local thermodynamic equilibrium are posed. The concentration, surface tension and dynamic viscosity are assumed to be temperature dependent. All details with regard to the problem statement could be found in [11, 12] for the case of the lubrication approximation.

3 Development of the mathematical model

In the [11] the lubrication theory has been employed to derive reduced set of equations, so the film aspect ratio has been taken asymptotically small $\varepsilon = H_0/l << 1$. And considering inertial terms to be negligible, the system of equations with boundary conditions has been rewritten in dimensionless form. And afterwards has been reduced to a solution of five governing equations: for the film thickness, temperature distribution in liquid and gas, vapor concentration in the gas and gas pressure.

Let us assume here that inertial terms in the motion equations are not negligible and should be taken into account. To derive the full statement mathematical model, let us perform the change of variables in the liquid and gas phases as follows:

$$\xi = z/H, u^1 = uH, v^1 = vH, w^1 = w - u\xi H_x - v\xi H_y, \tag{1}$$

$$\eta = \frac{\omega - 1}{H_C - H} z - \frac{H\omega - H_C}{H_C - H}, \quad u_g^1 = \frac{H_C - H}{\omega - 1} u_g,$$
$$v_g^1 = \frac{H_C - H}{\omega - 1} v_g, w_g^1 = w_g - \frac{H_C - z}{H_C - H} H_x u_g - \frac{H_C - z}{H_C - H} H_y v_g \tag{2}$$

Where $\omega = H_C / H_0$. This change of variables leads to simplification of the problem: spatial variables become nondependent on time and having flat free border and the kinematic interfacial conditions is simplified and becomes linear. Then the governing equations with boundary conditions could be rewritten in the layer $\xi \in (0, 1)$, $\eta \in (1, \omega)$, $-\infty < x, y < \infty$. The continuity equations in the liquid and in the gas retain their form in new variables but kinematic condition at the free interface simplifies. To rewrite governing equations with boundary conditions in the dimensionless form let us define scales for the liquid velocity, longitudinal length and temperature as follows:

$$U = \frac{\mu_0}{\rho H_0}, l = \left(\frac{\sigma_0 H_0^2}{\rho U^2}\right)^{1/3}, \mu_0 = \mu(T_0), [T] = \frac{Q_0 H_0}{\kappa S_0} \text{ or } [T] = \max_{t,x,y}(T_1(t,x,y) - T_0). \tag{3}$$

Here Q_0/S_0 is the average heat flux at the heater, S_0 the heating area. System of equations with boundary conditions has been rewritten in dimensionless form with the following dimensionless criteria of similarity and coefficients:

$$\mathbf{A} = \frac{g \cos\alpha H_0^2}{U^2 l}, \quad \mathbf{C} = \frac{g \sin\alpha H_0}{U^2}, \quad \mathbf{Ma} = \frac{\sigma_T [T] H_0}{\mu_0 U l}, \quad \mathbf{L} = \frac{\lambda \rho_g D}{\kappa [T]},$$
$$\mathbf{Sh} = \frac{\mu_{g0}}{\rho_g D}, \quad \mathbf{Cr} = \frac{\sigma_T [T]}{\sigma_0}, \quad \mathbf{Ca}^{-1} = \frac{\mu_0 U l^2}{\sigma_0 H_0^2}, \quad \mathbf{R} = \frac{D l}{H_0^2 U}, \quad \mathbf{Pr} = \frac{c_p \mu_0}{\kappa}, \tag{4}$$
$$\mathbf{Pr_g} = \frac{c_{pg} \mu_{g0}}{\kappa_g}, \quad \mathbf{P} = \frac{\rho_g D l^2}{\sigma_0 H_0^3}, k_1 = \frac{\mu_{0g}}{\mu_0}, k_2 = \frac{\kappa_g}{\kappa}, k_3 = \frac{\rho_g}{\rho}.$$

Where α - channel inclination angle, κ - thermal conductivity , λ - latent heat of vaporization, μ - liquid dynamic viscosity, ρ - liquid density, σ - surface tension, D - diffusion coefficient, c_p - specific heat of the liquid. Afterwards dimensionless system of equations with boundary conditions has been reduced to a solution of five governing equations: for the film thickness, temperature distribution in liquid and gas, vapor concentration in gas phase and gas pressure. Numerical solution of the problem is implemented by the finite difference method. The alternating directions implicit (ADI) method is used to solve the system of grid equations with boundary conditions. To solve the problem on each fractional grid step the Thomas algorithm is used. Channel height average pressure in liquid and gas has been calculated using the following formulas:

$$\tilde{p}(x,y,t) = \int_0^1 p(x,y,\xi,t) d\xi, \quad \tilde{p}_g(x,y,t) = (1/(\omega - 1)) \int_1^\omega p_g(x,y,\xi,t) d\xi. \tag{5}$$

This leads to a simplification of the numerical solution of the problem. Material constants correspond to water and nitrogen, inclination angle is equal to 0° [14].

4 Results

Calculations have been performed for the constant channel height H_C = 500 μm, α=0, heater size is equal to 3 x 6 mm^2 (length x width) and the heat flux q = 0.5 W/cm^2. The initial temperature is equal to 30°C. Nondimensional evaporation rate from the gas-liquid interface, $\mathbf{J}_{total}$, is calculated using formula $\mathbf{J}_{total} = \int_{\Lambda} \frac{1}{h(x,y)} \frac{\partial C}{\partial \xi}\Big|_{\xi=1} dxdy$, here Λ is the area of calculations, C - mass fraction of moisture in the gas phase. $\mathbf{J}_{total}$ is proportional by physical meaning to the value $(C_0 D\rho_g l^2)/H_0$ [kg/s]. Reynolds numbers $\mathbf{Re} = Q\rho/\mu_0$, $\mathbf{Re}_g = Q_g \rho_g / \mu_{0g}$ characterize flow intensity in the liquid and gas, here Q, Q_g – fluid and gas flow rates per unit width, correspondingly. All values presented in the graphs are non-dimensional, namely $t = \hat{t}l/U$, $T = T_0 + [T]\hat{T}$, $H = H_0 h$. Influence of the liquid and gas Reynolds numbers values on the main process characteristics including the interface deformations, maximum temperature and evaporation has been investigated numerically. Reynolds numbers in the liquid and gas are varied from 2 to 25 and from 15 to 200, correspondingly, Fig. 2-3.

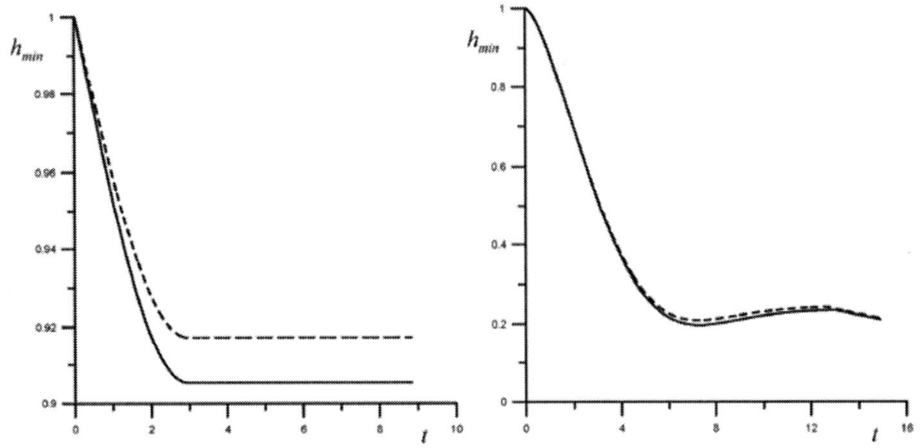

(1) (2)

Figure 2. Minimum film thickness versus time. Dashed lines – full Navier-Stokes equations, solid lines - lubrication approximation. (1) - Re = 25 and $\mathrm{Re_g}$ = 200; (2) - Re = 2,5 and $\mathrm{Re_g}$ = 20.

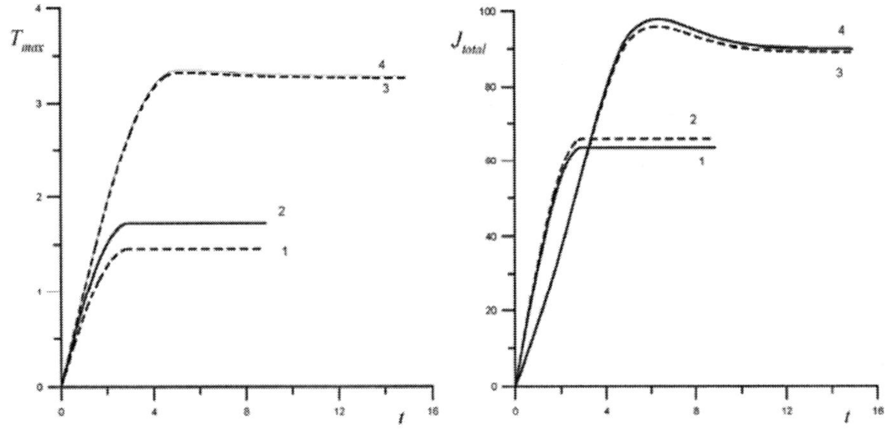

(1) (2)

Figure 3. (1) - Maximum dimensionless temperature at the gas -liquid interface. (2) - Evaporation rate from the gas-liquid interface. Dashed lines – full Navier-Stokes equations, solid lines - lubrication approximation. (1) - Re = 25 and $\mathrm{Re_g}$ = 200; (2) - Re = 2,5 and $\mathrm{Re_g}$ = 20.

In addition, comparison of the numerical results obtained using the model based on the full Navier-Stokes equations and using the simplified model developed in the framework of the thin layer approximation has been performed. The comparison shows that at low values of Reynolds numbers, simplified model well describes all the main characteristics of the gas and liquid motion. As one can see in Fig. 2 and 3 the difference between the calculated parameters for both models does not exceed 4%. Lines 1 and 2 correspond to the Reynolds numbers $Re = 25$ and $Re_g = 200$ as well as lines 3 and 4 to the $Re = 2{,}5$ and $Re_g = 20$. With the gas Reynolds number significant increase differences between numerical results starts to grow.

5 Conclusion

New full statement three-dimensional mathematical model of joint motion of a viscous liquid film and gas in a microchannel at local heating developed by taking into account the heat transfer by flows, evaporation and heat transfer at the gas-liquid interface is derived. The model is based on the full system of Navier-Stokes equations, taking into account the inertial terms in motion equations. Comparison of the numerical results obtained using the model based on the full Navier-Stokes equations and using the simplified model developed in the framework of the thin layer approximation is performed. The comparison shows that at low Reynolds numbers, simplified model well describes all the main characteristics of the gas and liquid motion. With the gas Reynolds numbers significant increase difference between numerical results starts to grow.

Acknowledgements

The study was financially supported by the Russian Science Foundation, project no. 14_19_01755.

References

1. J. P. Burelbach, S.G. Bankoff, S. H. Davis, Journal of Fluid Mechanics, **195** (1988)
2. R.V. Craster, O.K. Matar, Reviews of Modern Physics, **81** (2009)
3. E.A. Chinnov and O.A. Kabov, Microgravity Science and Technology, **XIX**-3/4 (2007)
4. O.A. Kabov, Yu.V. Lyulin, I.V. Marchuk and D.V. Zaitsev, International Journal of Heat and Fluid Flow, **28**, 103 (2007)
5. D.V. Zaitsev, D.A. Rodionov and O.A. Kabov, Technical Physics letters, **35**, 680 (2009)
6. A. Oron, S. H. Davis, S. G. Bankoff, Reviews of Modern Physics, **69** (1997)
7. J.M. Skotheim, U. Thiele, B. Scheid, Journal of Fluid Mechanics, **475** (2003)
8. V.S. Ajaev, E.Ya. Gatapova, O.A. Kabov, Physical Review E, **84**, 4 (2011)
9. A.M.Frank, O. A. Kabov, Physics of Fluids, **18** (2006)
10. S. Kalliadasis, A. Kiyashko and E.A. Demekhin, Journal of Fluid Mechanics, **475** (2003)
11. E.Ya. Gatapova, O.A. Kabov, International Journal of Heat and Mass Transfer, **51**, 19-20 (2008)
12. Yu. Kabova, V.V. Kuznetsov, O. Kabov, T. Gambaryan-Roisman, P. Stephan, International Journal of Heat and Mass Transfer **68** (2014)
13. Yu. Kabova, V.V. Kuznetsov, O. Kabov, Interfacial Phenomena and Heat Transfer **2**,1 (2014)
14. *Tables of physical units. Handbook.* (Ed.: Acad. I.K. Kikoin. M.: Atomizdat, 1976)

MATEC Web of Conferences **84**, 00021 (2016) DOI: 10.1051/matecconf/20168400021
International Symposium IPHT 2016

Liquid rivulet moved by a gas flow in a minichannel

Vladimir V. Kuznetsov[1] and Yulia O. Kabova[2,3,a]

[1]*Lavrentyev Institute of Hydrodynamics, SB RAS, 630090, 15 Lavrentyev prosp., Novosibirsk, Russia*
[2]*Institute of Thermophysics, SB RAS, 630090, 1 Lavrentyev prosp., Novosibirsk, Russia*
[3]*Novosibirsk State University of Architecture and Civil Engineering, 630008, 113 Leningradskaya Str., Novosibirsk, Russia*

Abstract. The mathematical model of steady-state flow of incompressible and nonisothermal liquid rivulet moved by a gas in an inclined minichannel was developed. Action of van der Waals and capillary forces, as well as gravity and tangential stress at the gas-liquid interface are taking into account. Numerically it is found out that rivulet formation gives a significant disturbing effect on the gas velocity distribution. It is shown that at the rivulet profile clearly visible three-phase contact line and region with practically constant inclination could be seen near the thin liquid film areas. The angle of this inclination could be considered as a contact angle. Region with practically constant derivative of the function h(y) with respect to y, near the contact line confirms this statement.

1 Introduction

Rivulet flows with and without phase-change are abundant in nature and play an important role in our daily life. Fast development in semiconductor technology, including shrinking feature size, increasing transistor density and faster circuit speeds have resulted in greater non-uniformity of on-chip power dissipation, generating localized submillimeter hot spots often exceeding 1 kW/cm^2 in heat flux. Rivulets driven by gravity or forced by gas/vapor flow in narrow channels (100–300 μm) are assumed to provide very high heat transfer intensity especially near the contact line area of the fluid on the heated surface [1, 2, 3, 4, 5]. Rivulet flows represent a promising approach to thermal management of advanced semiconductor devices with high power dissipation, in particularly for space applications: energy production, electronic cooling devices, life support systems and waste water treatment for long duration space exploration missions [6]. Understanding of fundamental aspects of rivulet dynamics might provide a way to control these flows. In [7], a model of a rivulet flow down an inclined substrate was proposed under assumption that the free-surface curvature of the rivulet is constant. In [8] a rivulet flow down an inclined cylinder was considered. The rivulet and gas flow in minichannel were theoretically investigated under conditions of variable gravity in [9]. The mathematical model describes steady-state isothermal rivulets with prescribed contact angle. In general it was found that the flow dynamics at microgravity differs significantly from that at normal gravity. Profile of the rivulet and gas-liquid interface deformations are affected significantly by the flow rates and gravity.

[a] Corresponding author: kabova@itp.nsc.ru

2 Problem statement

Let us consider an incompressible and nonisothermal liquid rivulet moved by a gas flow in an inclined minichannel with rectangular cross-section of certain width l and height d, Fig. 1. The liquid and gas motions are assumed to be laminar and steady state and temperature distribution on the bottom wall of the channel is assumed to be permanent along the entire length of the rivulet. The film motion is determined by the shear stress, induced by the gas flow on the free surface, and gravity.

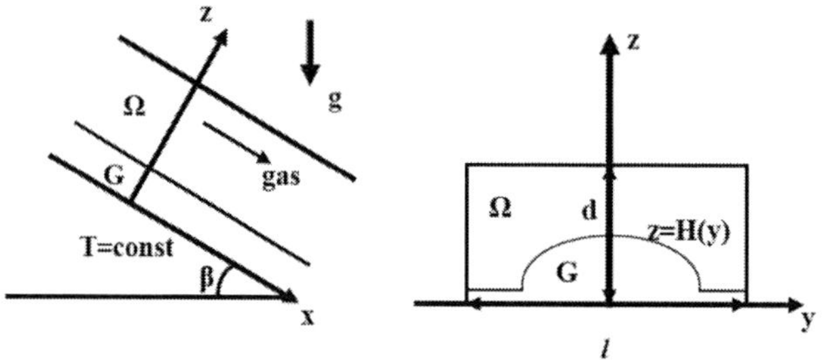

Figure 1. A sketch of the physical model.

Let the liquid and the gas occupy the following areas: $G=\{-\infty<x<\infty,-l/2<y<l/2,0<z<H(y)\}$ and $\Omega=\{-\infty<x<\infty,-l/2<y<l/2,H(y)<z<d\}$, correspondingly. The flows in the liquid and gas phases are described by the Navier–Stokes equations for incompressible fluid. Thus, in the areas G and Ω we have the following equations:

$$\rho(\boldsymbol{v}\cdot\nabla)\boldsymbol{v}=-\nabla p+\rho\boldsymbol{g}+\mu\Delta\boldsymbol{v},\nabla\cdot\boldsymbol{v}=0 \quad (1)$$

$$\rho_g(\boldsymbol{v}_g\cdot\nabla)\boldsymbol{v}_g=-\nabla p_g+\rho_g\boldsymbol{g}+\mu_g\Delta\boldsymbol{v}_g,\ \nabla\cdot\boldsymbol{v}_g=0. \quad (2)$$

Here, **v** and $\mathbf{v}_g$ are the velocity vectors; p and p_g are the pressure distributions in the liquid and the gas $\boldsymbol{g}=(|\boldsymbol{g}|\sin\beta,0,|\boldsymbol{g}|\cos\beta)$ is the free fall acceleration, ρ,ρ_g - densities of the liquid and the gas, μ,μ_g - dynamic viscosities of the liquid and the gas, β - inclination angle of the channel to the horizon. Boundary conditions for system (1)–(2) will take the form:

at $z=0$ and $y=\pm l/2,\ z<H(\pm l/2)$

$$\boldsymbol{v}=0, \quad (3)$$

at $z=d$ and $y=\pm l/2,\ z>H(\pm l/2)$

$$\boldsymbol{v}_g=0, \quad (4)$$

and at $z=H(y)$

$$\boldsymbol{v}=\boldsymbol{v}_g, \quad (5)$$

$$(P-P_g)\boldsymbol{n}=2\sigma K\boldsymbol{n}-\frac{A}{6\pi h^3}\boldsymbol{n}, \quad (6)$$

where σ is the surface tension coefficient, K is the mean curvature of the film surface $z=H(y)$, $P_{ij}=-p\delta_{ij}+\mu(\partial v_i/\partial x_j+\partial v_j/\partial x_i)$, $P_{gij}=-p_g\delta_{ij}+\mu_g(\partial v_{gi}/\partial x_j+\partial v_{gj}/\partial x_i)$ (δ_{ij} – the Kronecker symbols), $\boldsymbol{n}=\left(0,-H'/\sqrt{1+H'^2},1/\sqrt{1+H'^2}\right)$ - normal vector to the surface $z=H(y)$, A - Hamaker constant. Disjoining pressure is taken into account in (6). Modeling of the rivulet form and main features of the flow near the three-phase contact lines should take into account the thin liquid film appearing in the vicinity of the contact line, where the action of the van der Waals forces is significant [10]. Let us assume that the bottom wall of the channel is smooth, and thin liquid film covers it all. The height of this film is fast decreasing moving off the rivulet and has no effect on the results obtained.

3 Development of the mathematical model

The lubrication-type mathematical model based on the predominant velocity direction approach could be built. Let us assume that solution of the problem (1)–(6) could be found out in the following form: $\mathbf{v} = (v(y, z), 0, 0)$, $\mathbf{v}_g = (v_g(y, z), 0, 0)$, $\rho_g \ll \rho$. Accordingly, the gravity does not affect significantly the gas flow, which is induced by a constant longitudinal pressure gradient λ. The continuity equations in (1) and (2) are satisfied exactly. Accordingly, system (1)–(6) could be rewritten as follows

$$\frac{\partial^2 v}{\partial y^2} + \frac{\partial^2 v}{\partial z^2} = \frac{p_x - \rho g \sin\beta}{\mu}, \tag{7}$$

$$p_y = 0, \tag{8}$$

$$-p_z - \rho g \cos\beta = 0, \tag{9}$$

$$\frac{\partial^2 v_g}{\partial y^2} + \frac{\partial^2 v_g}{\partial z^2} = -\frac{\lambda}{\mu_g}, \tag{10}$$

$$p_g = p_0 - \lambda x, \ p_0 = const. \tag{11}$$

Here p_0 is the pressure in the gas at $x = 0$. Conditions (3), (4) and the velocity continuity condition (5) save their form. And projection of the dynamic condition (6) on the x axis gives the equation

$$\mu\left(-H'\frac{\partial v}{\partial y} + \frac{\partial v}{\partial z}\right) = \mu_g\left(-H'\frac{\partial v_g}{\partial y} + \frac{\partial v_g}{\partial z}\right), \tag{12}$$

the projection of the dynamic condition on the normal **n** taking into account (11) gives:

$$p - p_0 + \lambda x = 2\sigma K + \frac{A}{6\pi h^3}. \tag{13}$$

Conditions (8) and (9) are satisfied if we assume that in the liquid phase

$$p = p_1 - \lambda x + \rho g z \cos\beta, \quad p_1 = const. \tag{14}$$

Here p_1 – pressure at the liquid – solid interface. Then, from condition (13) and the definition of the mean curvature we obtain equation for the function $z = H(y)$:

$$\frac{H''}{(1+H'^2)^{3/2}} - \frac{\rho g \cos\beta}{\sigma} H + C_1 = \frac{A}{6\pi h^3 \sigma}, \tag{15}$$

$$H(0) = b. \tag{16}$$

Constant $C_1 = p_1 - p_0$ should be calculated. Hence, the problem is reduced to the determination of the function $z = H(y)$, which specifies the shape of the rivulet, the velocity components in the liquid and gas phases $v(y, z)$, $v_g(y, z)$, and the constant C_1, i.e. the system of Eqs. (7), (10), (15) with conditions (3)–(5), (12), and (16) should be solved numerically. Due to the one-dimensionality of the vectors **v** and $\mathbf{v}_g$, each of the conditions (3)–(5) is equivalent to the one scalar equality, and the pressure distributions in the liquid and the gas are given by the explicit formulas (11), (14). For the numerical solution of the problem formulated above, it is necessary to specify two constants, namely, the rivulet maximum height b and the longitudinal pressure gradient λ. If they are known, then after the solution of the problem the liquid and gas flow rates Q and Q_g could be found out by the formulas

$$Q=\iint\limits_{G_0} v\,dydz\,,\; Q_g=\iint\limits_{\Omega_0} v_g\,dydz\,, \tag{17}$$

where G_0 and Ω_0 are the cross-sections of areas G and Ω by the plane x = const. This problem is known as a "direct problem". In the simpler form, this problem was solved in [11]. The velocity field in the gas was not calculated, and the action of the gas flow on the liquid was replaced by the presence of a given constant shear stress directed along the x axis at the interface. But so called "inverse problem" is much more widely used for practical applications, because it allows calculating the values of b and λ from a condition of the correspondence of the flow rates calculated using formulas (17) after the solution of the direct problem with the prescribed flow rates. In the present paper, we assume that gas and liquids Reynolds numbers, $\mathbf{Re}=Q\rho/\mu$, $\mathbf{Re}_g=Q_g\rho_g/\mu_g$, are the given parameters and shape of the liquid rivulet should be calculated, so the inverse problem is solved. The numerical solution was conducted as follows. We assume that for some arbitrary b_0 and λ_0 the solution of the direct problem was obtained, and hence the functions $Q_0(b_0, \lambda_0)$ and $Q_{0g}(b_0, \lambda_0)$ specifying the values of the flow rates are known. Then the values of unknown b and λ could be found from the solution of the system

$$Q_0(b_0,\lambda_0)=Q,\quad Q_{0g}(b_0,\lambda_0)=Q_g\,. \tag{18}$$

Here values Q and Q_g are prescribed. Eqs. (19) were solved using the modified Newton's method. On each iteration step, while solving the direct problem, the system (15)–(16) was solved to find the function $z = H(y)$ and then Eqs. (7) and (10), with boundary conditions (3)–(5) and (12) were solved to find the velocity fields **v**, **v***g*. The latter problem was solved by a finite-difference method using fractional steps with respect to a fictitious time introduced for the calculations.

4 Results

In the numerical calculations, the material constants for the gas phase correspond to the nitrogen, and for the liquid to the FC-72: μ = 0.735×10^{-3} kg/m·s, μg = 17.75×10^{-6} kg/m·s, ρ = 1687.8 kg/m^3, ρg=1.2505 kg/m^3, and σ = 12.57×10^{-3} N/m, β=0, d=1,4 mm, l=4 cm, T=20 °C, A=10^{-20} m^{2}kg/s^2.

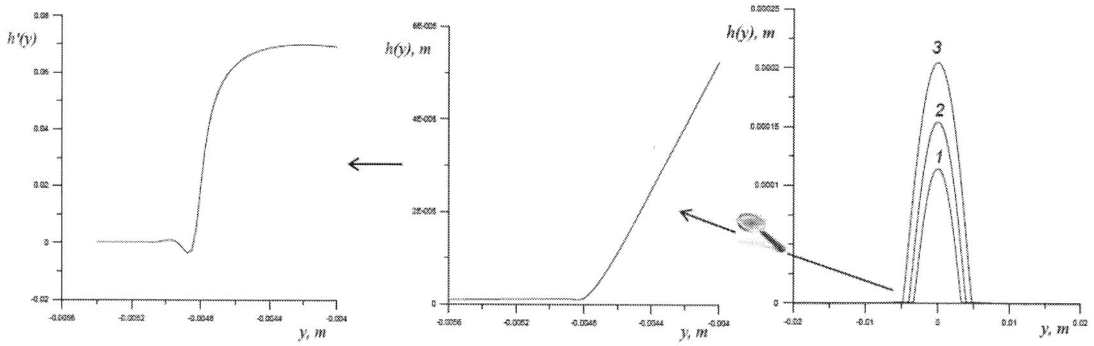

(3) (2) (1)

Figure 2. (1) - Rivulet profiles in cross sections perpendicular to the flow. Re =8. 1 - Re_g = 400; 2 - Re_g = 200; 3 - Re_g = 100. (2) - Thin liquid film near the contact line and contact angle. Re =8, Re_g = 100. (3) - Derivative of the function h(y) with respect to y, near the contact line. Re =8, Re_g = 100.

Figure 2, (1) shows the rivulet profiles in cross sections perpendicular to the flow at different gas Reynolds numbers. One can see that decreasing of gas velocity leads to the important change in the rivulet profile. Formally it is assumed that rivulet and thin liquid film fully covers the bottom wall of the channel, but at the rivulet profile clearly visible three-phase contact line and region with practically constant inclination could be seen near the thin liquid film areas, Fig.2, (2), and the angle of this inclination could be considered as a contact angle. Region with practically constant derivative of the function h(y) with respect to y, near the contact line confirms this statement, Fig. 2, (3). Figure 3 shows velocity field in cross section of the channel. The rivulet flow region is clearly visible because the velocity in the liquid is smaller than in the gas. One can see that rivulet formation gives a significant disturbing effect on the gas velocity distribution.

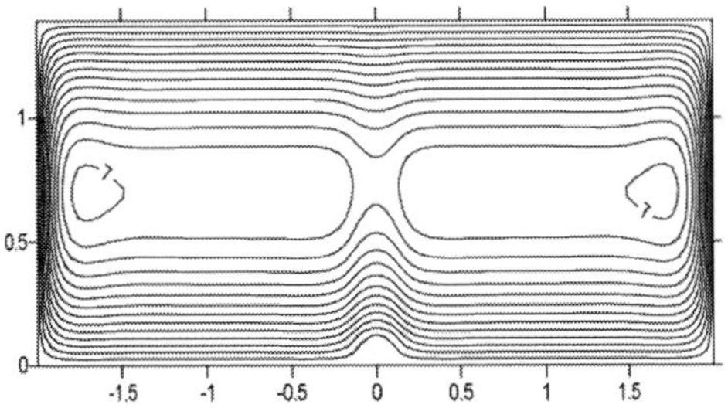

Figure 3. Velocity field in cross section (m/s). Re = 2, $Re_g = 20$.

5 Conclusion

For an incompressible and nonisothermal liquid rivulet moved by a gas flow in an inclined minichannel the lubrication-type mathematical model based on the predominant velocity direction approach was built. Action of the van der Waals and capillary forces, as well as gravity and tangential stress at the gas-liquid interface are taking into account. Numerically it is found out that rivulet formation gives a significant disturbing effect on the gas velocity distribution. It is shown that at the rivulet profile clearly visible three-phase contact line and region with practically constant inclination could be seen near the thin liquid film areas, and the angle of this inclination could be considered as a contact angle. Region with practically constant derivative of the function h(y) with respect to y, near the contact line confirms this statement.

Acknowledgements

The study was financially supported by the Russian Science Foundation, project no. 14_19_01755. The numerical algorithm was elaborated under the support of RFBR (No 14-08-00163).

References

1. P. Stephan, C. Brandt, Proc. I Int. Conf. on Microchannels and Minichannels (ed. S.G. Kandlikar), Rochester, NY, 24–25 April (2003)
2. A.Ye. Rednikov, P. Colinet, Langmuir **27**, 5 (2011)
3. V. S. Ajaev, E. Ya. Gatapova, O. A. Kabov, Advances in Colloid and Interface Science **228**, (2016)
4. A.L. Karchevsky, I.V. Marchuk and O.A. Kabov, Applied Mathematical Modeling **40,** (2016)
5. D.V. Zaitsev, A.M. Lozano, H. Auracher, O.A. Kabov, Microgravity science and technology, **19**, 71 (2007)
6. V.V.Cheverda, A.Glushchuk, P.Queeckers, S.B.Chikov, O.A.Kabov, Microgravity Science and Technology **25** (2013)
7. G.D. Towell and L.B. Rothfeld, AIChE Journal, **12,** 5 (1966)

8. S.V. Alekseenko, P.I. Geshev, and P.A. Kuibin, Doklady Physics, **42**, 5 (1997)
9. M.V. Bartashevich, O.A. Kabov, V.V. Kuznetsov, Fluid Dynamics, **45**, 6 (2010)
10. B.V. Deriagin, N.V. Churaev, V.M. Muller (1985, M.: Nauka)
11. T.G. Myers, H.X. Liang, and B. Wetton, International Journal of Non-linear Mechanics **39** (2004)

MATEC Web of Conferences **84**, 00022 (2016) DOI: 10.1051/matecconf/20168400022
International Symposium IPHT 2016

Evaporation dynamics and Marangoni number estimation for sessile picoliter liquid drop of binary mixture solution

Peter Lebedev-Stepanov [1,2,a], Alexander Kobelev[2] and Sergey Efimov[2]

[1]*Photochemistry Center of Russian Academy of Sciences , 119421, Moscow, Russia*
[2]*National Research Nuclear University "MEPhI", 115409, Moscow, Russia*

Abstract. We propose the evaporation model of picoliter sessile drop of binary solvent mixture (with infinitely soluble in each other components) based on Hu and Larson solution for single solvent sessile drop and Raoult law for saturated vapor density of components of binary mixture in wide range of undimensional molar binary concentration of the components. Concentration Marangoni number estimation for such a system is also considered for prediction of liquid flows structure for further applications in dissipative particle dynamics in binary mixture evaporating drop.

1 Introduction

The evaporating picoliter liquid sessile drop of axis-symmetric shape is an interesting object of theoretical and experimental investigations due to its important fundamental and practice applications. The evaporation of sessile drop of binary solvent mixture deposited by inkjet device on a flat substrate is considered. In case of picoliter drop, the typical thermal Marangoni number is small ($M_{a\mathrm{T}}$<10), and thermocapillary convection can be neglected, unlike of the concentration driven convection.

The thermal Marangoni number is determined by $M_{a\mathrm{T}} = -\dfrac{c_V \Delta T H}{\nu\kappa}\dfrac{d\sigma}{dT}$, where c_V is a specific heat of liquid, ν is a kinematic viscosity, κ is a thermal conductivity, $\dfrac{d\sigma}{dT}$ is the rate of the change of the surface tension with the temperature, ΔT is the temperature difference between the substrate plane and the apex of the drop, and h is a drop height. To obtain the value of ΔT we have been supposed [1] the estimation $\Delta T \propto \dfrac{\Delta H \rho_S D h}{\kappa R}(1-\chi)$, where ΔH denotes an enthalpy (heat) of evaporation, ρ_S is a saturated vapor mass density, D is diffusion coefficient of vapor in the surrounding air, χ is relative asymptotic value of the vapor concentration in the atmosphere (relative humidity of air in case of considering the water evaporation), and R is a drop radius. Analogously, concentration Marangoni

[a] Corresponding author : petrls@yandex.ru

number can be estimated by expression $Ma_C = \frac{\Delta c H}{D_L \eta}\left|\frac{\partial \sigma}{\partial c}\right|$, where Δc is normalized (relative) concentration difference of mixture component between the substrate plane and the apex, η is characteristic dynamical viscosity coefficient in liquid, σ is surface tension, D_L is a diffusion coefficient of water molecules in liquid mixture solution, n_{S0} and n_{L0} are concentrations of saturate water vapor and liquid vapor respectively, H is a height of the drop. Our estimation has shown that the value of Ma_C reaches more than 10^4 even for a picoliter binary mixture droplet, and the Marangoni instability really takes place. Recently, the concentration Marangoni driven migration of particles in such a system was experimentally investigated [2]. There was shown that the polystyrene nanoparticles were moved by toroidal vortex flow acted inside a drop, and the direction of this flow depends on solvent components types (ethanol/water, isopropanol/water, metoxypropanol/water), ratio of mixture, particle size and environmental conditions.

To estimate Marangoni number, we have developed novel evaporation model of picoliter sessile drop of binary solvent mixture (with infinitely soluble in each other components) based on Hu and Larson [3] solution for single solvent sessile drop and Raoult law for saturated vapor density of components of binary mixture in wide range of undimensional molar binary concentration of the components. The application of this model also includes the boundary conditions determination for Marangoni flows calculation to use in computer dissipative particle dynamics in picoliter drop of binary solvent mixture [1].

2 Model

Let's consider the sessile droplet of binary mixture solution with infinite mutual solubility of components in mixture at given external conditions.

There are two characteristic times in this system: drop evaporation time, $t_e \propto \frac{L^2 n_L}{D(n_S - n_\infty)}$, and relaxation time, $t_r \propto \frac{L^2}{D_L}$, where *L* is characteristic size of drop; *D*, D_L are the diffusion coefficient of molecule in air and in liquid respectively. Obviously, non-depending on *L* ratio $\frac{t_r}{t_e} = \frac{D(n_S - n_\infty)}{D_L n_L}$ (n_s, n_L, n_∞ are the concentrations of saturated vapor, liquid and vapor far from the drop respectively) is the criterion of concentration relaxation rate into drop. If $\frac{t_r}{t_e} << 1$ then *c=const* inside the liquid drop is good assumption, due to there is fast relaxation of concentration of binary components in the liquid drop of solution. With account of these approximations we obtain the next equations of evaporation dynamics

$$\frac{d\phi}{dt} = -\frac{2\phi(1+\cos\phi)^2}{\pi R^2 \sin\phi}\left[K_1(c-h) + K_2(1-h)\right], \quad (1)$$

$$\frac{dc}{dt} = \frac{2\phi\sin^2\phi}{\pi R^2\left(\frac{2}{3} - \cos\phi + \frac{\cos^3\phi}{3}\right)}\left[K_1(c-h)(c-1) + K_2 c(1-c)\right], \quad (2)$$

where ϕ is contact anngle of the droplet, *h* is relative humidity (if the first component of mixture is water, as we mean here), *c* is the undimensional water concentration, so that (1-*c*) is a concentration of the second component of the mixture, $K_1 = 2D_1 \frac{n_{S10}}{n_{L10}}$, $K_2 = 2D_2 \frac{n_{S20}}{n_{L20}}$.

To estimate the Marangoni number in such a system, we obtain the next equation

$$Ma_C = \frac{w(h-\bar{c})H}{D_L \eta} \left| \frac{\partial \sigma}{\partial c} \right|, \quad (3)$$

where $w = \frac{Dn_{S0}}{D_L n_{L0}}$, $w \ll 1$ (D_L is a diffusion coefficient of water molecules in liquid mixture solution, n_{S0} and n_{L0} are concentrations of saturate water vapor and liquid vapor respectively, H is a height of the drop.

3 Results

Let us consider the water/mthoxipropanol binary mixture solution sessile drop (Table 1) with initial mix molar proportion of 50% by mass deposited onto glass. The initial parameters of drop are: contact angle $\phi = 30^0$, radius of contact line, R=50 um; volume calculated from given ϕ and R by formula $V = \pi \left(\frac{R}{\sin\phi} \right)^3 \left(\frac{2}{3} - \cos\phi + \frac{\cos^3\phi}{3} \right)$ is 1308 um^3.

Table 1. Parameters of individual components of binary mixture at 20°C

Fluid	ρ /kgm^{-3}	P / kPa	n_S / m^{-3}	n_L / m^{-3}	D/ m^{2}c^{-1}	K/ m^{2}c^{-1}	D_L m^{2}c^{-1}	η /Pa s
Water , H_2O	1000	2.3	$5.7 \cdot 10^{23}$	$3.3 \cdot 10^{28}$	$2.5 \cdot 10^{-5}$	$8.6 \cdot 10^{-10}$	$2.5 \cdot 10^{-9}$	10^{-3}
Methoxypropanol, MP, $C_4O_2H_{10}$	920	1.2	$3.0 \cdot 10^{23}$	$6.2 \cdot 10^{27}$	$0.5 \cdot 10^{-5}$	$4.8 \cdot 10^{-10}$	$0.5 \cdot 10^{-9}$	1.9 10^{-3}

The calculation results obtaining by formulas (1)-(2) are shown at the Fig.1. Evaporation of water-metoxipropanol binary mixture drop with pinned radius of 0.05 mm, initial contact angle 45° and relative humidity 0.9 is shown at Fig. 2.

From these results (Fig.3) it is clear that it is possible that when a drop of a binary solution will initially increase in size if $K_1(c-h) + K_2(1-c) < 0$.

It is interesting to compare the theoretical model results with experimental data. However, as far as we know, the experimental data on the dynamics of evaporation of the picoliter binary solvent mixture droplet are not currently available.

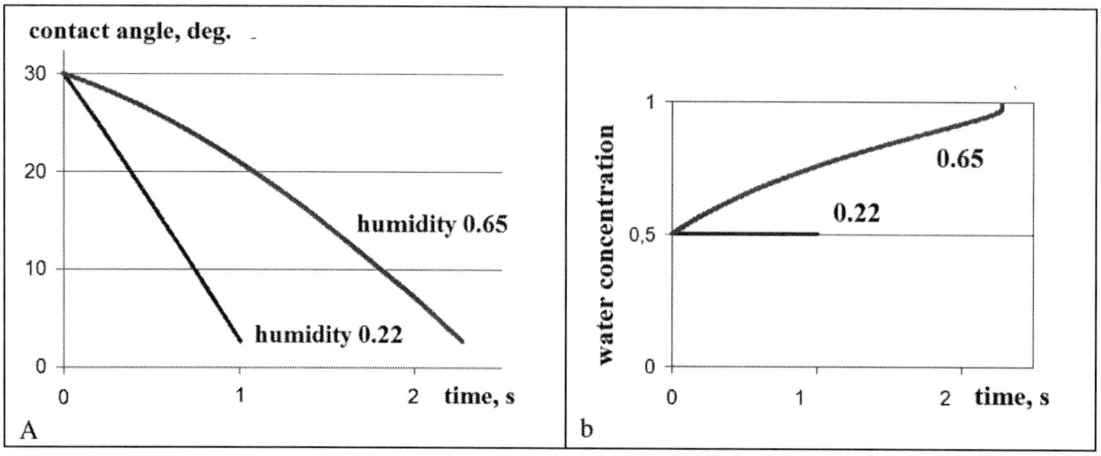

Figure 1. a – contact angle evolution during evaporation of water-MP droplet (initial molar concentration is 0.5) with pinning, b – water undimensional concentration evolution in the same case.

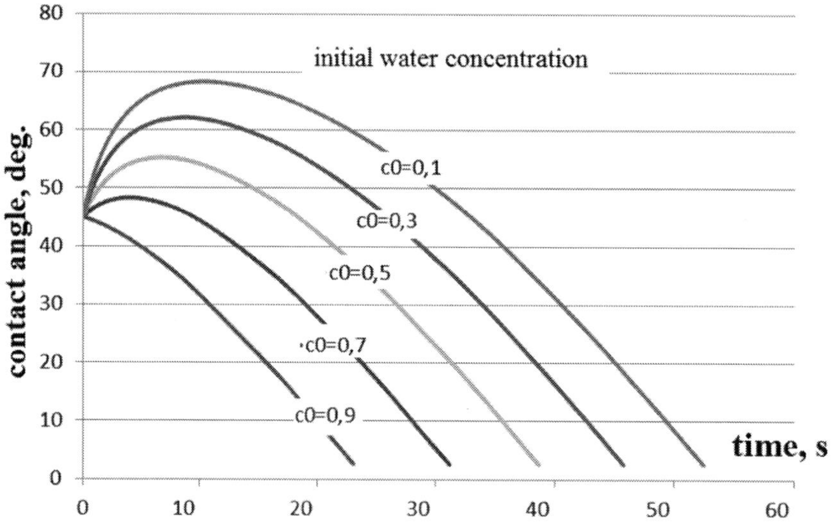

Figure 2. Evaporation of water-metoxipropanol binary mixture drop with pinned radius of 0.05 mm, initial contact angle 45° and relative humidity 0.9 (calculation result).

Authors thank B.S. Olga Sobanina for her help in testing of the model. This work was supported by the Russian Foundation for Basic Research, project no. 15-03-08050.

References

1. A. Mecke, I. Lee, J.R. Baker jr., M.M. Banaszak Holl, B.G. Orr, Eur. Phys. J. E **14**, 7 P. Lebedev-Stepanov, K. Vlasov, Colloids and Surfaces A: Physicochem. Eng. Aspects, **432**, (2013).
2. E. L. Talbot, A. Berson, L. Yang, C. D. Bain, 1-st Int. Workshop on Wetting and evaporation: droplets of pure and complex fluids. Marseilles, France, Book of Abstracts, 169 (2013).
3. H. Hu, R.G. Larson, J. Phys. Chem. B **106**, 1334 (2002).

MATEC Web of Conferences **84**, 00023 (2016) DOI: 10.1051/matecconf/20168400023
International Symposium IPHT 2016

Vapor condensation on curvilinear fins with condensate suction from the interfin space

Yuriy Lyulin[1,2,a], Irina Tueva[1,3] and Igor Marchuk[1,4]

[1] *Institute of Thermophysics, 1, Lavrentiev Ave., 630090, Novosibirsk, Russia*
[2] *Tomsk Polytechnic University, 30, Lenina Ave, Tomsk, 634050, Russia*
[3] *Novosibirsk State University, 2, Pirogov Ave., 630090, Novosibirsk, Russia*
[4] *Novosibirsk State Agrarian University, 160, Dobrolyubov Ave., 630039, Novosibirsk, Russia*

Abstract. Vapor condensation of the pure ethanol on the curvilinear fin was studied theoretically. Optimized profile of the fin surface providing maximum condensate flow rate has been calculated. New experimental rig and condenser with the optimized fin profile have been designed and created. Numerical simulation of the average heat transfer coefficient in depends of temperature drop and interfin space has been performed. It was found that the value of the flooding level of the interfin space significantly affects the intensity of condensation as a whole, however, it does not affect significantly the process of condensation on the upper convex part of the fin surface.

1 Introduction

Vapor condensation on finned surfaces is widely used in technique applications. Numerous experimental studies and theoretical models are presented in the literature. Gregorig [1] first derived a theoretical model for condensation on curvilinear fin with account of surface tension. The fin was described by equation for a spiral with variable direction of rotation. Subsequent models for stationary vapour film condensation on parametric family of curvilinear fins with power dependence of the curvature have been presented in [2-5]. Some experimental results related to the film wise vapour condensation on curvilinear fins are presented in [6, 7]. Description of various theoretical models and experimental results on film wise condensation are presented in several books and reviews [8-12]. The widespread validation method of the theoretical models is based on comparison between measurement of average heat transfer coefficient and one calculated by the model. But in most articles the flooding of the interfin space are not considered. The flooding of the interfin space in condenser leads to decreasing of intensive condensation area and reduces significantly the intensity of condensation. The problem of removing condensate from the interfin space is particularly important at microgravity conditions. One solution to this problem would be to use suction of the condensate from the interfin grooves.

The main goals of the work are development of a new experimental rig and numerical study of film condensation on the curvilinear fins with the suction of condensate from the interfin space.

---

[a] Corresponding author: lyulin@itp.nsc.ru

2 Experimental setup

For the studies of vapor condensation in the condenser with curvilinear fins a new experimental rig was created, see Fig. 1. The experimental rig consists of the following components: test cell (1); high precision liquid pump (2); evaporator (3); liquid flow meter (4); personal computer (5); temperature and pressure sensors (6); vapor flow controller (7); pressure controller (8); vacuum pump (9); control and data acquisition system (10); heat insulated box (11); heater fan (12); thermostat (13); liquid container (14); valves (15-18). Pure vapor of the working liquid prepared in the evaporator at temperature and pressure maintained. Vapor from the evaporator is supplied to the test cell due to the pressure drop provided by the vacuum pump. The pressure of the test cell is defined by temperature of the condenser. The pressure controller jointly with the vacuum pump is used for the fine adjustment of the pressure. A mass flow controller regulates the vapor flow rate at the inlet of test cell. At the outlet of the test cell the vapor is partially pumped out to avoid accumulating of the nondensables. The test cell with all sensors is placed inside of the thermoinsulating box to avoid the unwanted condensation that may appear inside the sensors and the connecting tubing. The temperature of the air inside the thermoinsulating box is maintained at 1-3 C° above the saturation temperate of the working vapor using the heating fan. Since presence any cold part in the experimental rig (except the condenser) can lead to an unwanted condensation. This unwanted condensation may appear inside the sensors and the connecting tubing. As a result the unwanted condensation causes the pressure and the flow fluctuations, which can terminate the experiment. Condenser temperature is maintained by cooling water. The condensed liquid is pumped from the test cell to liquid container using the high precision liquid pump.

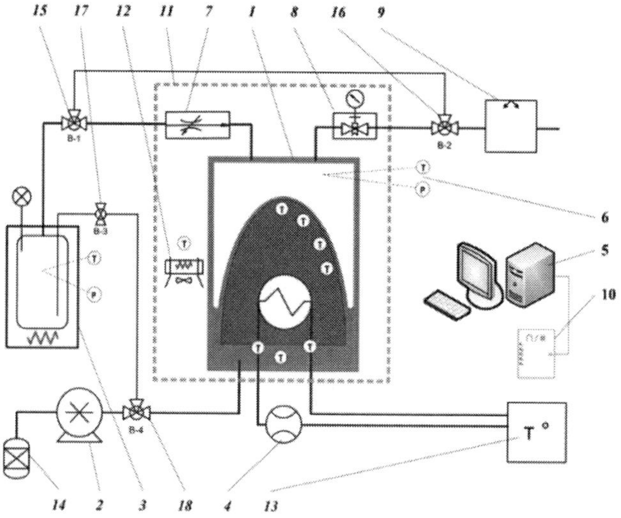

Figure 1. Scheme of the experimental rig.

The most important part of the experimental rig is the test cell. The design and photo of the test cell with a longitudinal fin are shown in Fig. 2. The test cell consists of the condenser in shape of the longitudinal fin, vapor chamber composed of transparent walls and distribution cambers for cooling water. The condenser fin is made of highly thermally conductive material (Cu). Shape of fin for experimental test cell is calculated using the theoretical model and optimized to obtain the maximum condensate flow rate from the fin surface for given geometric parameters. Surface profile of the fin is shown in Fig. 3. The given generatrix curve of the fin surface is a solution of the corresponding variational problem presented in [4, 7]. The curvature of the generated curve of the fin is defined as a function of curve length s by formula: $\kappa(s) = (as+b)^{-1/2} - c$. The formula constants of defined fin have the following values: a = 6.769519e + 000, b = 4.339554e-005, c = -2.762646e + 001, the curvature at the fin top - 1 mm^{-1}, the length of the generatrix curve S = 60 mm. Full rotation of the curve is equal to 90 °. The height of the fin is 54 mm, width 30 mm and length 200 mm. Cooling water from thermostat

DOI: 10.1051/matecconf/20168400023

controls the temperature of the condenser. The cooling water is pumping through the distribution chambers and special channels in the fin body with a diameter of 5 mm. The distribution chambers are fitted at the housing face of test cell and have a water temperature measurement. Attachment of the distribution chambers with the condenser is made with the help of two bolts (M8 with 300 mm length) through housing face walls. Hermeticity of the test cell is achieved by using silicon sealant and rubber seal. The temperature of the condenser is measured with help of six thermocouples (OD = 0.5 mm) that installed by three pieces on each end of the fin. Thermocouples are installed in the fin body through the front sides and holes with a diameter of 1.5 mm. The implementation depth of the thermocouple tips is 100 mm and 30 mm for each side, correspondently. The holes are located on the top, in the middle and at the bottom of the fin at the distance of 5 mm from the surface of condensation. The housing of the test cell is made of transparent polycarbonate walls. The distance between the side walls and fin surface can be varied in the range of 0.5-1.5 mm. Vapor is supplied into the test cell through a stainless steel tube with inside diameter of 14 mm and length of 200 mm. The tube is placed at the distance of 20 mm over the top of the fin. Special holes with diameter of 2 mm with the step of 10 mm are located on the side of the tube providing uniform distribution of the vapor inside the test cell. The temperature of the vapor, condensed liquid and cooling water are measured by the thermocouples with diameter of 1.5 mm that embedded in the test cell using special fitting. Pressure sensor is installed in top cover. Fluid line for pumping out of the condensed liquid is connected from the bottom of test cell.

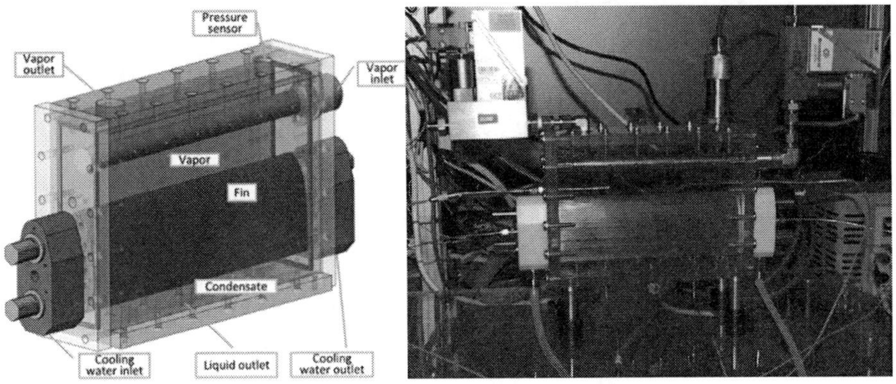

Figure 2. Design and photo of the test cell.

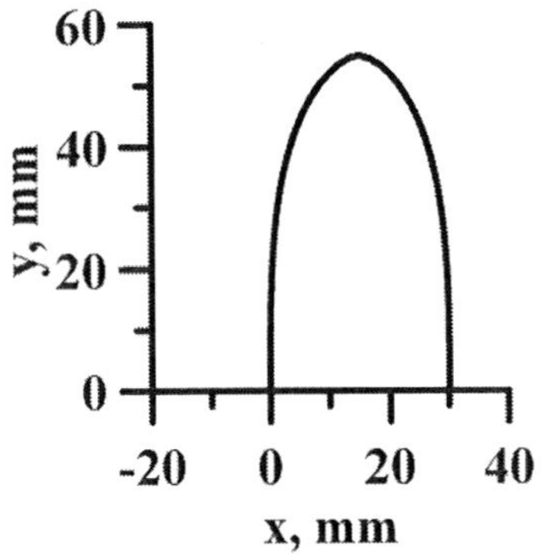

Figure 3. Calculated profile of the fin surface.

3 Numerical results

Numerical calculations of the vapor condensation of pure ethanol on the curvilinear fin have been performed using the theoretical model described in [4,7]. The temperature drop between the surface and vapor temperature varies from 0.2 K to 10 K. The dependence of the average heat transfer coefficient on the temperature drop for the atmospheric pressure is presented in Fig. 4. The heat transfer coefficients display the usual trend that expected for the condensation of the pure vapor on vertical wall. The heat transfer coefficient is reduced monotonically with the increase of the temperature drop. The values of the average heat transfer coefficient are in the range of 2.6-7.0 kW/ m^{2}K, Fig. 4. The distributions of the condensate film thickness for different values of the temperature drop are presented in Fig. 5. The minimum film thickness at top of the fin changes from 1.6 μm to 6.9 μm for the given temperature range. Dependence of the specific flow rate of condensed liquid along of the fin is shown in Fig. 6. It should be noted that the difference between the curves in Fig 6 is significantly greater than in the Fig 5. It is explained that the condensate flow rate is proportional to the third power of the condensate film thickness.

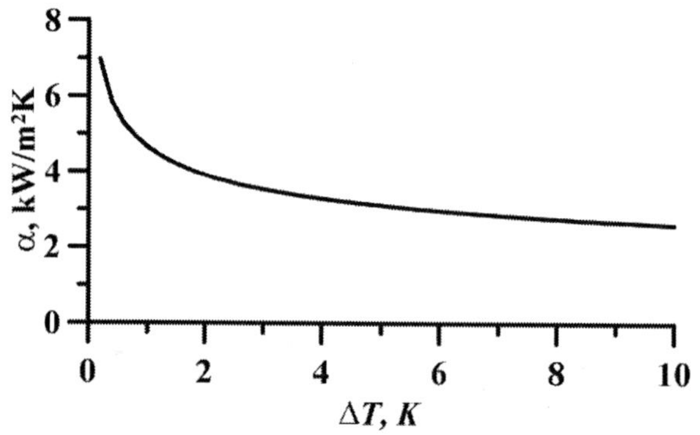

Figure 4. Dependence of the average heat transfer coefficient on the temperature difference.

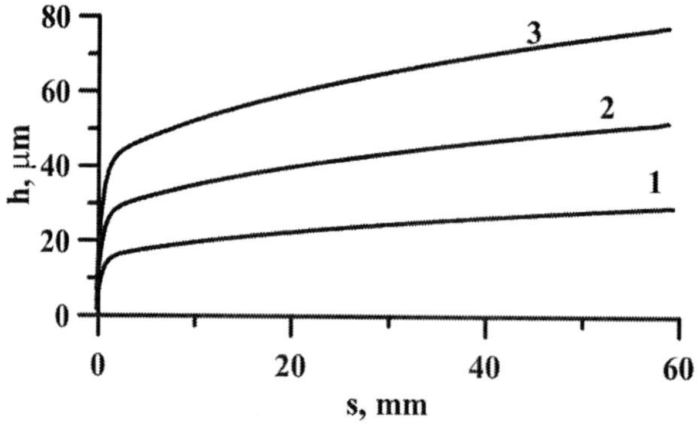

Figure 5. Condensate film thickness along the fin surface. 1- ΔT=0.2 K, h_{min} = 1.6 μm, α = 6.97 kW//m^{2}K; 1- ΔT=2 K, h_{min} = 3.8 μm, α = 3.91 kW//m^{2}K; 1- ΔT=10 K, h_{min} = 6.9 μm, α = 2.61 kW//m^{2}K.

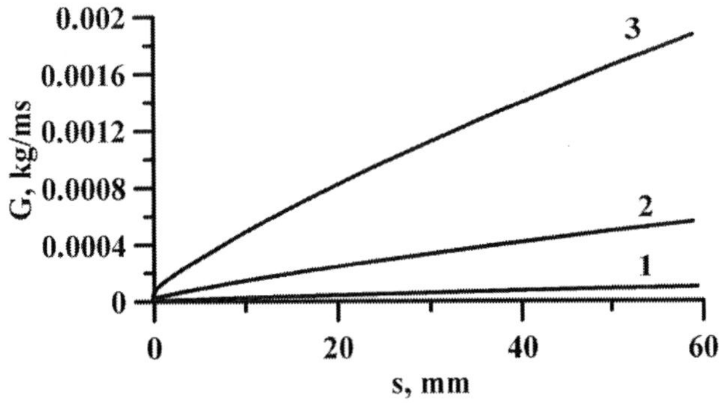

Figure 6. Mass flow rate of the condensate along the fin surface. 1- ΔT=0.2 K, 1- ΔT=2 K, 1- ΔT=10 K.

4 Conclusions

New experimental rig and numerical calculations for study of film vapor condensation on the curvilinear fins and numerical with the suction of condensate from the interfin space has been performed. Shape of fin for experimental test cell is calculated using the theoretical model and optimized to obtain the maximum condensate flow rate from the fin surface for given geometric parameters. The results of calculations of condensation of ethanol on the fin have been obtained for different values of the temperature difference between saturation temperature and fin surface temperature. The calculated heat transfer coefficient is in the range of 2.6-7.0 kW//m^{2}K. The value of the flooding level of the interfin grooves significantly affects the intensity of condensation as a whole, however, it does not affect the process of condensation on the upper convex part of the fin surface. For the calculation of operational parameters of the condenser it is possible take into account the total condensate flow only from the convex part of the condenser fins to determine its average performance.

The work was financially supported by the Russian Ministry of Education and Science (project identifier RFMEFI61614X0016).

References

1. R. Gregorig, Zeitschrift für angewandte Mathematik und Physik **Bd. 5,** 36 – 49 (1954)
2. C. Zener and A. Lavi, Journal of Heat Transfer **96**, 205-209 (1974)
3. T. Adamek, Warme - und Stoffubertragung **15**, 255 – 270 (1981)
4. I.V. Marchuk and O.A. Kabov, *Encyclopedia of Two-Phase Heat Transfer and Flow II: Special Topics and Applications,* **3,** 133-176 (2015)
5. I.V. Marchuk and O.A. Kabov, Doklady Physics, **61**, 19-23 (2016)
6. A. Glushchuk, I.V. Marchuk and O.A. Kabov, Journal Microgravity Science and Technology **23**, 65–74 (2011)
7. I.V. Marchuk, A.V. Gluschuk and O.A. Kabov, Technical Physics Letters **32(5)**, 388-391 (2006)
8. R.L. Webb, *Principles of Enhanced Heat Transfer*, 556 (1994)
9. J.R. Thome, *Engineering Data Book III* (2008)
10. V.G. Rifert, H.F. Smirnov, *Condensation Heat Transfer Enhancement*, 371 (2004)
11. J. Rose, Enhanced condensation heat transfer, **49 (3)**, 626-635 (2006)
12. A. Cavallini, G. Censi, D. Del Col, L. Doretti, G.A. Longo, L. Rossetto, C. Zilio, International Journal of Refrigeration **26**, 373-392 (2003)

MATEC Web of Conferences **84**, 00024 (2016) DOI: 10.1051/matecconf/20168400024
International Symposium IPHT 2016

Criterion for dry spot development in isothermal liquid film on a horizontal substrate

Leonid Maltsev[1], and Oleg Kabov[1,2,a]

[1]*Kutateladze Institute of Thermophysics SB RAS, Novosibirsk, Russia*
[2] *National Research Tomsk Polytechnic University, Tomsk, Russia*

Abstract. The paper proposes the criterion for development of dry spots in isothermal liquid films on a horizontal substrate and the formulas for gravity and surface tension forces applied at a given contact angle in the plane of the substrate on the liquid rim element surrounding a dry spot. The balance of these forces determines further pattern of initial small dry spot – whether it will disappear or develop into a large spot.

1 Introduction

Studies of rupture of thin liquid films on solid surfaces are important for modeling of multiphase flows in microfluidic devices, heat exchange systems, mining industry, and for biomedical applications such as dynamics of the tear film in the eye [1]. When the interface approaches the wall, the film can rupture, resulting in the formation of a three-phase contact line. The overall dynamics of many types of multiphase flows encountered in various applications depends on these local phenomena [2]. In cooling systems based on thin-film flows driven by either gravity or shear stresses at the liquid-gas interface, formation of dry spots results in significant reduction of heat flux from the heated wall, as shown experimentally in [3]. Liquid films, flowing under gravity over a localized heater on vertical or inclined flat plates, rupture at sufficiently high heat fluxes generated by the heater [4, 5].

An important aspect of the problem of the liquid film rupture is the substrate wettability effect, [6]. In most theoretical models describing the isothermal rupture of a liquid film, the equilibrium contact angle appears as a major parameter determining the critical film thickness [7]. Experiments on the rupture of the liquid film in the absence of heating qualitatively confirm the dependence of the critical thickness of the film on the contact angle [8]. The equilibrium contact angle is used as basic parameter in many models of rupture of the heated liquid film [9]. However, this is inconsistent with some experimental works [3]. It should be noted that submicron liquid film formed on the interface of dry spots can make a major contribution to the total heat and mass transfer due to intensive evaporation [10, 11, 12]. In some cases, rupture of the film is accompanied by formation of ultrafine residual liquid film, which exists for a very limited time [5,13].

Current research provides a theoretical analysis of the effect of the equilibrium contact angle, surface tension and liquid weight on the critical size of dry spots in isothermal liquid film of a given thickness on a horizontal flat substrate.

2 Analysis of the forces acting on the liquid rim surrounding the dry spot

Let the motionless liquid film rest on a horizontal substrate under the action of gravity. Assume that at some point, a dry spot, having the shape of a circle, appears in the film. The dry spot is usually surrounded by a liquid rim. Figure 1 shows a general view of a dry spot surrounded by a rim, and Fig. 2 presents a cross section of the rim by a plane passing through the center of the dry spot.

Each element of the rim cut off by central angle $\delta\varphi$ is subjected to the action of the gravity and the force caused by surface tension on a curved surface of the rim. Let us write expressions for these forces acting within the plane of the plate. We introduce the following notation: ρ is the fluid density, r_0 is the radius of the dry spot, g is the acceleration of gravity, and h is the thickness of the liquid film.

Gravity causes static pressure in the film:

$$p = \rho g z \tag{1}$$

The total action of pressure forces on the rim element in the plane of the substrate is equal to

$$P = \frac{1}{2}\rho g h^2 r_1 \delta\varphi \tag{2}$$

and is directed towards the center of the rim.

Capillary force N acting on the rim element in the plane of the substrate can be represented in the following form,

$$N = \sigma r_0 \delta\varphi \int_{(L)} (k_1 + k_2) cos\psi dS \tag{3}$$

Here σ is the surface tension of the liquid, k_1 and k_2 are the main curvatures of the rim surface at the points of contour L of the rim cross section.

According to Meunier's theorem [14], the curvature associated with the axial symmetry of the rim, can be written as

$$k_1 = \frac{\cos\psi}{r} \tag{4}$$

where r is the distance from a given point of the contour L to the axis of symmetry of the rim.

The second curvature related to the shape of the contour L, can be expressed as

$$k_2 = \frac{d\vartheta}{dS} \tag{5}$$

Where $\vartheta = \frac{\pi}{2} - \psi$ is the angle between the substrate and the tangent to the contour L.

Consequently,

$$N = \sigma\left[-\int_{(L)} \frac{\cos^2\psi}{r(\psi)} R(\psi) d\psi + \int_{(L)} \sin\vartheta \, d\vartheta\right] r_0 \delta\varphi \tag{6}$$

here $R(\psi)$ is the radius of the local curvature of the contour L. The contribution to the total force from the surface tension at the solid-liquid interface can be shown to be zero.

It is natural to assume that the contour L is a smooth curve with a continuous change (not necessarily monotonic) of the angle ϑ from $-\theta_0$ to 0, where θ_0 is contact angle. Then the second integral in equation (6) does not depend on the shape of L and is equal to (1- cos θ_0), so that equation (6) takes the form

$$N = \sigma(-K + 1 - \cos\theta_0) r_0 \delta\varphi \tag{7}$$

Here $K = \int_{(L)} \frac{\cos^2\psi}{r(\psi)} R(\psi) d\psi$

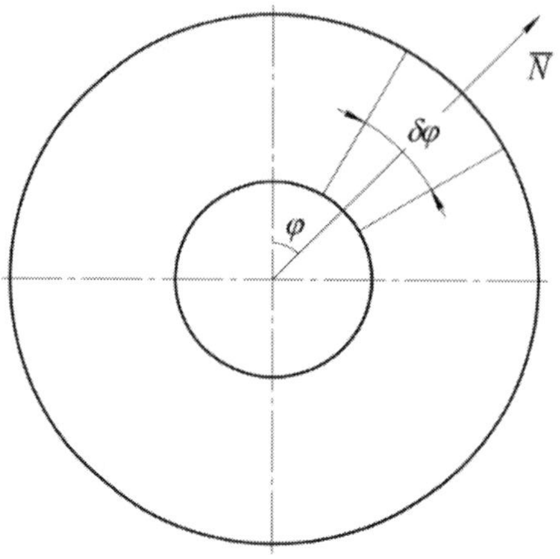

Figure 1. Schematic diagram of the liquid flow.

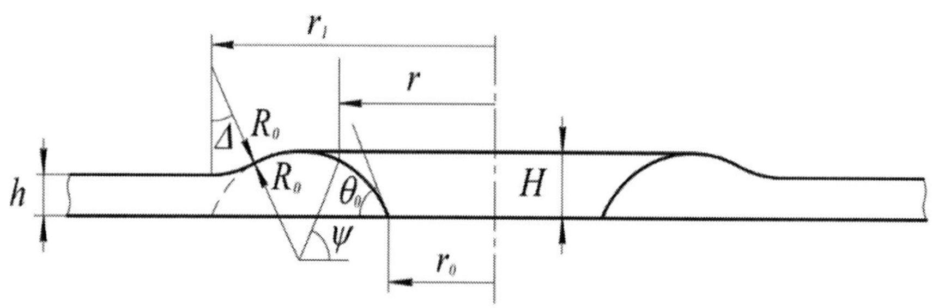

Figure 2. The cross-section of the rim surrounding the dry spot.

3 Condition of the critical state of a dry spot in the film

Under the forces acting on the rim elements, the rim begins to extend symmetrically, absorbing the liquid film, or conversely shrink. We neglect the liquid flow within the rim assuming that each element of the rim moves along the corresponding radius of the rim integrally. Then the behavior of a small dry spot in the film is determined by the sign of the total force acting on the rim element towards the radius of the dry spot. The spot disappears if this force is directed toward the center of the rim, and grows in size if the force is directed from the centre.

Hence, the equality

$$P = N \tag{8}$$

is the condition for the equilibrium state of the rim. Substituting the expressions for summands in this equality, we obtain the equation for determining the critical parameters of dry spot in the liquid film

$$\frac{1}{2}\rho g h^2 r_1 - \sigma(1 - \cos\theta_0 - K)r_0 = 0 \tag{9}$$

or in a dimensionless form

$$\frac{r_1}{2r_0 Bo} - (1 - \cos\theta_0 - K) = 0 \tag{10}$$

where $Bo = \sigma/\rho g h^2$ is the Bond number.

In the overwhelming number of works dealing with the modeling of dry spots and rivulets flowing along the interfaces of dry spots, the interfaces of the cross section of the rim or rivulet are approximated by circular arcs. Let us assume that the curved interface of the rim cross section is composed of two arcs of a circle with a radius R_0 (Fig. 2).

Thus, we believe that the cross-section of the free interface of the film in the area of a dry spot has the shape of an arc of a circle with a certain radius R_0, which is adjacent to the plate substrate at an angle θ_0. The most visible characteristic of the initial perturbation, obviously, is the rim height H, which is associated with the radius R_0 by the relationship

$$H = R_0(1 - \cos\theta_0) \quad (11)$$

Let assume the following scenario for the initial dry spot formation. A gas (steam) bubble appears in the liquid film directly on the wall. It is separated from the atmosphere by a thin film membrane. At some point, the membrane collapses and a round dry spot without a rim appears in the film.

In this case $H = h$ and the cross section of the rim is a half segment of a circle with a radius R_0 that is adjacent to the substrate at an angle equal to θ_0. Then

$$R_0 = \frac{h}{(1-\cos\theta_0)}, \qquad r_1 = r_0 + R_0 \sin\theta_0 \quad (12)$$

After the cross-sectional shape of the rim is known, the K integral in the expression for N (7) takes a definite value

$$K = K_0\left(\frac{\pi}{2} - \theta_0, \frac{\pi}{2}, \frac{r_0}{h}(1 - cos\theta_0) + sin\theta_0\right), \quad (13)$$

where $K_0(\psi_1, \psi_2, C) = \int_{\psi_1}^{\psi_2} \frac{\cos^2\psi}{C - \cos\psi}$.

Substituting all the necessary values in equation (10), we obtain the equation for determining the critical parameters of an initial perturbation in the following form

$$\frac{1}{2Bo}\left(1 + \frac{h}{r_0}\frac{sin\theta_0}{1-cos\theta_0}\right) - (1 - cos\theta_0 - K) = 0 \quad (14)$$

Solving equation (14) with respect to r_0/h, we see that for a given geometry of the free interface of the rim, critical value of the dry spot radius is determined by the liquid film thickness, Bond number and contact angle. Figure 3 shows the dependences of r_0/h versus (θ_0) for three values of Bond number. For clarity, Table 1 shows film thicknesses for water at various Bond numbers selected for the calculations.

Table 1.

g=9.8 m/s, ρ=1000 kg/m^3, σ=0.073 N/m			
Bo	3	10	100
h, mm	1.56	0.86	0.27

4 Conclusions

Critical ratios of the dry spot radius to the liquid film thickness are determined by Bond number and contact angle. In this case, at the contact angles greater than θ_0 = 70–80 degrees, the effect of Bond number on the ratio r_0/h is quite small, and the values of r_0/h are also small. However, at moderate and small values of θ_0, the role of the Bond number is extremely important, and the values of r_0/h become substantially greater than unity, i.e. for small values of θ_0 even large dry spots tend to disappear.

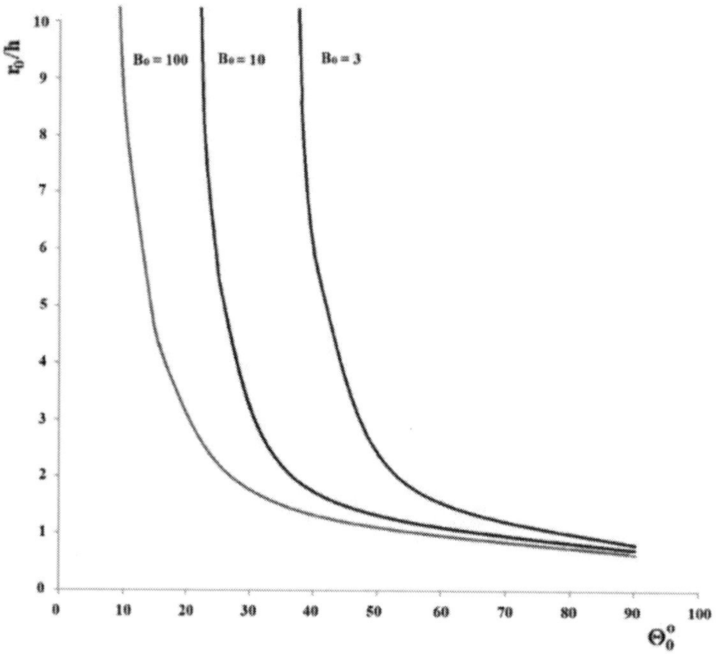

Figure 3. The calculation results of critical size of dry spot in the liquid film depending on the contact angle and Bond number.

We gratefully acknowledge the support from the Ministry of Education and Science of the Russian Federation (Agreement No. 14.613.21.0011, project identifier RFMEFI61314X0011)

References

1. Ajaev V.S., Interfacial Phenomena and Heat Transfer, **1**(1), 81–92 (2013).
2. Ajaev V.S., Gatapova, E.Ya., Kabov O.A., Advances in Colloid and Interface Science, **228**, 92-104 (2016).
3. Zaitsev D.V., Kabov O.A., Cheverda V.V., and Bufetov N.S., High temperature, **42**, 450 (2004).
4. Zaitsev D.V., Kabov, O.A., Microgravity science and technology, **19**, 174 (2007).
5. Zaitsev D.V., Rodionov D.A., Kabov O.A., Microgravity science and technology, **19**, 100 (2007).
6. Zaitsev D.V., Kirichenko D.P., Kabov O.A., Technical Physics Letters, **41**, 551 (2015).
7. El-Genk M.S., Saber H.H., International Journal of Heat and Mass Transfer, **44**, 2809 (2001).
8. Kadoura M., Chandra S., Experiments in Fluids, **54**, 1465 (2013).
9. El-Genk M.S., Saber H.H., J. Heat Transfer, **124**, 39 (2002).
10. Potash M. and Wayner Jr., Int. J. Heat and Mass Transfer, **15**, 1851 (1972).
11. Stephan P.C. and Busse C.A., Int. J. Heat Mass Transfer, **35**, 383-391 (1992).
12. Karchevsky A.L., Marchuk I.V. and Kabov O.A., Applied Mathematical Modeling, **40**, 1029–1037, (2016).
13. Lyulin Yu. V., Spesivtsev S. E., Marchuk I. V., and Kabov O. A., Technical Physics Letters, **41**, 1034–1037 (2015).
14. Korn G., Korn T. *Handbook of Mathematics* (Nauka, Moscow, 720 p., 1968).

MATEC Web of Conferences **84**, 00025 (2016) DOI: 10.1051/matecconf/20168400025
International Symposium IPHT 2016

The investigation of boiling crisis of nanofluids

Andrey Minakov[1,2,a], Maxim Pryazhnikov[1], Vladimir Zhigarev[1] and Valery Rudyak[3]

[1]*Siberian Federal University*
[2]*Institute of Thermophysics SB RAS*
[3]*Novosibirsk State University of Civil Engineering*

Abstract. Saturated boiling of nanofluids on a cylindrical heater with different diameters is experimentally studied. Studied nanofluids were prepared using distilled water and different metal oxides nanoparticles. The volume concentration of the nanoparticles was changed from 0.05 to 1%. It has been measured that the critical heat flux for nanofluids was much higher than for water. A strong dependence of CHF on the material and size of the nanoparticles and duration of boiling and size of heater was shown.

1 Introduction

Studies of the last two decades have shown that nanofluids have unusual transfer properties. In particular, small adds of nanoparticles into carrying liquid can considerably increase its heat conductivity and viscosity [1–4]. This stimulated many thermophysical applications of nanofluids, particularly, aimed at intensification of heat exchange. It has been found that nanofluids have really enhanced coefficients of heat transfer (see, for example, [5, 6] and literature cited there in). The further intention to increase this coefficient stimulated the study of heat transfer of nanofluids during boiling. These works have been performed quite intensely during the last decade. Nevertheless, the obtained results are rather contradictory. For example, in [7] it is noted that adding of nanoparticles does not change heat transport considerably, with an even decrease in the heat transfer coefficient during boiling been seen in [8]. Conversely, in [9–11], this coefficient rose.

In this paper an experimental study of saturated boiling of nanofluids on cylindrical heaters micro sizes. The aim of this study was to study the effect of nanoparticles on critical heat flux (CHF) during boiling. And also study the effect of the size of the heater. As the heater used nichrome wire of various diameters. Heater diameter ranged from 100 microns to 300 microns.

2 Experimental apparatus and procedure

In this work, the effect of nanoparticles on critical density of heat flow during boiling was studied. Studies of thermophysical properties of nanofluids have shown that they generally depend not only on the volume concentration of particles, but also on their sizes [3, 12, 13]. Therefore, in this work, nanofluids with considerably different nanoparticle sizes were considered. In all cases, the carrying liquid was distilled water and diamond nanoparticles, iron oxide (III) and SiO_2 nanoparticles with volume concentration from 0.05 to 1% were used. To prepare nanoparticles, a standard two stage process was used. After adding some quantity of the nanopowder to water, a vessel with the nanoliquid was put for half an hour in a UZDNA - ultrasonic disperser to decompose nanoparticle aggregates. All particle sizes given in the paper this is the sizes of conglomerates. Measurement of conglomerates distribution in the fluid in terms of their size was carried out by means of the CPS Disk Centrifuge DC24000. The particle size measurement carried out in the liquid. Therefore, the size of the conglomerates are taken into account automatically.

The scheme of the experimental setup scheme was described in [14-15]. The studied liquid was placed in a sealed glass flask 8 cm in diameter. The liquid volume in the flask was 300 mL. A nichrome wire with diameter d = 100, 200, 300 micron, length L = 34 mm fixed by copper cables was put as a heater in the vessel with the liquid. The flask with the studied liquid was sealed; therefore, the formed condensate dropped into the flask, thus sustaining saturation conditions in the working chamber. To control the liquid temperature in the flask, a chromel–copel thermocouple was used, which was connected to TMP200 temperature potentiometer . The end of the thermocouple was leveled with the nichrome heater at 2 cm distance from it. The reservoir with the studied liquid was placed into water bath in which, with the help of electric heater, a constant temperature was maintained, which was about 0.5° lower than the boiling temperature.

3 Results and discussion

The critical heat load of the nanofluids changes considerably with an increase in the volume concentration of nanoparticles. The dependence of relative thermal load q/q_w (q_w is the critical density of the heat flow for water) on the volume concentration of nanoparticles is shown in Fig. 1a. The here displayed values of heat flow critical density were obtained by averaging over five independent experiments. The critical density of heat for water was q_w = 1.32 MW/m^2.

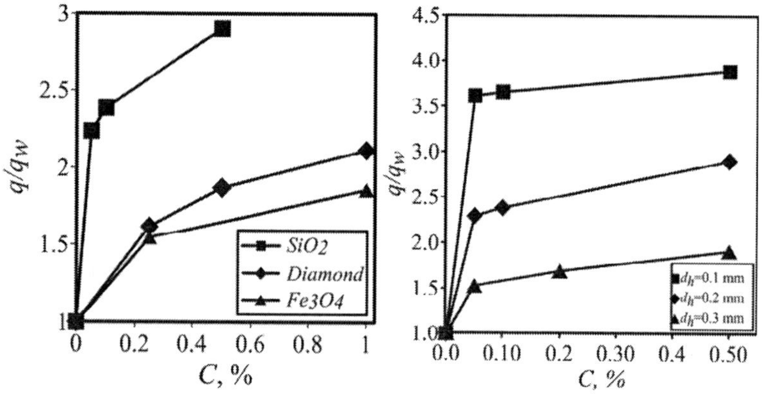

Figure 1. The dependence of critical heat flux density on the concentrations for different materials of nanoparticles (a), and for different diameter (b) of nichrome heater.

Analysis of heat flux density behavior during boiling of nanofluids shows that the use of nanoparticles brings about a considerable increase in the critical thermal load. Especially for nanoparticle SiO_2. As early as at a volume particle concentration of 0.25%, the critical density of the heat flux increases by more than 100% and continues to rise with further increase in the concentration of nanoparticles. Hence, use of nanofluids can significantly increase the value of the thermal load as compared to a conventional heat carrier and, consequently, this can increase the energy efficiency of setups in which the main process is boiling.

It was further investigated the influence of the diameter of the heater on the value of the critical heat flux at the boiling. These studies were conducted for SiO_2 nanofluids, because it showed the best results in the previous experiments (see Fig.1b). For these studies, a series of experiments with different diameters of the heaters was performed. As a heater used nichrome wire diameter 100, 200, 300 microns. The results of these measurements are shown in Fig.1b. It is seen that with decreasing diameter of the heater ratio increases dramatically for all concentrations.

DOI: 10.1051/matecconf/20168400025

Also we investigated the influence of the diameter of the nanoparticle on the value of the critical heat flux at the boiling. These studies were conducted for SiO_2 nanofluids and heater diameter 200 microns. The particle diameters ranged from 25 nm to 100 nm. The results of these measurements are shown in Figure 2. It is evident that with decreasing particle diameter ratio is reduced. We think that this is due to the deposition of particles on the wall heater. During deposition of particles on the surface of the formed roughened surface (see Fig.3). With increase of the particle diameter size of the heater surface roughness increases. It is known that rougher surfaces of the heat flux density higher than smooth.

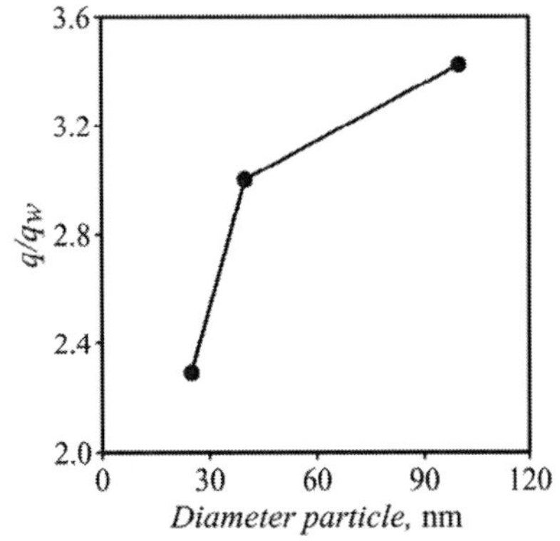

Figure 2. The dependence of heat flux density on the nanoparticles diameters (a)

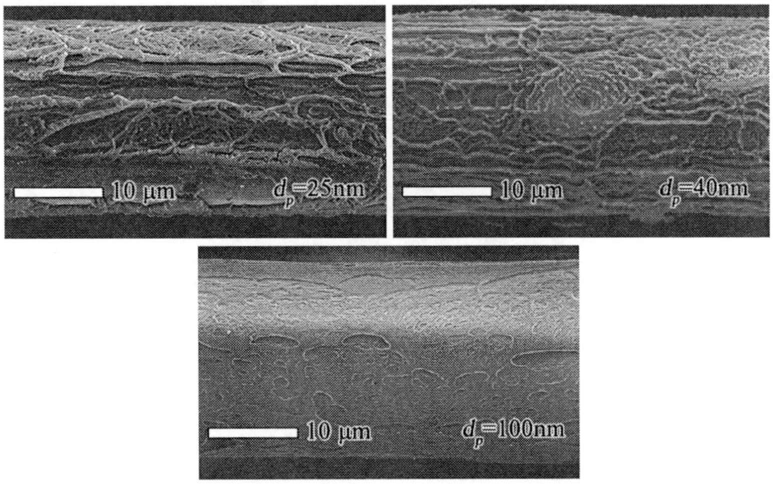

Figure 3. SEM images of nichrome wire surface after boiling for 0.05% SiO_2 concentration.

4 Conclusions

Boiling characteristics and CHF enhancement according to material, particle size, heater dimensions, and concentrations were investigated by pool boiling experiments of nanofluids. CHF is significantly enhanced in nanofluids (more than 3 times). CHF is enhanced even at very low concentrations. It is shown that there is a strong dependence CHF on the material and size of the nanoparticles. With increasing of the particle diameter the CHF increases. This is due to the dimensions of the roughness formed during the deposition of the particles. As the size of the heater flow is reduced. This is due to the influence of microflows around roughen the surface of the heater. It can assume that the surface particle interaction between the nanoparticles and the heating surface is the factor that causes an appreciable increase or decrease in CHF, more than what can be accounted for by the fluid property variation.

This work was partially financially supported by the Russian Science Foundation (Contract № 14-19-00312).

References

1. S. Das, S. U. S. Choi, W. Yu, T. Pradeep, *Nanofluids Science and Tecnology* (Wiley Interscience, New Jersey, 2007).
2. X.Q. Wang, A. S. Mujumbar, International J. Thermal Sciences **46**, 1 (2007).
3. V.Y. Rudyak, , A.V. Minakov, M.S. Smetanina, M.I. Pryazhnikov. Doklady Akademii Nauk. **467**(3), 1-3 (2016).
4. V. Ya. Rudyak, Advances in Nanoparticles **2**, 266 (2013).
5. V. I. Terekhov, S. V. Kalinina, V. V. Lemanov, Teplofiz. Aeromech. No. **2**, 173 (2010).
6. D. V. Guzei, A. V. Minakov, V. Ya. Rudyak, A. A. Dekterev, Techn. Phys. Lett. **40**, 203, (2014).
7. S. M. Kwark, R. Kumar, G. Moreno, J. Yoo, S. M. You, Int. J. Heat Mass Transfer **53** (5–6), 972 (2009).
8. S. K. Das, G. Narayan, and Baby A. K. Prakash, J. Nanopart. Res. **10** (7), 1099 (2008).
9. J. S. Coursey and J. Kim, Int. J. Heat Fluid Flow, **29** (6), 1577 (2008).
10. M. H. Shi, M. Q. Shuai, Z. Q. Chen, Q. Li, Y. Xuan, J. Enhanced Heat Transfer **14** (3), 223 (2007).
11. B. S. Fokin, M. Ya. Belen'kii, V. I. Al'myashev, V. B. Khabenskii, O. V. Al'myasheva, V. V. Gusarov, Pis'ma Zh. Tekh. Fiz. **35** (10), 1 (2009).
12. E. V. Timofeeva, D. S. Smith, W. Yu, D. M. France, D. Singh, J. L. Routbo, Nanotecnology **21**, 215703 (2010).
13. A.V. Minakov, A.S. Lobasov, D.V. Guzei, M.I. Pryazhnikov, V.Ya Rudyak. ATE. **88**,*140-148 (2015).*
14. A.V. Minakov, A.S. Lobasov, V.Y. Rudyak, D.V. Guzei, M.I. Pryazhnikov. Tech. Phys. Let. **40** (7), 562-564 (2014)
15. M.I. Pryazhnikov, A.V. Minakov, V.Y. Rudyak. Tech. Phys. Let. **41** (9), 891-893 (2015)

Effect of the Kapitza temperature jump on thermal processes in nanofluids

Sergey Novopashin, Marina Serebryakova[a] and Gennadiy Sukhinin

Kutateladze Institute of Thermophysics, Lavrentyev Ave., 1, 630090 Novosibirsk, Russian Federation

Abstract. Two analytical solutions describing thermal processes in a nanofluid based on spherical nanoparticles taking into account the Kapitza temperature jump on a particle-fluid boundary were found. In the first solution the thermal conductivity of nanofluids was found with the help of Maxwell approach. The second solution describes stationary heat exchange between a spherical particle and fluid in two different conditions. A dimensionless criterion characterizing the effect of the Kapitza temperature jump on thermal processes in nanofluids has been obtained in both solutions.

1 Introduction

A temperature jump at an interface between two media was detected by Kapitza [1] in experiments on the thermal conductivity of superfluid helium. This phenomenon is often referred to as the Kapitza interfacial thermal resistance (KITR). Since the KITR is low, it is observed in systems having small dimensions or a high thermal conductivity. Until recently, it has been thought that the Kapitza temperature jump at room temperature can be disregarded [2]. However, the development of nanotechnologies initiated investigations of this phenomenon in a wide temperature range and for various pairs of solids [3, 4]. The effect of the KITR on the thermal conductivity of solid composite materials was theoretically studied in [5]. The KITP at room temperature can vary in the range of $G = (2 \times 10^{-10} - 2 \times 10^{-7})$ m^2 K/W [4]. The corresponding characteristic scale for copper (thermal conductivity is $\kappa \approx 400$ W/(m K)) is evaluated: $L = G\kappa \approx 10^{-7}$–$10^{-4}$ m. The KITR at a fluid–solid interface was studied in [6]. The KITR can range from 3×10^{-9} to 10^{-8} m^2 K/W when the surface of the solid is lyophilic or lyophobic to the given fluid. The thermal conductivity of fluids is much lower, e.g., the thermal conductivity of water is $\kappa \approx 0.6$ W/(m K). The characteristic scale at which the contribution of the KITR should be taken into account for, e.g., water is estimated as $L = G\kappa \approx (2 \times 10^{-9} - 6 \times 10^{-9})$ m. The characteristic distance between particles in suspensions with nanoparticles (nanofluids [7]) with a characteristic volume content of about 1% is larger than the diameter of a particle by a factor of 3–5. Consequently, the Kapitza temperature jump can affect thermal processes in nanofluids. We indicate two thermal problems for which the presence of the Kapitza temperature jump can be important. The first problem is the effect of the KITR on the thermal conductivity of nanofluids. The second problem is heat transfer between a particle and a medium at the release of heat in nanoparticles, e.g., at hyperthermia with the heating of magnetic nanoparticles by an ac magnetic

[a]*e-mail: serebriakova.marina@gmail.com*

field [10, 11] or at a phase transition in the material of a nanoparticle used for the intensification of heat transfer [12, 13].

2 Thermal conductivity of nanofluids

Let us consider a linear distribution of temperatures at infinity (constant gradient of temperature): $\vec{\nabla} T = \vec{A} = const$ [14]. Indices p and f will be used to label the values of a particle and fluids. Thus, the coefficients of thermal conductivity of these regions are κ_p and κ_f, respectively. Let us assume that the radius of the particle is a. Applying standard boundary conditions, temperature equality and thermal flow continuity on contact surfaces, we find the distribution of temperatures using the following equations:

$$T_p(\vec{r}) = C_p \vec{A}\cdot\vec{r}, \ \ (r<a); \ \ T_f(\vec{r}) = \left(1 + C_p / r^3\right)\vec{A}\cdot\vec{r} \ \ (r \ge a) \tag{1}$$

Constants C_p and C_f are calculated with the help of conditions on the particle boundary:

$$T_p(a) - T_f(a) = Gq, \ \ q = -\kappa_p \frac{\partial T_p}{\partial r} = -\kappa_f \frac{\partial T_f}{\partial r} \ \ (r=a) \tag{2}$$

Here G is the Kapitza Interfacial Thermal Resistance (KITP). Calculations give temperature distributions:

$$T_p(\vec{r}) = \frac{3\kappa_f}{\kappa_p + 2\kappa_f(1+\kappa_p G/a)}\vec{A}\cdot\vec{r}; \ \ T_f(\vec{r}) = \left(1 + \frac{\kappa_f(1+\kappa_p G/a) - \kappa_p}{\kappa_p + 2\kappa_f(1+\kappa_p G/a)}\cdot\frac{a^3}{r^3}\right)\vec{A}\cdot\vec{r} \tag{3}$$

Calculating heat conductivity of nanofluids we will use Maxwell approach [9]. Let nanofluid ocupy a spherical region with radius R in the same base fluid. In this case temperature distribution in a base fluid in the accordance with (3) at $G = 0$ will be as follows:

$$T_f(\vec{r}) = \left(1 + \frac{\kappa_f - \kappa_{np}}{\kappa_{nf} + 2\kappa_f}\cdot\frac{a^3}{r^3}\right)\vec{A}\cdot\vec{r}, \tag{4}$$

where κ_{nf} is unknown heat conductivity. On the other hand, temperature distribution could be found as a sum of n spherical particles. Substituting $na^3 = \varphi R^3$, where φ is volume fraction of nanoparticles, and introducing dimensionless paremeter $\gamma = G\kappa_f/a$, we get a final dependence:

$$\frac{\kappa_{nf}}{\kappa_f} = \frac{1 + 2\kappa_f/\kappa_p - 2\varphi\left(\kappa_f/\kappa_p - 1\right) + 2\gamma(1-\varphi)}{1 + 2\kappa_f/\kappa_p + \varphi\left(\kappa_f/\kappa_p - 1\right) + \gamma(2+\varphi)} \tag{5}$$

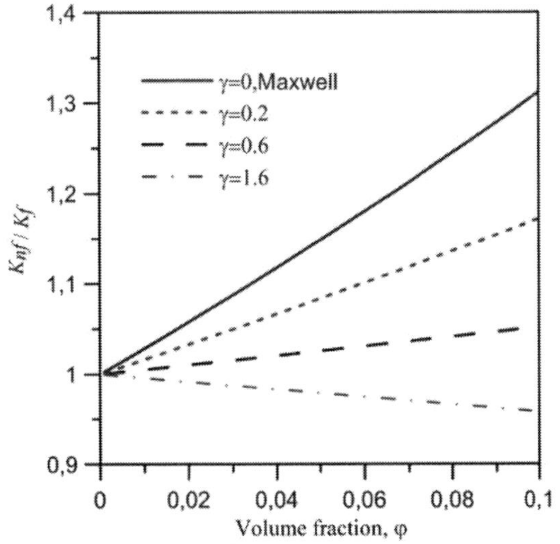

Figure1. The influence of KITP on thermal conductivity for a different value of parameter γ ($\kappa_p >> \kappa_f$).

The analysis of the formulae (5) shows that for any value of $\gamma > 0$ the KITP results in decreasing heat conductivity of nonofluids with respect to Maxwell's prediction. For practically important case $\kappa_p >> \kappa_f$, one gets $\kappa_{nf} = \kappa_f\left(1+2\varphi+2\gamma(1-\varphi)\right)/\left(1-\varphi+\gamma(2+\varphi)\right)$. Figure 1 shows the influence of KITP for different value of parameter γ. It is interesting to note that for the condition $\kappa_p >> \kappa_f$ and $\gamma = 1$, $\kappa_{nf} \approx \kappa_f$.

3 Heat exchange between a particle and a fluid

We solve a steady-state inhomogeneous heat conduction equation taking into account the Kapitza temperature jump at the boundary of a spherical particle of radius $\boldsymbol{a}$ under two different conditions. First, a heat source Q = const inside the spherical particle and temperature at infinity $\boldsymbol{T_\infty}$ is given. Second, the temperature of the particle $\boldsymbol{T_p}$ and the temperature at infinity $\boldsymbol{T_\infty}$ are given (this case corresponds to phase transition in a particle at temperature $\boldsymbol{T_p}$). Boundary conditions are the same as (2). The solution for both cases are obtained analytically. For the first case we have:

$$T_f(r)-T_\infty = \frac{q}{6\kappa_f}(a^2-r^2)+\frac{a^2 q}{3\kappa_f}\left(1+\frac{G\kappa_f}{a}\right), \qquad (r < a)$$

$$T_f(r)-T_\infty = \frac{a^3 q}{3\kappa_f}\cdot\frac{1}{r}, \qquad (r > a) \qquad (6)$$

For the second case the temperature distribution in fluid is given below:

$$T_f(r)-T_\infty = \frac{(T_p-T_\infty)}{(G\kappa_f/a+1)}\frac{a}{r}\,. \qquad (7)$$

In both cases considered above, the effect of the KITR is characterized by the dimensionless parameter $\gamma = G\kappa_f/a$ expressed in terms of the interfacial resistance, radius of the particle, and thermal conductivity of the fluid. Figure 2 shows the dependences of the temperature of the particle at a given thermal flux and the thermal flux at a given temperature difference between the particle and infinity as a function of the parameter γ. The temperature and thermal flux are normalized to the values in the absence of the Kapitza temperature jump.

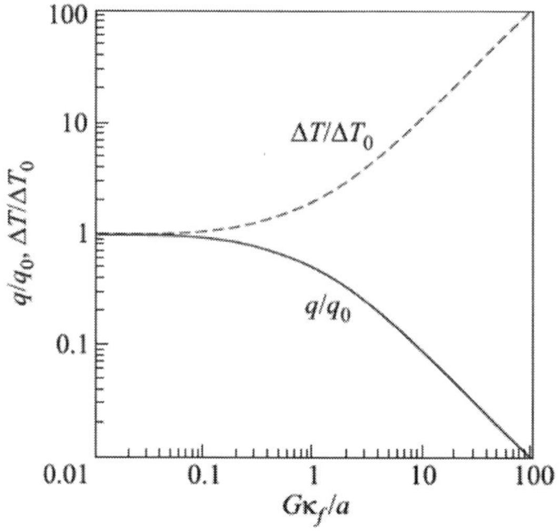

Figure2. Thermal flux (at given temperatures) and the temperature at the boundary of the particle at a given thermal flux versus the Kapitza temperature jump.

4 Conclusions

As calculations show, thermal processes in nanofluids can be characterized by the dimensionless parameter $\gamma = G\kappa_f / a$ expressed in terms of interfacial resistance, the radius of the particle, and thermal conductivity of the fluid. The general character of this parameter makes it possible to treat it as a similarity parameter when describing thermal processes in nanofluids with allowance for the Kapitza temperature jump.

The analysis of formulae (5) shows that the thermal conductivity of a nanofluid could be less than the thermal conductivity of a base fluid even when the thermal conductivity of particles is considerably higher than the thermal conductivity of the fluid (see Figure 1).

According to Figure 2, the Kapitza temperature jump should be taken into account when the parameter $\gamma > 1$. It is worth noting that nanoparticles used both in hyperthermia and at the intensification of heat transfer owing to the heat of a phase transition are particles in shell (core-shell). In this case, there are two contact surfaces and two Kapitza temperature jumps correspondingly. Furthermore, the finiteness of the shell also contributes to additional thermal resistance. For this reason, the total contribution can be significant when solving particular problems.

Let us make an important remark with respect to thermal processes in nanofluids and KITP. To get stable nanofluid with respect to coagulation of lyophobic nanoparticles different surfactants are used. Molecules of the surfactant create some shell around particles. As it is shown in [15], the contribution of this shell with thickness δ and thermal conductivity κ_s on thermal processes is the same as the value of the Kapitza temperature jump with parameter $G = \delta/\kappa_s$.

Acknowledgement

This work was supported by the Russian Science Foundation; Project no. 14-19-01379.

References

1. P. L. Kapitza, Zh. Eksp. Teor. Fiz. **11**, 1 (1941).
2. I. M. Khalatnikov, *Theory of Superfluidity* (Nauka, Moscow, 1971) [in Russian].
3. D. G. Cahill, W. K. Ford, K. E. Goodson, G. D. Mahan, A. Majumdar, H. J. Maris, R. Merlin, and S. R. Phillpot, J. Appl. Phys. **93**, 793 (2003).
4. H.-K. Lyeo and D. G. Cahill, Phys. Rev. B **73**, 144301 (2006).
5. C.-W. Nan, R. Birringer, D. R. Clarke, and H. Gleiter, J. Appl. Phys. **81**, 6692 (1997).
6. Z. Ge, D. G. Cahill, and P. V. Braun, Phys. Rev. Lett. **96**, 186101 (2006).
7. S. U. S. Choi, in Developments and Applications of Non-Newtonian Flows, Fluids Engineering Division FED, Vol. 231, Materials Division MD, Vol. 66 (ASME, New York, 1995), 99.
8. J. Buongiorno, D. C. Venerus, N. Prabhat, Th. McKrell, J. Townsend, R. Christianson, Y. V. Tolmachev, P. Keblinski, L.-W. Hu, J. L. Alvarado, I. Ch. Bang, S. W. Bishnoi, M. Bonetti, F. Botz, A. Cecere, et al., J. Appl. Phys. **106**, 094312 (2009).
9. J. C. Maxwell, *A Treatise on Electricity and Magnetism*, 2nd ed. (Clarendon, Oxford, UK, 1881).
10. S. Laurent, S. Dutz, U. O. Hafeli, and M. Mahmoudi, Adv. Colloid Interface Sci. **166**, 8 (2011).
11. I. Sharifi, H. Shokrollahi, and S. Amiri, J. Magn. Magn. Mater. **324**, 903 (2012).
12. D. A. Scott, A. Lampureux, and B. R. Baliga, ASME J. Heat Transfer **135**, 011013 (2013).
13. W. Wu, H. Bostanci, L. C. Chow, Y. Hong, S. J. Ding, M. Su, and J. P. Kizito, ASME J. Heat Transfer **135**, 052202 (2013).
14. Landau L.D., Lifshitz E.M. *Fluid Mechanics*, (Pergamon Press, 1987).
15. Sukhinin G.I., Serebryakova M.A. and Novopashin S.A. J. Heat Transfer (2016).

MATEC Web of Conferences 84, 00027 (2016) DOI: 10.1051/matecconf/20168400027
International Symposium IPHT 2016

Study of heat transfer in the bubble meniscus with an array of temperature micro-sensors

Evgeniy Orlik[1,2,a], Elena Bykovskaya[1] and Haruhiko Ohta[3]

[1] *Institute of Thermophysics, 1, Lavrentiev Ave., 630090, Novosibirsk, Russia*
[2] *Novosibirsk State Agricultural University, 630039, Novosibirsk, Russia*
[3] *Kyushu University, 744 Motooka, Nishi-ku, Fukuoka 813-0044, Japan*

Abstract. Experimental setup has been designed and manufactured to study the evaporation processes in the bubble meniscus. New method for accurate measurements of local heat transfer has been proposed in collaboration with Kyushu University. Two side transparent coatings (thin film uniform heater and an array of temperature micro-sensors) are developed on sapphire substrates. Application of such substrate allows one to investigate heat transfer in the vicinity of the contact line. The shadow technique is used to define shape and contact angles of bubble.

1 Introduction

In recent years, progress in electronic technology resulted in increasing heat dissipation from semiconductor chips. Thus there is a strong demand for thermal management systems with higher performance compared with the existing ordinary convective air or single-phase liquid systems [1, 2]. The cooling by phase change attracts much attention for micro-electronics and space applications because of potentially high efficiency at low mass flow consumption [3, 4, 5]. It is known that heat transfer increases by reducing thickness of liquid film. The experimental work shows that heat transfer near the contact line is three times higher compared to average for sessile drop on electrically heated thin foil [6]. Some experimental works related to vapor bubble and its meniscus can be founded in the literature [7]. The goal of the present experiment is to study the boiling and condensation of vapor in the bubble, and also the process of evaporation at the meniscus, rupture of liquid layer, dynamics of receding contact line, under the control of the heating from the substrate. These studies are performed in collaboration with Kyushu University, where a unique technology of manufacturing micro-sensors (micro-heaters) have been developed that provides high accuracy of heat transfer measurements [8]. Micro-sensors create relief on surface, which may significantly affect the behavior of the contact line and the evaporation. Analysis of the effect of patterned and structured surfaces on contact line and its dynamic has been reviewed in [9].

2 Experimental setup

Experimental setup consists of three elements (Fig. 1): a transparent test cell with inner size of $60\times60\times8mm^3$, a pressure damper connected to the test cell and a transparent cover box with air fan electrically heated to control surround temperature. A pressure damper is mounted on a platform which

[a] Corresponding author: orlik.evgeniy@gmail.com

 DOI: 10.1051/matecconf/20168400027

allows changing hydrostatic pressure and equipped with Peltier elements controlling the temperature in a liquid.

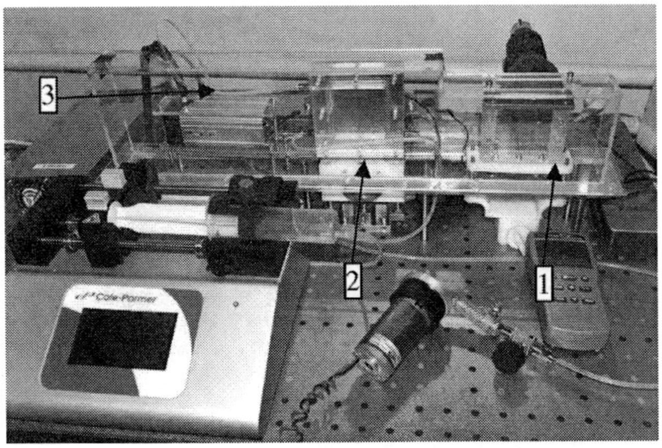

Figure 1. Photograph of experimental setup: 1 –test sell, 2 – pressure damper, 3 – transparent cover box.

The bubble shape is visualized from the top and from the side with the help of the schlieren technique and shadow technique with resolution of 6 µm/pix respectively, (as shown in Fig. 2 and Fig. 3). The images from the side (Fig. 4) are processed in the Drop Shape Analysis software by KRÜSS using different methods including Young-Laplace method [10]. The result of the contact angle hysteresis (CAH) measurements of water on the uniform transparent heater (Fig. 5) is shown in Fig. 6. CAH on uniform transparent heater more than two times less compare surface without coating (15° compare 40° correspondingly). Using of fast cameras and infrared scanner is also expected.

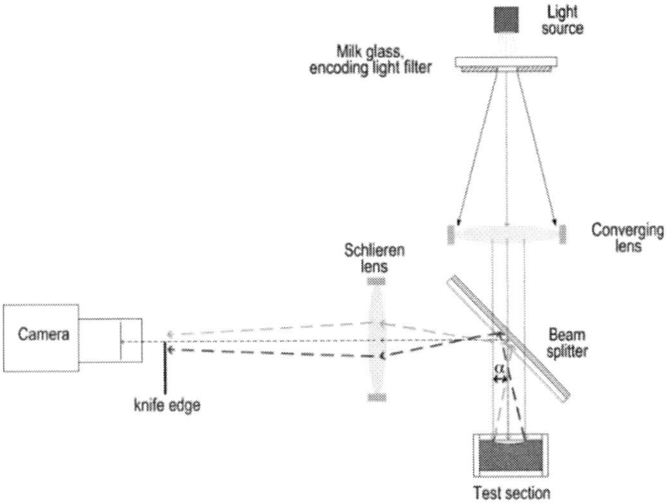

Figure 2. Schlieren method from the top.

 DOI: 10.1051/matecconf/20168400027

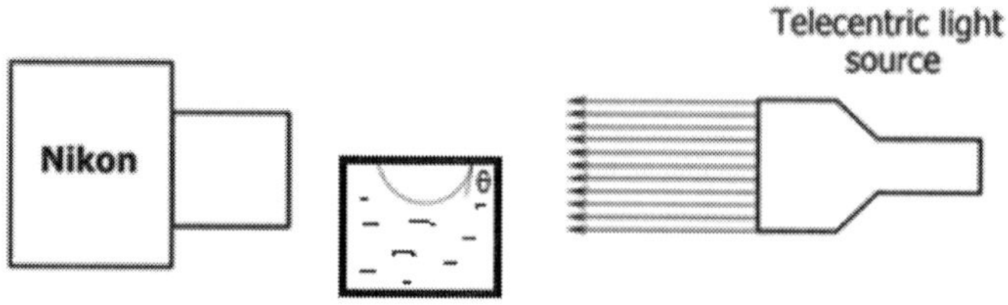

Figure 3. Shadow technique with optical resolution of 6μm/pix from the side.

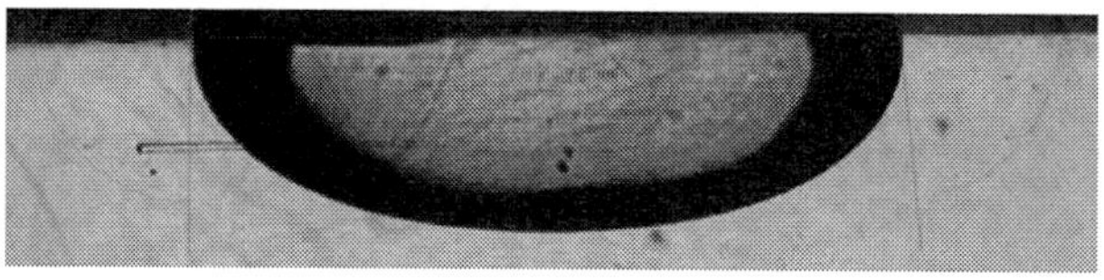

Figure 4. Photograph of the bubble in water on the transparent heater.

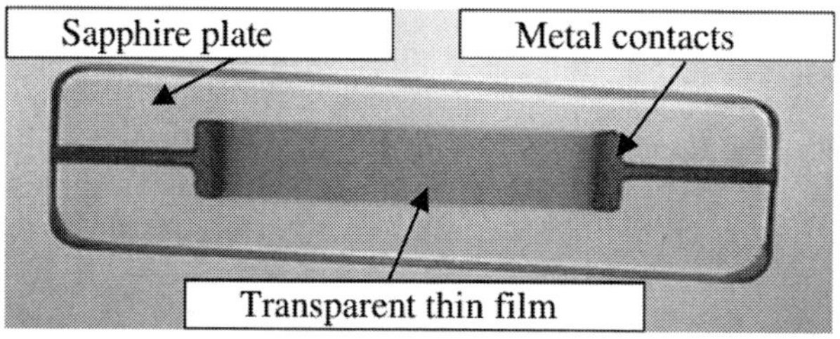

Figure 5. Photograph of sapphire substrate with thin film heater.

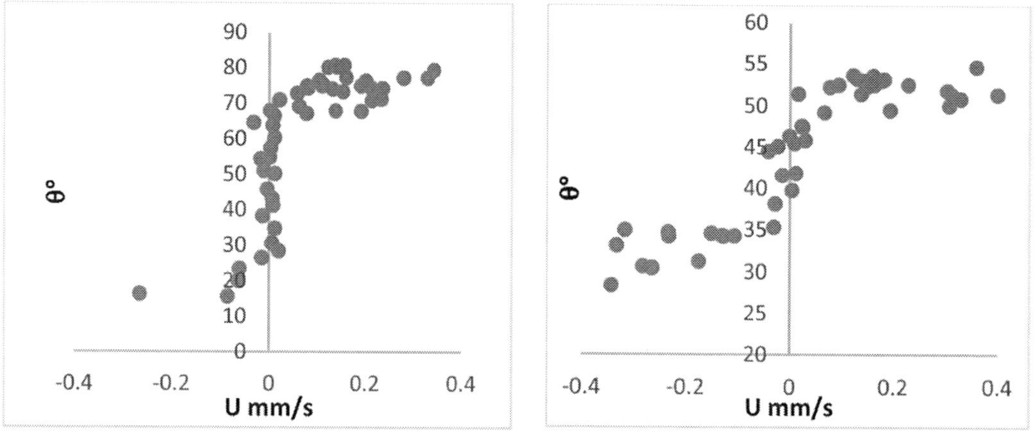

Figure 6. Contact angle hysteresis of water on the sapphire (left) and transparent heater (right).

Two types of substrates are used: 1) 3 mm thick sapphire plate with a uniform transparent thin film heater (Fig. 5) made of indium tin oxide (ITO). Thickness of the ITO layer is 1.1 micron, which has been coated by 450 nm silicon dioxide (SiO_2) layer to prevent electrolysis. Thin film heater stay in direct contact with liquid and it is considered that heat losses are negligible. 2) Substrate for local heat transfer measurements using temperature platinum micro-sensors is shown in Fig 7. The surface area of a single sensor is $3\times0.5mm^2$. Distance between sensors is 0.1 mm. The temperature platinum micro-sensors stay in contact with a liquid and measure local surface temperature. Heat flux is known from electrical power, hence local heat transfer coefficient can be found from temperature surface and heat flux.

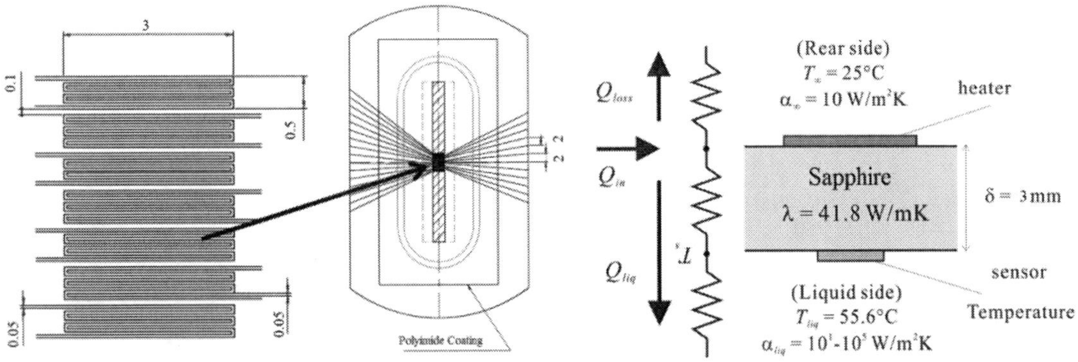

Figure7. Scheme of substrate with an array of temperature micro-sensors for heat transfer measurements.

3 Conclusions

Manufactured experimental is used to perform wide investigations of vapor bubble meniscus. A unique technology of manufacturing micro-sensors (micro-heaters) has been developed that provide high accuracy of heat transfer measurements. Transparent substrate and test cell allows observed bubble from all sides using different optical techniques including infrared diagnostics.

We gratefully acknowledge the support from the Ministry of Education and Science of Russia (Project identifier RFMEFI61314X0011).

References

1. Bar-Cohen A., Sherwood G., Hodes M., and Solbreken G., IEEE Transactions on Components, Hybrids, and Manufacturing Technology, **18**, 3, pp. 502–509, (1995).
2. Mahajan R., Chiu C., Chrysler G., Proceedings of the IEEE, **94**, 8, 1476–1486, (2006).
3. Kabov O.A., Chinnov E.A., High temperatures, **39**, 703-713 (2001).
4. Iorio C.S., Goncharova O.N. and Kabov O.A., Micrograviy sci. technol., **21**, 313–319 (2009).
5. Tomoki Hirokawa, Haruhiko Ohta and Oleg A. Kabov, Interfacial Phenomena and Heat Transfer, **3**, 303–317, (2015).
6. I. Marchuk, A. Karchevsky, A. Surtaev, O. Kabov, International Journal of Aerospace Engineering, **2015**, 1 (2015).
7. An Zou, Ashish Chanana, Amit Agrawal, Peter C. Wayner, Jr. & Shalabh, C. Maroo1, Scientific Reports, **6**, 20240 (2016)
8. O. Kawanami, H. Ohta, O. Kabov, Y. Sakata, Y. Kotani, Y. Asada, T. Nagayasu, Y.Shinmoto, S. Chikov, P. Queeckers, J.Straub, Microgravity sci. technol., **21**, 3–8, (2009).
9. V.S. Ajaev, E.Ya. Gatapova, O.A. Kabov, Advances in Colloid and Interface Science, **228**, 92–104, (2016).
10. I.V. Marchuk, V.V. Cheverda, P.A. Strizhak, O.A. Kabov, Thermophysics and Aeromechanics **22**, 297–303, (2015).

The effect of temperature and gas Reynolds number on evaporation of a sessile liquid drop in mini-channel

Evgeniy Orlik[1,2,a], Elena Bykovskaya[1] and Bo-Feng Bai[3]

[1] *Institute of Thermophysics, 1, Lavrentiev Ave., 630090, Novosibirsk, Russia*
[2] *Novosibirsk State Agricultural University, 630039, Novosibirsk, Russia*
[3] *State key lab of Multiphase Flow in Power Engineering, Xi'an Jiaotong University, 710049, Xi'an, China*

Abstract. Experimental setup has been designed and manufactured to study the evaporation processes of liquid drop under blowing gas in mini-channel. The height of channel can be varied from 3 to 20 mm. Substrates are removable and its surface temperature is kept to constant value. The shadow method is main measurement technique. Series of experiments with 100 μl water drop on polished stainless still substrate are carried out in channel with 9 mm height. Dependences of evaporating rate for different range of temperatures and gas Reynolds numbers are obtained.

1 Introduction

Drop is a fairly simple physical object containing different complex phenomena. Evaporation of sessile liquid droplet is a key point of many processes such as painting, coating, humidification, the agriculture irrigation. Experimental studies show that the wettability and contact angle hysteresis have a significant effect on the drop evaporation rate [1, 2] and thermocapillary film rupture [3]. It is known that surfactants affect the wettability, hence there is a need that properties of test liquid be carefully characterized. Also the contact angle hysteresis has an effect on the spreading of the drop on a solid surface under gravity, which is shown in [4]. Effect of gravity on evaporation has to be considered in space applications. In recent paper [5] it is shown that Marangoni effect (flow driven by surface tension gradient) modifies the drop interface shape and impacts on evaporation. Deep understanding of evaporation requires knowledge what is happening inside droplet. Experimental works and numerical simulations show that evaporation leads to convective flow inside droplet with appearing vortex structures [6]. Such experimental investigation requires development of new optical methods, one of them based on Micro-PIV. Gas action on evaporation of drop is also known and studied [7], but mostly all investigations have been carried out in open ambient air, where conditions are not very controlled. Dynamics of drop on surface by a shearing air in channel have been studied in [8], it is shown strong influence of contact angle hysteresis and drop sizes on the dynamics. This experimental work presents results of evaporation of sessile drop on a heated substrate in mini channel under blowing air laminar flow.

[a] Corresponding author: orlik.evgeniy@gmail.com

2 Experimental setup

Experimental setup with changeable height of channel from 3 to 20 mm is shown in Fig. 1. The goal is to study the effect of the substrate temperature and Reynolds number of air flow on the evaporation rate of a sessile drop. Distilled deionized nano-filtered water was used as the working liquid which is obtained from Milli-Q system. Initial drop volume was slightly varied and was around 100 µl. Polished stainless steel substrate was used in the experiments. The substrate temperature was maintained at 25.0, 50.0 and 70.0 ±0.25 °C using Peltier elements and PR-59 controller. The gas temperature was held at 25 °C and Reynolds number was varied from 0 to 2000 using Bronkhorst flow regulator. Flow rate in outlet was measured with Bronkhorst flow meter. Developed laminar flow was obtained due to long inlet length which is 10 times of channel height. The channel was equipped with optical glasses from top and side. The drop shape was visualized from the top and from the side with the help of the shadow technique with resolution of 6 µm/pix (as shown in Fig. 2 and 3). The images from the side were processed with the Drop Shape Analysis software by KRÜSS using different methods including Young-Laplace method [9]. The advancing static contact angle was θ=85±6°.

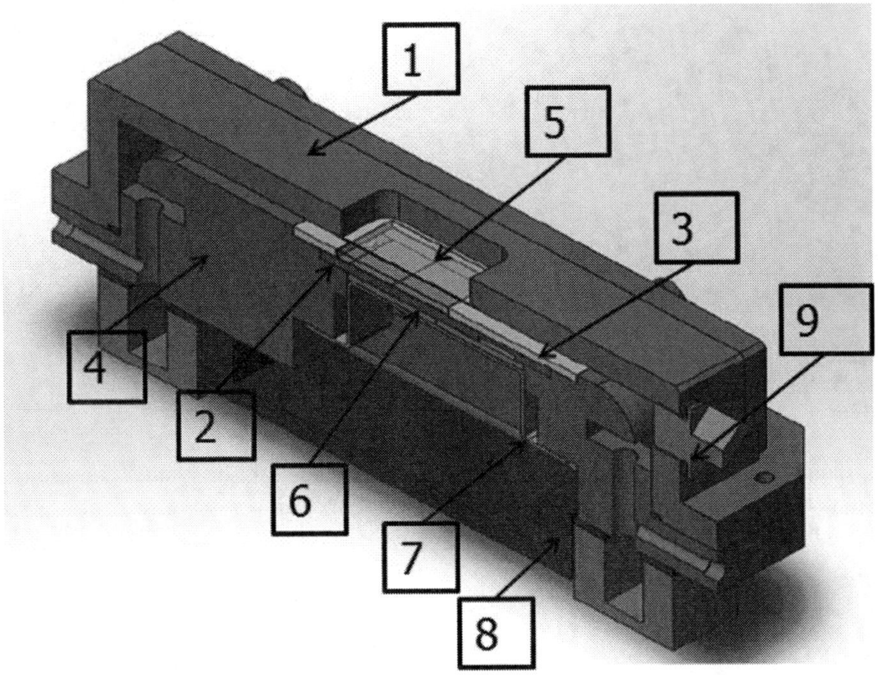

Figure 1. Experimental setup. 1 – cover, 2 – copper insert, 3 – stainless steel substrate, 5 – optical window, 6 – Peltier element, 7 – heat exchanger, 8 – support, 9 – port for syringe.

 DOI: 10.1051/matecconf/20168400028

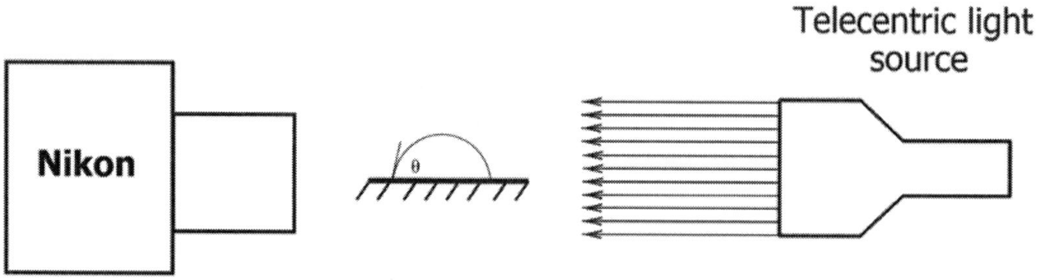

Figure 2. Schematic of the shadow technique with optical resolution of 6μm/pix.

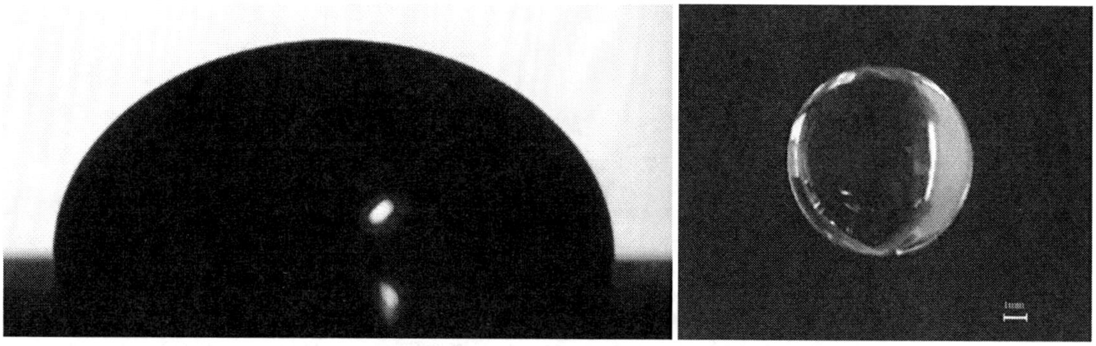

Figure 3. Photographs of 100 μl water drop on polished stainless still substrate (side and top view).

3 Results

First experimental results have been obtained in channel with height of 9 mm. The life time of liquid droplet on unheated substrate depending on Reynolds number is shown in Fig. 4. Evaporation rate which is inversely to life time is proportional to Reynolds number at power 0.285.

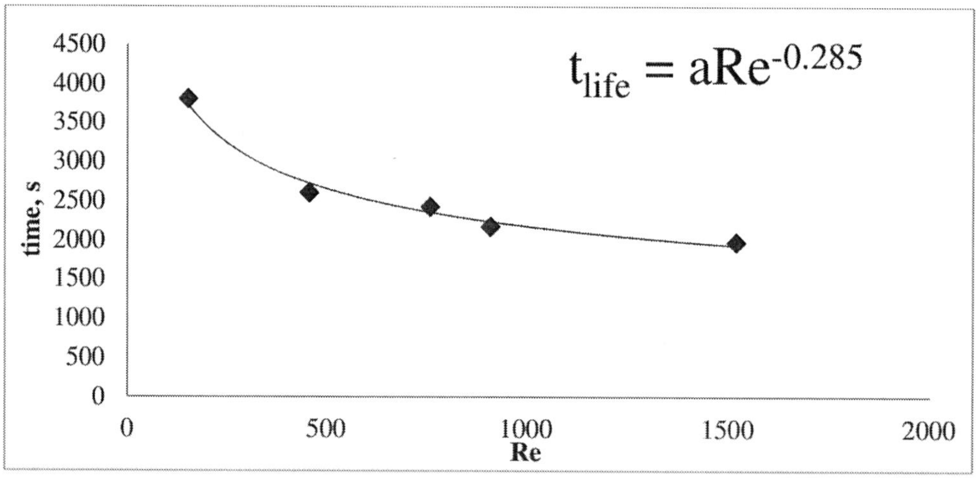

Figure4. Evaporation time of water drop vs. gas Reynolds number. Temperature of substrate is 25 °C.

Influence of substrate temperature on evaporation rate is shown in Fig. 5. It is seen that the rate of the drop mass loss (evaporation rate) is almost constant with time. From analysis of the experimental data, we obtained that in the range of Re=1140-1900 the evaporation rate is defined by the substrate temperature as: $dV/dt \approx 0.042 + 0.004*(T_{substrate}-T_{25})$. Comparison of evaporation rates dV/dt depending on substrate temperature T and gas velocity U for different liquids is shown in Table 1, the data for Ethanol and HFE7100 taken from [10]. Mass evaporation rate Q in units of kg/(s m^2) has been measured in [10], but evaporation area in experiments [10] is constant, hence Q~dV/dt. From Table 1 one can see that the data are in qualitative agreement.

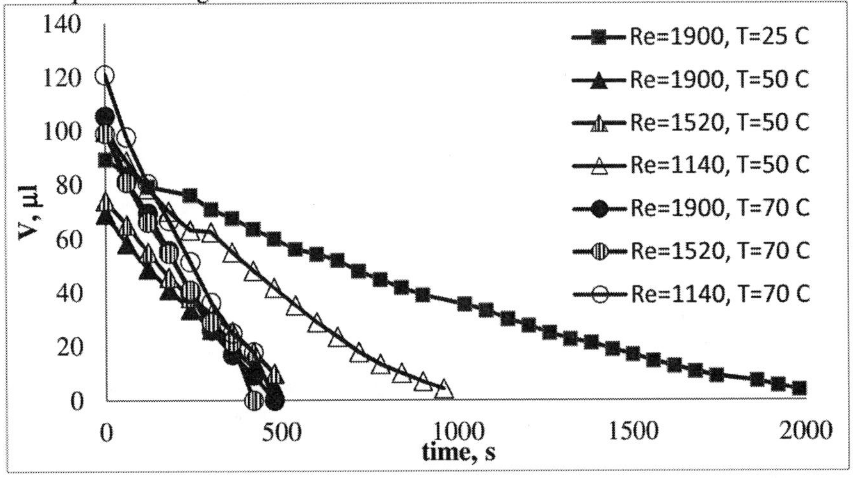

Figure 5. Change of drop volume with time for different substrate temperatures and gas Reynolds numbers.

Table 1.

$dV/dt \sim T^a$, $dV/dt \sim U^b$	Water sessile drop	Liquid surface (Ethanol –air)	Liquid surface (HFE7100-Nitrogen)
a	1	1.17	1.22
b	0.285	0.54	0.176

The work was financially supported by the Russian Ministry of Education and Science (Project identifier RFMEFI61315X0038).

References

1. E.Ya. Gatapova, A.A. Semenov, D.V. Zaitsev, O.A. Kabov, Colloids and Surfaces A: Physicochemical and Engineering Aspects, **441**, 776, (2014)
2. A.A. Semenov, D.V. Feoktistov, D.V. Zaitsev, G.V. Kuznetsov, O.A. Kabov, Thermophysics and Aeromechanics, **22**, 771, (2015)
3. D.V. Zaitsev, D.P. Kirichenko, and O.A. Kabov, Technical Physics Letters, **41**, 551, (2015)
4. O.A. Kabov, D.V. Zaitsev, Doklady Physics, **58**, 292, (2013)
5. Y. Tsoumpas, S. Dehaeck, A. Rednikov, P. Colinet, Langmuir, **31**, 13334, (2015)
6. P.J. S´aenz, K. Sefiane, J. Kim, O.K. Matar, P. Valluri, Journal of Fluid Mechanics **772**, 705–739, (2015)
7. L. Bin, R. Bennacer, A. Bouvet, Applied Thermal Engineering, **31**, 3792–3798, (2011)
8. J. Fan, M.C.T. Wilson, N. Kapur, Journal of Colloid and Interface Science, **356**, 286–292, (2011)
9. I.V. Marchuk, V.V. Cheverda, P.A. Strizhak, O.A. Kabov, Thermophysics and Aeromechanics **22**, 297–303, (2015)
10. Yu. V. Lyulin, D. V. Feoktistov, I. A. Afanas'ev, E. S. Chachilo, O. A. Kabov, G. V. Kuznetsov, Technical Physics Letters, **41**, 665–666, (2015)

Numerical modeling of flow structure and heat transfer in a mist turbulent flow downstream of a pipe sudden expansion

Maksim Pakhomov[a] and Viktor Terekhov

Kutateladze Institute of Thermophysics, Siberian Branch of the Russian Academy of Sciences, 630090, Novosibirsk, Russia

Abstract. Turbulent droplet-laden flow downstream of a sudden pipe expansion is numerically studied using Eulerian two-fluid model. The model is used to investigate the effect of droplet evaporation on the particle dispersion and on the gas phase turbulence modification. Turbulence suppression in the case of evaporating droplets is hardly observed near the wall, and the level of turbulence tends to the corresponding value for the single-phase flow regime. In the flow core, where evaporation is insignificant, a decrease in the level of gas turbulence (to 20% as compared to a single-phase flow) can be observed. The maximal effect of droplet evaporation is obtained in the wall region of the tube. A considerable increase in the maximal value of heat transfer on adding the evaporating droplets to the separated flow is shown (more than 2-folds as compared to the single-phase flow at a small value of droplet mass concentration of $M_{L1} \leq 0.05$). The addition of the solid non-evaporating particles causes a slight increase in the maximum value of heat transfer in the case of small particles and a decrease in heat transfer in the case of large particles.

1 Introduction

Two-phase droplet-laden separated flows are observed in many engineering and natural processes, such as cyclonic separation, flame stabilization in internal combustors, pneumatic transport, and many others. The separated flow is a typical two-dimensional shear flow consisting of several zones: main core flow, shear layer, recirculation, and flow relaxation regions. Each zone has specific features and typical length and time scales. The interactions between finely dispersed phase and turbulent gas-phase flows are very complex, and many of these interactions remain poorly understood [1]. Gas–droplet flows are usually inhomogeneous and anisotropic and often include flow separation and heat transfer between two-phase flow and the wall surface. A review of experimental and numerical works revealed that the effect of dispersed phase on the turbulence modification by particles in two-phase separated flows is an extremely complex phenomenon even in the case of gas–particle flows without phase changes.

The aim of the present short communication is to examine the effect of evaporating droplets on gas turbulence modification and heat transfer enhancement in sudden pipe expansion flow. The number of papers that have examined two-phase separated flows with evaporating droplets is very limited [2, 3]. These few studies do not provide sufficient information to evaluate all the factors

[a] Corresponding author : pakhomov@ngs.ru

affecting the flow structure, particle dispersion, and turbulence modification in sudden pipe expansion flows with evaporating droplets.

2 The mathematical model and numerical realization

The authors of the present paper used the Eulerian two-fluid approach for the modeling of dispersed phase, which treats the particulate phase as a continuous medium with properties analogous to those of a fluid [3]. This technique involves the solution of a second set of Navier–Stokes-like equations in addition to those of the carrier (gas) phase. A one-point probability density function (PDF) [4, 5] of the particle velocity for describing the action of small particles on turbulence is used. The governing mean and fluctuating equations for both phases are described in detail [3]. Gas–droplet turbulent flow is numerically predicted by the set of steady-state axisymmetrical Reynolds averaged Navier-Stokes (RANS) equations. Gas phase turbulence was modeled with the use of elliptic-blending second-moment closure [6]. Two-way coupling is achieved between dispersed and carrier phases in the mean and fluctuating transport. Particles' mean and fluctuating flow interactions are described by the two-way coupling model [3]. The volume fraction of the dispersed phase is lower $\Phi_1 < 10^{-4}$ and the droplets are fine ($d_1 < 100$ μm); therefore, the effects of inter-particle collisions are neglected when treating the hydrodynamic and heat and mass transfer processes in the two-phase flow [3]. Validation analysis of the developed model for both single-phase and two-phase flows in sudden pipe expansion was performed in [3].

The mean transport equations for both gas and dispersed phases and the second moment closure model are solved using a control volumes method on a staggered grid. The QUICK scheme [7] is used to approximate the convective terms, and the second-order accurate central difference scheme is adopted for the diffusion terms. The velocity correction is used to satisfy continuity through the SIMPLEC algorithm [8], which couples velocity and pressure.

The first cell is located at a distance $y_+ = yU_*/\nu = 0.3$–0.5 from the wall, where U_* is the friction velocity obtained for the flow in the inlet pipe. At least 10 control volumes have been generated to be able to resolve the mean velocity field and turbulence quantities in the viscosity-affected near-wall region ($y_+ < 10$). Grid sensitivity studies are carried out to determine the optimum grid resolution that gives the mesh-independent solution. For all numerical investigations performed in the study, a basic grid with 350×100 control volumes along the axial and radial directions is used. A more refined grid is applied in the recirculation region and in the zones of flow detachment and reattachment.

3 Numerical results and discussion

Numerical modeling is carried out for a monodisperse gas–droplet mixture in the initial cross-section, and then the droplet size is changed due to the evaporation process. The diameter of the pipe is $2R_1 =$ 20 mm before expansion and $2R_2 = 60$ mm behind the expansion, the expansion ratio is $\text{ER} = (R_2/R_1)^2$ = 9, and the step height is $H = 20$ mm. The mean-mass gas flow velocity before detachment is $U_{m1} =$ 10–30 m/s, and the Reynolds number for the gas phase varies within the range $\text{Re}_H = HU_{m1}/\nu = (1.33$–$4) \times 10^4$. The initial velocity of the dispersed phase is $U_{L1} = 0.8U_{m1}$. The Stokes number in the mean motion is Stk = 0.04–5, and the initial mass fraction changes within the range $M_{L1} = 0$–0.1. Computations are performed at a uniform wall heat flux density $q_W = 1$ kW/m^2. The case with the particles of $d_1 = 30$ μm, Stk = 0.4 is the basic one in this study.

All simulations are performed for both the evaporating and non-evaporating cases. In the case of non-evaporating particles, the predictions are carried out without consideration of the phase transition of the dispersed phase and with consideration of the thermal-physical properties of the two-phase flow. This case is artificial, but it allows us to analyze the effect of dispersed phase evaporation on the transfer processes and heat transfer in the separated two-phase flow.

The radial profiles of mass fraction of the dispersed phase are shown in Fig. 1a for the cases of evaporating droplets (continuous lines) and the non-evaporating particles (dashed line). A sharp decrease in the concentration of particles is observed due to their dispersion over the tube cross-section in the separation region. It should be noted that the low-inertia droplets at low Stokes numbers of Stk < 1 (d_1 < 50 μm) (lines *1* and *2*) are entrained well in the separated flow, and are available in almost the entire tube cross-section. The near-wall zone of the pipe (r/H > 1.2) is practically free from particles due to intense evaporation. The large droplets with Stk = 4 (d_1 = 100 μm) are not involved in the area of the recirculation flow but are mainly in the shear mixing layer and in the flow core. In the case of non-evaporating particles, their concentration in the wall region of the tube is higher than in the case of the evaporating droplets.

The profiles of the turbulent kinetic energy (TKE) in the two-phase separated flow along the pipe are shown in Fig. 1b, where k_0 is turbulence of the gas phase in the single-phase flow (without particles or droplets). In the two-phase flow, the TKE is suppressed by finely dispersed droplets in comparison with the single-phase one. This effect increases with the size of the dispersed phase, which is consistent with the data of studies on separated flow with solid particles [1, 2, 7]. Small particles (Stk < 1) are involved well in the turbulent motion of gas and take some energy away from the carrier medium. The turbulence suppression in the case of evaporating droplets is not observed in the wall zone and $k/k_0 \approx 1$ because this area is free of particles and the level of turbulence tends to the corresponding value for the single-phase flow regime. In the flow core, where evaporation is insignificant, a decrease in gas turbulence is observed (to 20%). It should be noted that the maximal effect of droplet evaporation is noticeable in the wall region of the pipe (the difference is up to 10%). An increase in TKE is observed and $k/k_0 \approx 1$ far from the cross-section of the flow detachment (x/H = 15) because most of the droplets have already evaporated.

The effect of the Stokes number (the initial size of droplets) on the magnitude of maximal Nusselt numbers is shown in Fig. 2. The local Nusselt number is calculated by the following relationship: $\mathrm{Nu} = q_W H/[\lambda(T_W - T_m)]$, where λ is the coefficient of heat conductivity of the gas flow, and T_W and T_m are the wall temperature and mean-mass temperature of the gas flow, respectively. The increase of the droplets' mass fraction intensifies the heat transfer significantly. This effect cannot be explained by flow turbulization on the addition of the dispersed phase. The main reason of the heat transfer intensification is the using of latent heat of droplets evaporation. This effect increases with an increase in the amount of dispersed phase. Two characteristic areas in the distribution of Nu_{max} in the investigated range of particle sizes should be noted. An increase in the initial droplet diameter has the more complex influence on heat transfer in the two-phase separated flow. Heat transfer enhancement is observed in the range of small particles, and then a drastic decrease occurs at Stokes numbers of Stk > 0.25. The dispersed phase is not involved in the separated motion for the largest droplets with diameter d_1 = 100 μm and Stk = 4. The heat transfer approximately corresponds to the value typical for the single-phase flow in the recirculation zone. The heat transfer rate increases due to the evaporating droplets in the near-wall zone behind the point of flow reattachment. This behavior of the maximal heat transfer is caused by the effect of various factors: more intense evaporation of small droplets, a decrease in the rate of their inertial precipitation, and weaker involvement of large particles in the separated flow. The increase in mass concentration of the dispersed phase causes a significant increase in heat transfer between the two-phase flow and wall as compared to the single-phase flow (Nu_{max} = 51). It should be noted the maximal increase in heat transfer in the gas–droplet flow occurs in the region of small particles, which intensively penetrate into the recirculation zone and thus reach the pipe wall.

This work is supported by the Russian Science Foundation (Project No. 14-19-00402).

References

1. J.R. Fessler, J.K. Eaton, J. Fluid Mech. **314** 97 (1999)
2. K. Hishida, T. Nagayasu, M. Maeda, Int. J. Heat Mass Transfer **38** 1773 (1995)
3. M.A. Pakhomov, V.I. Terekhov, Int. J. Heat Mass Transfer **66** 210 (2013)
4. I.V. Derevich, L.I. Zaichik, Fluid Dynamics **23** 722 (1988)
5. M.W. Reeks, Phys. Fluids A **3** 446 (1991)

DOI: 10.1051/matecconf/20168400029

6. A. Fadai-Ghotbi, R. Manceau, J. Boree, Flow, Turb. Combust. **81** 395 (2008)
7. B.P. Leonard, Computer Meth. Appl. Mech. Eng. **19** 59 (1979)
8. J.P. Van Doormaal, G.D. Raithby, Numerical Heat Transfer, pt. A **7** 147 (1984)

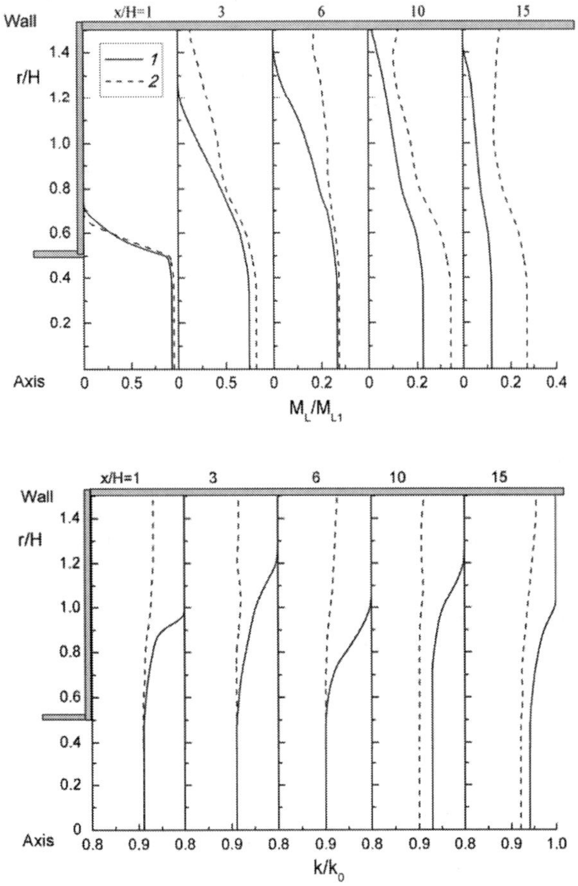

a

b

Figure 1. Radial droplets mass fraction distribution (a) and gas phase turbulence modulation (b) along the axial coordinate. *1* – evaporating droplets, *2* – non-evaporating particles. $M_{L1} = 0.05$, Stk = 0.4, $d_1 = 30$ μm.

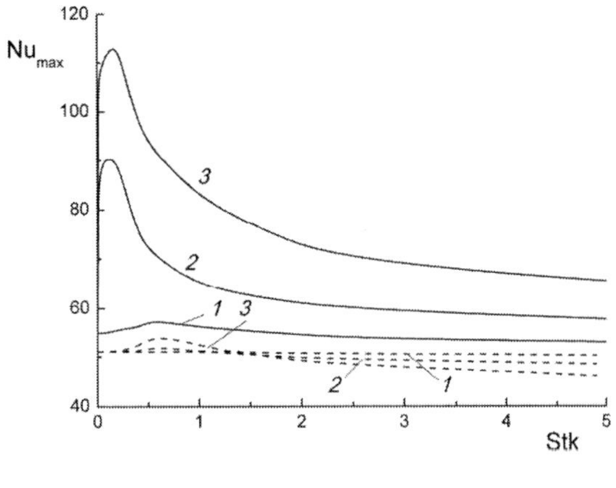

Figure 2. The effect of Stokes number on the variation of maximum Nusselt number in separated droplet-laden flow. *1* – $M_{L1} = 0.01$, *2* – 0.05, *3* – 0.1.

Characteristics of Layered Gels Formation by Additive Technologies

Boris Pokusaev, Sergey Karlov, Andrey Vyazmin[a], Kseniya Okhotnikova and Dmitry Nekrasov

Moscow State University of Mechanical Engineering, Environmental and Chemical Engineering Institute, ul. Staraya Basmannaya 21/4, Moscow, 105066 Russia

Abstract. In order to study the properties of the different density silica gels optical methods were used. It is shown that layers of gels spatially inhomogeneous and anisotropic, their optical density are significantly reduced near the free surface. It is found that on the surface of thin layers of silica gel there are convective structures. Features of multi-layered gel structures with different densities were studied. It is found that after the gels forming layers have poor adhesion to each other. This is due to the fact that the volume between different layers of gels filled with liquid. The presence of liquid in the interlayer volume is observed visually and confirmed by the light scattering of the laser beam. The thicknesses of the liquid boundaries between the layers of the gels were estimated.

1 Introduction

Gel is a dispersive system with liquid dispersing medium, and the dispersion phase makes up a spatial structured mesh due to intermolecular interaction in the contact sites. Gels are micro- or nanostructured media whose internal structure depends on the material and on the method of preparation. The characteristics of gel can change due to the internal processes [1, 2].

The promising application of gels is the use in regenerative medicine for growth of biological tissues, including the problem of culturing stem cells in vitro [3, 4]. This capillary network is intended for the delivery of nutrients to individual cells, and to remove products from the metabolism. It seems promising the idea that gels can used as a carrier during the formation of ordered bio-structured elements (based on the 3D printing) for additive medical technology. However, there are a number of factors that influence diffusion transport in gels and change mass-transfer laws [5, 6].

A large number of studies concerning the synthesis of different chemical nature gels and the study of its properties have been published. However, for use in technology gels have many common properties. For example, the most important properties that define the processes of mass transfer in the gels are nonstationarity and anisotropy due to the structure and behavior of the transfer medium.

For experimental study we used gels that are based on silicates. Such gels are well investigated (see, for example [7]), simply prepared and not toxic. Such gels are appropriate for additive technologies used for the creation of skeleton matrix (retaining biologically active gels, but transmissive to liquid nutrient) of bioreactors by 3D-printing. Gels of different density were investigated. The stock solution of sodium silicate was diluted with distilled water to the levels of density 1.04 – 1.08 g/cm^3. The gel-generating mixture was prepared by adding the hydrochloric acid.

[a] Corresponding author: av1958@list.ru

The aim of the investigation is to study of the fundamental properties of the gels layers by optical methods (for example, silica gels different density) in relation to its using in additive 3D technologies. It is important to obtain data on the properties of multilayer gels structures and to determine the effects on the properties of the boundary layers between the gels.

2 Nonuniformity and anisotropy of layered gel

For the investigation of the anisotropic properties of gel the spectroscopic method was used. A container for a gel has produced from special quartz glass. The experimental setup provides the instantaneous registration in spectral range of 200 – 1100 nm. The optical resolution is in average 1.0 nm. The spatial resolution is 1 mm. This setup allows us to investigate the intensity of the transmitted light through the gel as a function of the distance from the top surface. Thickness of wetting meniscus is about 1.5 mm. For this reason, measurements were made starting at a depth of 2 mm.

The intensity of the light passing through the gel, referred to the value corresponding to the transmittance intensity of the optical system without gel. Measurement processed for light wavelength of 500 nm, that corresponds to the minimum transmittance of the gels. The results of the measurements for gels with two different initial densities are shown in Fig. 1.

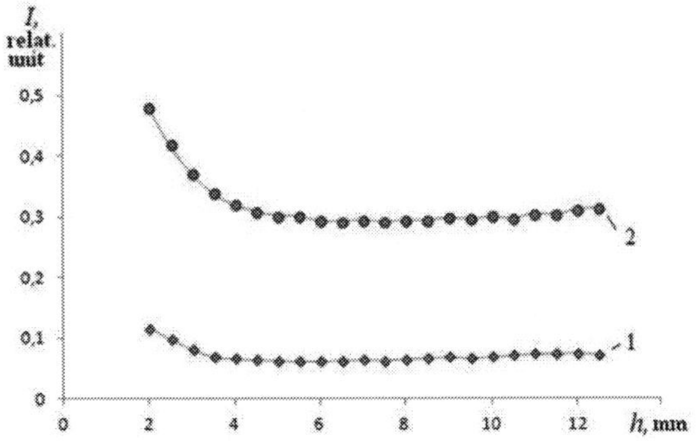

Figure 1. Intensities of transmitted light *I* (relat. units) vs. distance from the top surface of the gel layer *h* (mm) for different density: 1 - 1.08 g/cm^3; 2 - 1.04 g/cm^3. Measurements were carried out 1 hour after gel formation.

From the experimental data it can be concluded that gels of both densities are nonuniform in their thickness. The density of the surface layers of the gel can be less than their density inside the gel. The explanation of this fact, that describes the temporal dynamics of the process of syneresis, is the slow deposition of a microstructured phase in a liquid under the forces of gravity. The deposition rate is determined by the rate of filtration of the liquid phase from the lower layers of gel to the top.

The formation of silica gel from liquid media requires a few minutes and it is associated with chemical changes. As a result a structured phase is occurred; gel has different properties from original liquids. It is found that on the surface of thin layers of silica gel there are convective structures (Fig. 2). These structures make more complicated to depose the upper layers of the gel at bioprinting.

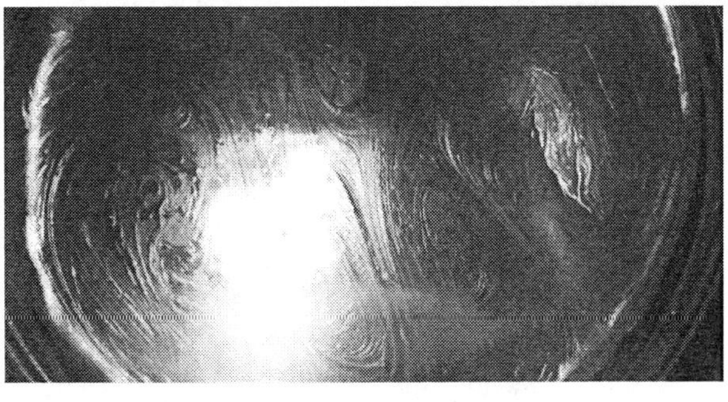

Figure 2. Photo of the convective structures appearing on the surface of a thin layer of silica gel during its formation.

3 Specific properties of multilayer gels

In order to create 3D additive technologies using gels, there is a need to check whether the properties of additivity are valid for multilayered gels. For this purpose it is required to determine which properties of multilayer gels differ from the homogeneous gel of the same chemical composition.

Gels are very promising for additive technologies, since they are easily form matrixes of different shapes and densities (Fig. 3). It is possible to apply a denser layer of gel on the less dense without any mixing corresponding to liquid media. Such bilayer systems are hydrodynamic stable. For layered gel structures presented on pictures the border between the layers (pointed by arrow) is clearly visible.

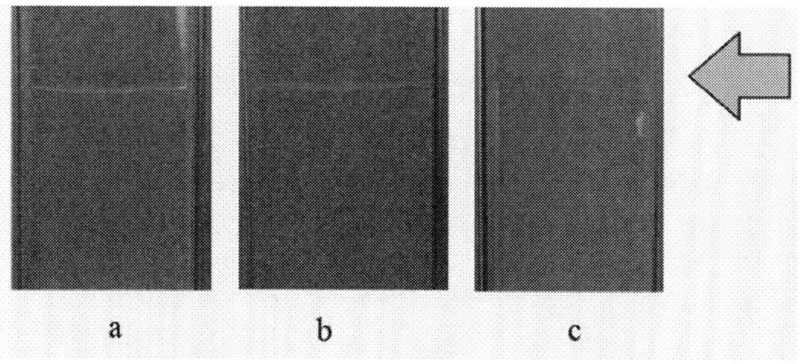

Figure 3. Photos of silica gel two-layer systems of different densities: a – both layers with density of 1.04 g/cm^3, b - both layers with 1.08 g/cm^3, c – the upper layer has the density of 1.08 g/cm^3; the lower has the density 1.04 g/cm^3. The arrow indicated the separation surface between these two layers.

A comparison of the spectrum of single-layered and double-layered gel structures of the same thickness did not show differences in the intensity of transmitted light, either in the form of spectrum. However, the presence of interphase boundaries may have an impact on the technological properties of gels. In Fig. 4 shows the mechanical properties of the gels after drying. On the left there is the initial lattice structure created from the fully formed gel, on the right – inadvertent mechanical disruption of this structure after the day as a result of the drying of the gel. Breaks the continuity of the patterns observed in the contact layers. This indicates poor adhesion between the layers of the gel.

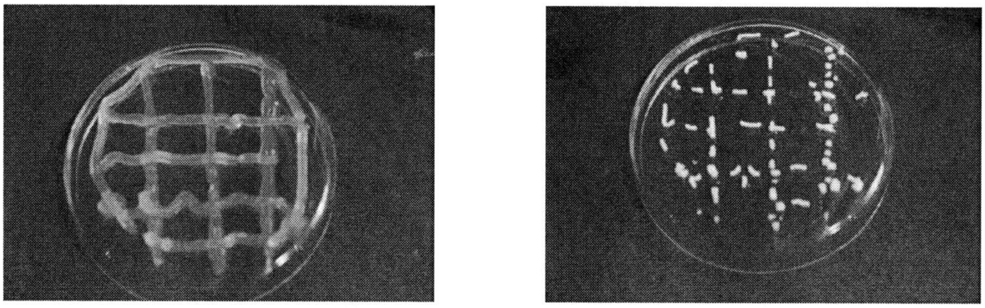

Figure 4. Photo of gel grating (left) on Petri dish, the same grating after drying of gel for 24 hours (right).

Important factor complicating the use of additive technologies to gels is their spontaneous densification. As a result of the pressure of the upper layers of the liquid is squeezed out of the lower layers forming the system drops on the perimeter of the upper layer (Fig. 5). Such a drop will obstruct the uniform application of the subsequent layers of the gel.

To make sure that between layers of the gel there is a thin layer of fluid, a study of impact of passing the green laser with wavelength 532 nm through it was performed. The main results are shown in Fig. 6. The passage of the beam through a homogeneous gel is uniform in all directions scattering of light, which fully corresponds to the scattering of light on the microinhomogeneities smaller wavelengths of light. When the ray go along the boundary layers of the gels it is clearly visible a lateral dispersion that occurs in a planar waveguide. This phenomenon is caused by the multiple reflections of light on the surfaces of the gels formed, that are limited by layer of liquid.

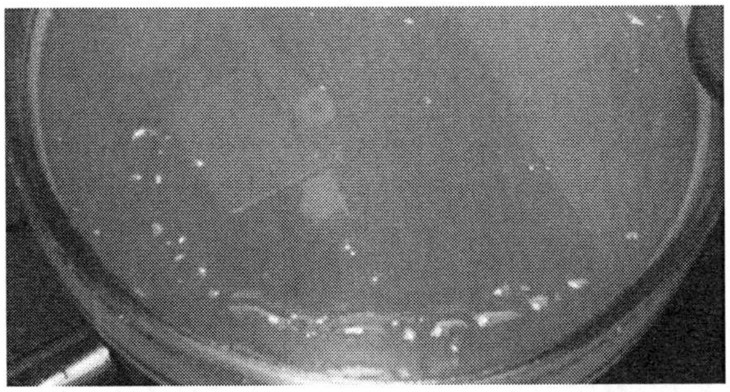

Figure 5. Photo of drops formed along the perimeter of upper layer during the formation gel two-layer matrix.

In order to estimate the thickness of the liquid layer between the gels a photographic method was used. Since the space between layers of the gel is a thin capillary, for receiving the contrast photos the inert colored dye was used that is well identified by spectral processing photo. The estimation showed that the thickness of the layer between gels is at least 0.5 mm and increases with aging of the gel.

Figure 6. Photo of characteristic scattering of laser beam of 532 nm passing through a homogeneous gel (left), the characteristic scattering of the laser beam passing through the boundary between the gels layers (right).

4 Conclusions

Layers of gels are not uniform in thickness. The density of gels decreases towards the free surface. This is true for gels of different density. On the free surface of the gel during its formation, the occurrence of spontaneous interfacial convection is possible.

Multilayered gels are different in properties from a single layered of the same thickness. Between the layers of gel a thin fluid-filled layer is formed. This layer reduces adhesion between the layers of gel, but can significantly increase mass transfer of gel matrix. These features of layered gel structures must be taken into account when using such materials in additive 3D printing technology.

Acknowledgement

This work was supported by the Russian Science Foundation (project no. 15-19-00177).

References

1. K. Kajiwara, Yo. Osada, *Gels handbook* (Elsevier, 2000)
2. M. Tokita, K. Nishinary, Prog. Coll. & Polymer Sci. **136**, (2009)
3. B. A. Westrin, A. Axelsson, Biotech. & Bioeng. **38**, 439 (1991).
4. J. A. Hendriks, L. Moroni, J. Riesle, J. R. De Wijn, C. A. Van Blitterswijk, Biomater. **34**, (2013)
5. B. Pokusaev, S. Karlov, A. Vyazmin, D. Nekrasov, Chem. Eng. Trans. **43**, 1681 (2015)
6. M. Lauffer, Biophys. J. **1**, 205 (1961)
7. N. A. Shabanova, P. D. Sarkisov, *Sol-gel technology*. (Binom. Moscow, Russia, 2012)

MATEC Web of Conferences **84**, 00031 (2016) DOI: 10.1051/matecconf/20168400031
International Symposium IPHT 2016

Unsteady Heat Transfer In a Gas Mixture

Alexey Polikarpov[1,2,a], Minh Tuan Ho[3] and Irina Graur[4]

[1] *Ural Federal University, Physics Department, Lenina str. 51, 620000, Ekaterinburg, Russia*
[2]*Kutateladze Institute of Thermophysics SB RAS, Lavrentyev Ave., Novosibirsk, 630090, Russia*
[3]*James Weir Fluids Laboratory, Department of Mechanical and Aerospace Engineering, University of Strathclyde, Glasgow G1 1XJ, UK*
[4]*Aix-Marseille University, rue Enrico Fermi 5, 13013, Marseille, France*

Abstract. The time-dependent problem of heat transfer between two parallel plates in a binary gas mixture was investigated on the basis of numerical solution of BGK type kinetic model equation with collision integral taken in the form of McCormack model. The time evolution of the normal heat flux was simulated and time needed to reach the steady state conditions was obtained.

1 Introduction

Heat and mass transfer phenomena in a gas near a heated or cooled surface play a key role in large number of technological processes. In addition, in most of these processes the transport phenomena are essentially time-dependent. Actual tendency of many technical devices miniaturization leads to increase of a ratio between the molecular mean free path and the characteristic scale of a device (this ratio is called Knudsen number). If the Knudsen number becomes larger than one a gas inside this device must be considered as rarefied and its behavior must be simulated on the basis of the Boltzmann equation.

The heat transfer in a rarefied gas mixture confined between two parallel plates was intensively studied during the last decades, see Refs. [1,2,3]. One of these works is devoted to the case of steady state regime [1]. The transient behaviour in collisionless regime was studied in Ref. [2], while the oscillatory boundary conditions were stated in Ref. [3]. Apparently, only a single work is devoted to the consideration of unsteady heat transfer in a binary gas mixture confined between concentric cylinders [4].

Here the fundamental problem of the transient heat transfer through a gas mixture between two parallel plates, induced by the instantaneous temperature rise and drop of the upper and lower plate, respectively, is simulated on the basis of the McCormack kinetic model [5].

2 Problem Statement

Let us consider two parallel infinite plates maintained at different temperatures. The temperature of the upper plate is larger than that of lower and the temperature difference between the plates is equal

[a] Corresponding author : Alexey/Polikarpov@urfu.ru

to ΔT. The plates are spaced at a distance H. The gap between the plates is filled by a rarefied gaseous mixture. The time evolution of the occurring heat flux is the goal of discussion.

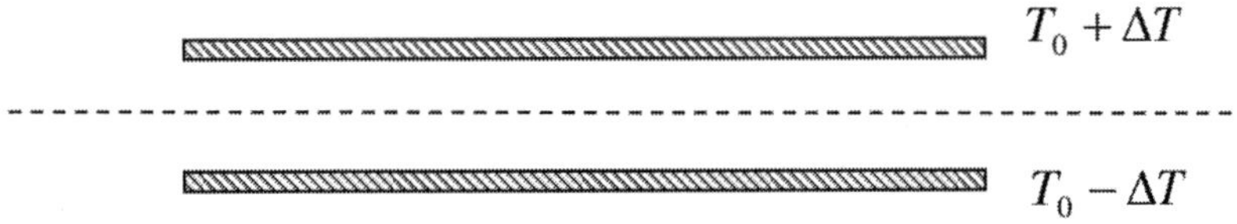

Figure 1. Task geometry

Initially a gas mixture between two parallel plates is in an equilibrium state. Then, the temperature of one plate increases instantaneously and the relaxation of this system to its new equilibrium state is observed. The time evolution of the heat flux is investigated in a wide range of governing parameters: Knudsen number, gas mixture composition and species concentration. The cases of the mixtures with similar (Ne-Ar) and different (He-Ar) molecular masses are considered. The time-dependent linearized kinetic equation for a binary gas mixture is solved numerically

$$\frac{\partial h_\alpha}{\partial t} + v_y \frac{\partial h_\alpha}{\partial y} = \sum_{\beta=1,2} L_{\alpha\beta} h_\beta, \alpha = 1,2, \tag{1}$$

where $L_{\alpha\beta} h$ is the linearized collision term. The McCormack model [5] is used here for the calculations of the collision term.

The discrete-velocity method is used to solve numerically the McCormack model equation. The discrete velocity set was taken according to roots of Gauss-Hermite polynomials. The second order spatial discretization with explicit time integration is used. The iteration procedure was used to calculate the unknown values of physical quantities, particularly the value of normal heat flux.

The governing parameters for this task are:

1. The molar concentration of the mixture, which is defined as

$$C = \frac{n_1}{n_1 + n_2} \tag{2}$$

n_α $(\alpha = 1,2)$ is the number density of specie α.

2. The rarefaction parameter which is defined as

$$\delta = \frac{pH}{\mu v_0} \tag{3}$$

Here p is the gas mixture pressure, v_0 is the most probable molecular speed of a gas mixture:

$$v_o = \sqrt{\frac{2kT_0}{m}} \tag{4}$$

$m = C_0 m_1 + (1 - C_0) m_2$ (m_1 is the mass of the lighter component), μ is the gas mixture viscosity, k is the Boltzmann constant.

The time evolution of normal heat flux is given in nondimensional form

$$q_y = \frac{q'_y}{p v_0} \frac{T_0}{\Delta T} \tag{5}$$

The upper superscript defines the dimensional value.

3 Results and Discussion

The dependence of the heat flux on the values of concentration parameter C_0 and rarefaction parameter δ are studied numerically. The time needed to reach the steady-state conditions is estimated. Figures 2 and 3 show the time evolution of heat flux in the center of the gap between the plates for three values of concentration C_0 and $\delta=10$ (near continuum regime). Figure 2 shows the case of quite dissimilar masses of species components (He-Ar), while Figure 3 illustrates the case of similar masses of species components (Ne-Ar).

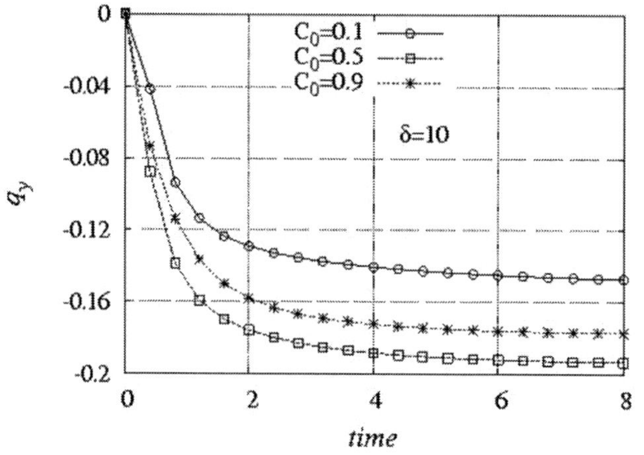

Figure 2. Time evolution of normal heat flux in the center of the gap for mixture He-Ar. Time unit on abscissa axis corresponds to one thousand discrete time steps Δt.

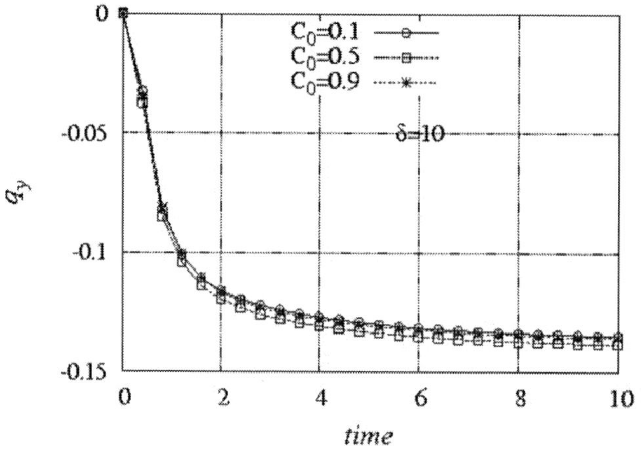

Figure 3. Time evolution of normal heat flux in the center of the gap for mixture Ne-Ar. Time unit on abscissa axis corresponds to one thousand discrete time steps Δt.

Due to sudden change of the temperatures of the both plates the value of a heat flux starts to decreases and it changes from «0» value to its steady state value. The results were obtained for three values of concentration C_0=0.1, 0.5, 0.9, for rarefaction parameter δ=10 and two types of a gas mixture. As it follows from Figures 2 for He-Ar mixture the normal heat flux relaxation speed (dq_y/dt) depends on the value of concentration parameter C_0, it reaches the maximum value for C_0=0.5 (Figure 2) and then drop for larger values of concentration parameter. The values of dimensional normal heat flux takes a maximum also for C_0=0.5. The time necessary to reach the steady state value has the order equivalent to 10 reference time. Figures 2,3 shows the considerable change of heat flux values happened in the earlier time interval less than two values of reference time (H/v_0) and then the curves approach the steady-state limit. The values of heat flux values for Ne-Ar mixture do not depend on the concentration parameter. Independently of the composition of the gas mixture the time needed to reach the steady state values has approximately the same order.

Acknowledgment

The study was supported by the Ministry of Education and Science of the Russian Federation (project identifier RFMEFI61614X0016).

References

1. Sone Y. J. Phys. Soc. Jpn., **20**(2):222 (1965).
2. D C Wadsworth, D A Erwin, and E Ph Muntz. J. of Fluid Mech., **248**(1):219, (1993)
3. A Manela and N. G. Hadjiconstantinou. Phys. Fluids, **20** (11):117104, (2008).
4. M Vargas, S Stefanov, and V Roussinov. International Journal of Heat and Mass Transfer, **59**:302, (2013).
5. McCormack F. J. Phys. Fluids **16**, 12. (1973).

Numerical Investigation of the Liquid Film Flows with Evaporation at Thermocapillary Interface

Ekaterina Rezanova [1, 2, a]

[1]*Altai State University, Department of Informatics, pr. Lenina, 61, 656049 Barnaul, Russia*
[2]*Kutateladze Institute of Thermophysics SB RAS, pr. Lavrentyeva, 1, 630090 Novosibirsk, Russia*

Abstract. Flows of the thin liquid layers on an inclined non-uniformly heated substrate are investigated numerically. The evaporation at the thermocapillary interface is taking into account. The Oberbeck-Boussinesq equations and the generalized kinematic, dynamic and energy conditions on a thermocapillary boundary are used for governing equations. The evolution equation, which determines the position of the interface, is obtained on the basis of the long-wave approximation of the equations for moderate Reynolds numbers. The numerical algorithm for solving of this evolution equation is presented. Comparison of the numerical results of flows of various liquids is presented.

1 Introduction

The effects such as gravity and thermocapillary forces, additional tangential stresses and some other factors may affect the flows in thin liquid layers. The influences of various effects on the flow character are investigated analytically, numerically and experimentally [1-6]. Evaporation at the interface should be taken into account in the modeling of such processes [1-3, 7-9]. A number of works are devoted to the numerical study of flow of evaporating thin liquid layers [2-4]. The boundary conditions should provide a fulfillment of the conservation laws on the interface [10-12].

The motion of a viscous incompressible liquid film along an inclined at an angle α impermeable substrate is investigated numerically. Accompanying flow of gas moves over a liquid layer. The two-dimensional mathematical model of the process is based on the Oberbeck-Boussinesq equations. The boundary conditions on thermocapillary interface are written with account of evaporation. Long-wave theory is applied to construct the nonlinear evolution equations for the liquid film thickness. This equation should incorporate the effects of gravity, viscosity, capillarity and thermocapillarity.

2 The problem statement

Let the interface position be determined by the equation $z=h(x,t)$. The equation defines the solid substrate has the form $z=0$, the x-axis coincides with the direction of fluid flow (see Fig.1). The gravity vector has the form $\mathbf{g}=(g_1, g_2)=(g \sin \alpha, - g \cos \alpha)$, $g = |\mathbf{g}|$.

[a] Corresponding author : katerezanova@mail.ru

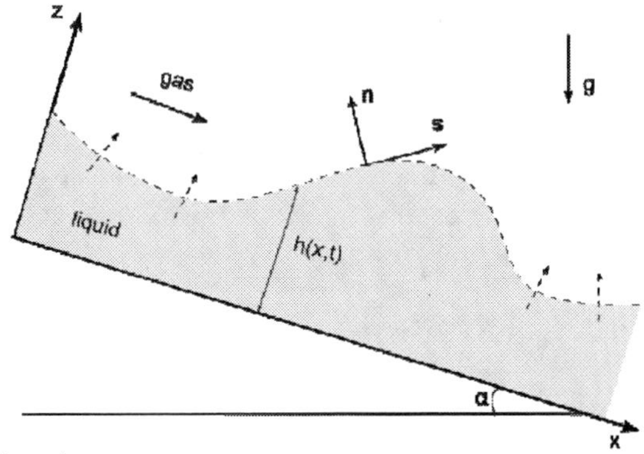

Figure 1. Geometry of flow domain

The one-sided mathematical model based on the Oberbeck-Boussinesq equations in the dimensionless form can be written as follows in the case, when the Reynolds number $Re=O(1)$ [7]:

$$\varepsilon^2\left(u_t+uu_x+wu_z\right)-\varepsilon^2 u_{xx}=u_{zz}-p'_x-\gamma_1 T\sin\alpha,$$

$$\varepsilon^4\left(w_t+uw_x+ww_z\right)-\varepsilon^4 w_{xx}-\varepsilon^2 w_{zz}=-p'_z+\gamma_2 T\cos\alpha,$$

$$u_x+w_z=0,$$

$$Pr\varepsilon^2\left(T_t+uT_x+Tw_z\right)-\varepsilon^2 T_{xx}=T_{zz}.$$

Here ε is the small parameter of the problem, $\varepsilon=d/l$ (d and l are the transverse and longitudinal characteristic lengths), $\mathbf{v}=(u, w)$ is the velocity vector; p' is the modified pressure ($p' = p - \gamma_1(\beta T_*)^{-1} x \sin\alpha + \gamma_2(\beta T_*)^{-1} z \cos\alpha$); T is the temperature; ρ – some relative value of the liquid density; $Pr=\nu/\chi$ is the Prandtl number; $\gamma_1=Gr/\varepsilon$; $\gamma_2=Gr$; $Gr=d^3 g\beta T_*/\nu^2$ is the Grashof number; the Reynolds number is $Re=u_* l/\nu$; ν, χ, β is the coefficients of the kinematic viscosity, thermal diffusivity and thermal expansion; T_* is the characteristic temperature drop; u_* is the characteristic longitudinal velocity of the liquid.

The generalized kinematic, dynamic and energy conditions [7, 12] can be written in the following dimensionless form after simplifications:

$$-\varepsilon(h_t+h_x u-w)=EJ_{ev},$$

$$u_z=\alpha_\tau \tau(x,t)-\alpha_{Ma}\left(T_x+h_x T_z\right),\quad p=p^g-\alpha_{Ca}h_{xx}\left(1-\alpha_\sigma T\right),$$

$$-T_z=\bar{\beta}_3 J_{ev}+\bar{\beta}_6 h_{xx} J_{ev}.$$

Here J_{ev} is the local mass flux at the thermocapillary interface, $E=\kappa T/(\lambda_U\, \rho\nu)$ is the evaporation parameter, κ is the heat conductivity coefficient, λ_U is the latent heat of vaporization, $\alpha_\sigma=MaCa/Pr$, $Ma=\sigma_T T_* l/(\rho\nu\chi)$ is the Marangoni number, $Ca=u_*\rho\nu/\sigma_0$ is the capillary number, σ_T is the temperature coefficient of surface tension, σ_0 is the value of the coefficient of surface tension σ at the some value of temperature (we assume the linear dependence of the surface tension on temperature $\sigma=\sigma_0-\sigma_T(T-T_0)$), $\alpha_{Ma}=Ma\varepsilon/Pr$, $\alpha_{Ca}=\varepsilon/Ca$, $\alpha_\tau=\bar{\rho}\bar{\nu}\bar{\upsilon}\varepsilon/\bar{h}$, $\bar{\rho},\bar{\nu}$ – ratio of the density and the coefficient of kinematic viscosity of the gas and liquid, $\bar{\upsilon}=u_*^g/u_*$, $\bar{h}$ – ratio of the characteristic scale of gas layer to l, $\bar{\beta}_3=\varepsilon$, $\bar{\beta}_6=\left(1-1/\bar{\rho}\right)\varepsilon/(\bar{U}Ca)$, $\bar{U}=\lambda_U/u_*^2$. The local mass flux of vapor at the interface J_{ev} is determined by the Hertz-Knudsen equation [2]: $J_{ev}=\alpha_J T|_{z=h(x,t)}$, where $\alpha_J=\alpha\rho_s\lambda_U T_*/J_*\left(M/\left(2\pi R_g T_s^3\right)\right)^{1/2}$, α is an accommodation coefficient, ρ_s is the vapor density, J_* – is the characteristic value of the vapor mass flux, M is the molecular weight, R_g is the universal gas constant, T_s – saturated vapor temperature.

The no-slip conditions for velocity and the non-uniform heating are given on the solid plate :

$$u|_{z=0}=0,\qquad w|_{z=0}=0.$$

3 The thin layer equation: scheme of the numerical solution

The problem of finding of the unknown functions (longitudinal and transverse velocities, pressure and temperature) will be solved after obtaining the layer thickness h. Let us consider the problem statement for the zero-th order terms of the expansion in powers of ε of the input functions. The following evolution equation determines the interface position (a consequence of the kinematic condition) [7]:

$$h_t + h_x\left\{\frac{h^4}{24}\gamma_2\cos\alpha A_x + \frac{h^3}{6}(\gamma_2\cos\alpha(\Theta_0)_x + \gamma_1\sin\alpha A) + \frac{h^2}{2}((C_0)_x + \gamma_1\sin\alpha\Theta_0) + C_1 h\right\} +$$
$$+\left\{\frac{h^5}{120}\gamma_2\cos\alpha A_{xx} + \frac{h^4}{24}(\gamma_2\cos\alpha(\Theta_0)_{xx} + \gamma_1\sin\alpha A_x) + \right. \quad (1)$$
$$\left. + \frac{h^3}{6}((C_0)_{xx} + (\Theta_0)_x\gamma_1\sin\alpha) + \frac{h^2}{2}(C_1)_x\right\} + \frac{E\alpha_J}{\varepsilon}\{Ah + \Theta_0\} = 0.$$

Here the function Θ_0 defines the heating of the solid substrate, functions A, C_0 and C_1 are determined from the following relations:

$$A = \frac{(\bar\beta_3\alpha_J + \bar\beta_6 h_{xx}\alpha_J)\Theta_0}{1 - \bar\beta_3\alpha_J - \bar\beta_6 h_{xx}\alpha_J h},$$

$$C_0(x,t) = p^g - \alpha_{Ca}h_{xx}\left(1 - \alpha_\sigma\Theta^0\right) - \gamma_2\cos\alpha(A\frac{h^2}{2} + \Theta_0 h) - \frac{\gamma_1}{Bu}x\sin\alpha + \gamma_2 h\cos\alpha, \quad (2)$$

$$C_1(x,t) = \alpha_\tau\tau(x,t) - \alpha_{Ma}\tilde\Theta - (C_0)_x h - \gamma_1\sin\alpha h(A\frac{h^2}{2} + \Theta_0 h) - \gamma_2\cos\alpha(A_x\frac{h^3}{6} + (\Theta_0)_x\frac{h}{2}). \quad (3)$$

Here $\tilde\Theta = A_x h + (\Theta_0)_x + h_x A$, $\Theta^0 = Ah + \Theta_0$.

The periodic problem of finding the function h satisfying the equation (1) on the interval [-L; L] is considered. The following periodic conditions on the boundaries x=-L, x=L are fulfilled:

$$h|_{x=-L} = h|_{x=L},\quad h_x|_{x=-L} = h_x|_{x=L},\quad h_{xx}|_{x=-L} = h_{xx}|_{x=L},\quad h_{xxx}|_{x=-L} = h_{xxx}|_{x=L}.$$

After substitution of expressions (2) and (3) the equation (1) can be rewritten as follows:

$$h_t + A_4 h_{xxxx} + A_3 h_{xxx} + A_2 h_{xx} + A_1 h + D = 0. \quad (4)$$

The coefficients A_4, A_3, A_2, A_1, D are the functions which depend on A, h, Θ_0 and their derivatives. Equation (4) is solved numerically by using the following implicit finite-difference scheme [13]:

$$\frac{h^{k+1} - h^k}{\tau} + A_4^k h_{xxxx}^{k+1} + A_3^k h_{xxx}^{k+1} + A_2^k h_{xx}^{k+1} + A_1^k h^{k+1} + D^k = 0. \quad (5)$$

The scheme (5) is solved with use of the finite-difference analogues of the second-order approximation for all derivatives with respect to x. The uniform finite-difference mesh x_1, x_2, ..., x_{N+1} ($x_n = -L + (n-1)\Delta x$, $\Delta x = 2L/N$) is used for implementation of the scheme (5).

The scheme leads to a system of linear algebraic equations which can be solved by using the five-point Thomas algorithm and Thomas algorithm with parameter. The unknown value of the liquid layer thickness h at x=-L, x=L is chosen as the parameter.

Periodic movement of ethanol and HFE-7100 is studied numerically. The initial position of the thermocapillary boundary is given by the equation $h_0 = 1 - \delta_1\cos kx$, where δ_1=0.01, $k=\pi/2$. Nonuniformly heating of the substrate is determined using the function $\Theta_0 = 1 + \delta_0\cos k_1x\cdot\cos k_2t$, where δ_0=0.25, $k_1=\pi/2$, k_2=2.

Figure 2 presents the process of evaporation of the thin layer for the two liquids: HFE-7100 and ethanol in the case of low gravity ($g=9.81\cdot10^{-2}$ m/s). The angle of inclination of the substrate α is equal to $\pi/6$. Numerical investigation shows that the character of change of layer thickness is similar for different fluids but the quantitative results differ and the difference increases over time.

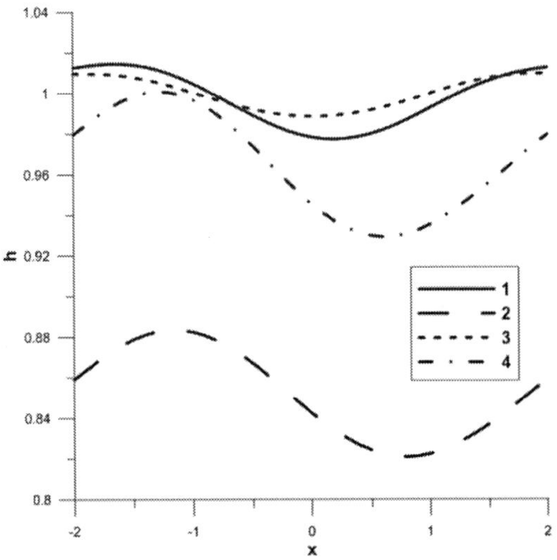

Figure 2. The behaviour of the liquid layer thickness, the case of low gravity (g=9.81·10^{-2} m/s): 1 – "HFE-7100-nitrogen" system, t=10^{-4}, 2 – "HFE-7100-nitrogen" system, t=10^{-2}, 3 – "ethanol-nitrogen" system, t=10^{-4}, 4 – "ethanol-nitrogen" system, t=10^{-2}.

Acknowledgments. The research has been supported by the Russian Ministry of Education and Science (project indicator RFMEFI61314X0011).

References

1. A. Oron, S.H. Davis, S.G. Bankoff, Reviews of Modern Physics, **69** (**3**), 931-980 (1997)
2. S. Miladinova, S. Slavtchev, G. Lebon, J.-C. Legros, In: Proc. 6th Workshop on Transport Phenomena in Two-Phase Flow, Bourgas, 121-128 (2001)
3. Yu.O. Kabova, V.V. Kuznetsov, O.A. Kabov, International Journal of Heat and Mass Transfer, **55**, 1271–1278 (2011)
4. Yu.O. Kabova, V.V. Kuznetsov, O.A. Kabov, T. Gambaryan-Roisman, P. Stephan, International Journal of Heat and Mass Transfer, **68**, 527-541 (2014)
5. Yu.V. Lyulin, O.A. Kabov, Int. J. of Heat and Mass Transfer, **70**, 599-609 (2014)
6. R.V. Craster, O.K. Matar, Review of Modern Physics, **81(3)**, 1131-1198 (2009)
7. O.N. Goncharova, E.V. Rezanova, Izvestia AltGU, **85** (**1/1**), 70-74 (2015)
8. O.N. Goncharova, E.V. Rezanova, Ya.A. Tarasov, Izvestia AltGU, **81** (**1/1**), 47-52 (2014)
9. O.N. Goncharova, E.V. Rezanova, Izvestia AltGU, **81** (**1/2**), 21-25 (2014)
10. C.S. Iorio, O.N. Goncharova, O.A. Kabov, Microgravity Sci. Technol, **21**(**1**), 313-320 (2009)
11. C.S. Iorio, O.N. Goncharova, O.A. Kabov, Computational Thermal Sci., **3**(**4**), 333-342 (2011)
12. O.N. Goncharova, Izvestia AltGU, **73** (**1/2**), 12-18 (2012)
13. E.V. Rezanova, Izvestia AltGU, **89** (**1/1**), 168-172 (2016)

Two-phase flow patterns in horizontal rectangular minichannel

Fedor Ron'shin[1,a]

[1]*Kutateladze Institute of Thermophysics, 630090, Lavrentiev ave. 1, Novosibirsk, Russia*

Abstract. The two-phase flow in a short horizontal channel of rectangular cross-section of 1×19 mm^2 has been studied experimentally. Five conventional two-phase flow patterns have been detected (bubble, churn, stratified, annular and jet) and transitions between them have been determined. It is shown that a change in the width of the horizontal channels has a substantial effect on the boundaries between the flow regimes.

1 Introduction

In general, there is a rapid development of miniaturized heat exchangers. Recent studies show that the heat exchange systems with mini- and microscales are more energy efficient than the macro-systems with the channels greater than 1 mm. The trend towards a decrease in characteristic sizes of devices in various technical fields determines development of studies on hydrodynamic and heat transfer in mini- and microchannels. With decreasing of the channel height the ratio of surface area to channel volume increases inversely proportional to its minimal transverse size, and this causes high heat transfer intensity in the microsystems. In application to the processes in reacting systems, for instance, burning of combustible liquids in small channels, heat transfer intensification can either improve or deteriorate the situation [1].

There are a significant number of researches on the two-phase flow in mini- and microchannels. The review articles on two-phase flow in channels of different configurations can be found in [2-4]. Analysis of the researches shows that the structure of two-phase flow is generally influenced by the following parameters: channel geometry and dimensions, structure of the inlet section, fluid properties such as viscosity and surface tension. It is shown that in many studies round channels have been used, although rectangular channels have greater prospects for using in thermal stabilization systems. Flow patterns in the horizontal rectangular channel height below 1 mm are quite different from the classical two-phase flow regimes for horizontal channels with height more than 2 mm [4]. In the works of [5, 6] it is shown that the regimes of gas-liquid flow in the minichannels significantly depend on the conditions of phase input to the channel. In [7, 8] it is discussed the mechanisms influencing on the formation of patterns in mini- and microchannels. In the [9] there is an attempt to summarize various works on the study of two-phase flow in microchannels. It is shown that different authors use different criteria and distinguish different patterns of two-phase flow; thence, it is difficult to perform a comparison with other works.

[a] Corresponding author : f.ronshin@gmail.com

The aim of present work is to study the patterns of two-phase flow and the boundaries between them in the short (length 80 mm) rectangular channel cross-section of 1×19 mm^2, and compare it with the minichannel with cross-section of 1×10 mm^2.

2 Experimental setup and technique of measurement

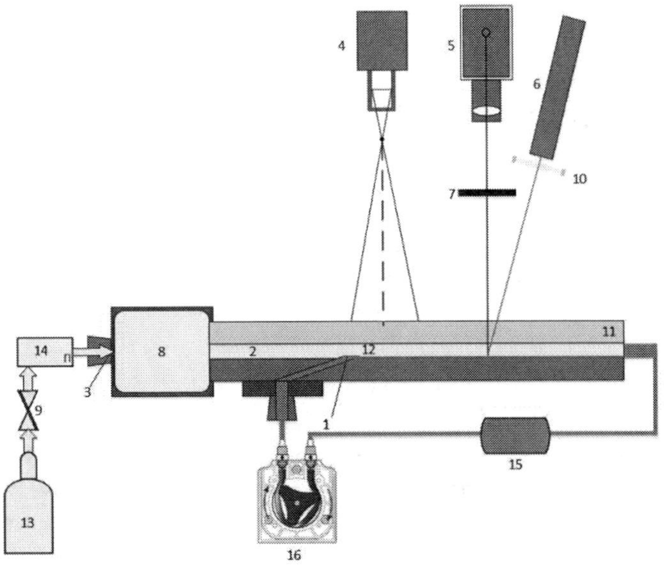

Figure 1. Schematic diagram of the experimental arrangement.

The schematic diagram of the experimental setup is demonstrated in figure 1. The basic element of the test cells is a flat plate made of stainless steel with liquid nozzle (1). The stainless steel plate is placed on the textolite basement. The frame with the stainless steel plate is covered by optical window (11), so that rectangular minichannel (2) is created by the special metallic inserts with height 1 mm.

The working liquid (pure distilled water) and the dry nitrogen from the gas bottle (13) have been used in the experiment. The working gas is supplied from the standard gas bottle (~ 150 bars) (13). Initial gas pressure is set at 5 bars by pressure regulator (9) in order to provide a sufficient gas flow rate through the gas-controllers. A gas flow rate is regulated with the help of mass flow Bronkhost controller (14). Gas flow rate is varied from 0.5 l/minute to 75 l/minute. The working gas is supplied through the gas inlet (3) to the gas chamber (8). Water is supplied to the test cell by the Ismatec peristaltic pump (16) from a liquid container (15). The liquid flow rate is regulated in the range between 1 ml/min and 50 ml/min. The water film is supplied from the liquid nozzle (1) of the height 50 μm. The film, formed in the nozzle (1), is driven along the channel by the moving gas. The mixture of vapor, gas and liquid and residual vapor is evacuated to the liquid container (15). The gas is evacuated to the atmosphere. All parameters have been recorded by the data acquisition systems, developed on the LabView software platform.

High-speed LIF (Laser Induced Fluorescence) technique has been used to study the two-phase flow in the mini-channel (1 mm height). This technique is described in more details in [10]. Distilled water with fluorophore Rhodamin 6G has been used for this optical method. The liquid-gas interaction in the channel is monitored in the region (12) by the digital video camera (4) and LIF technique. The "Aries" laser (6) with wavelength 532 nm and power of 50 mW is used to initiate fluorescence. The laser beam is unwrapped into the line (which is oriented orthogonally to the flow direction) by the cylindrical lens (10) which is placed at the distance 52 mm from the liquid entrance (1). The high-speed (up to 2.1 kHz) digital camera VS-Ld-751 (5) is used to register light from the liquid. The step low-pass filter (7) is placed in front of the lens of the camera to cut off laser emission which is reflected from the gas-liquid interface.

3 Results and discussion

The flow patterns in the rectangular minichannel with cross-section of 1x19 mm have been studied experimentally. The flow regime map has been created for the investigated channel (Fig. 1). The superficial gas and liquid velocities defined as the flow rates divided by the channel cross section area have been used as coordinates of this map. Following two-phase flow regimes have been registered: bubble, churn, annular, stratified and jet.

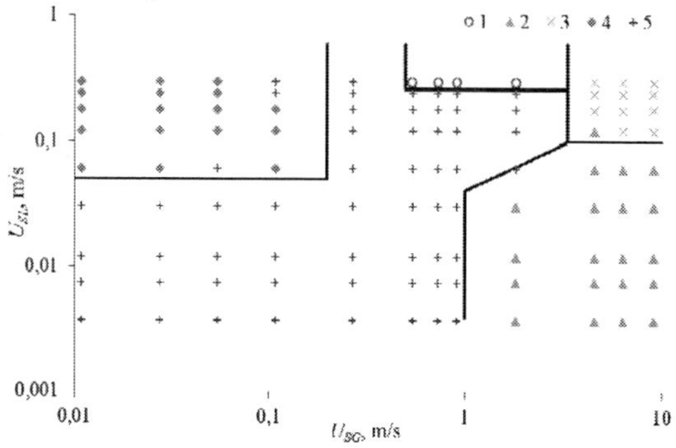

Figure 2. Map of two-phase flow in the minichannel with cross section of 1×19mm^2.
Flow regimes: (1) churn; (2) stratified; (3) annular; (4) bubble; (5) jet.

At very low superficial liquid velocities, gas moves in the central part of the channel, while the bulk of the liquid moves along the periphery of the side walls. Disturbances on the surface of the liquid are not detected. Stationary jet flow pattern is observed at low superficial liquid and gas velocities, when the gas flow occupied no more than half channel cross-section. The jet flow regime (U_{SG} = 0.027 m/s, U_{SL} = 0.03 m/s) is demonstrated in figure 3a. On the one of the sides of the channel, one can see the movement of the bubbles. This is transitional pattern to the bubble ones. By increasing the superficial liquid velocity, liquid pulsations occur in the sides of the channel. By increasing the superficial gas velocity (Fig. 3b), when the gas begins to occupy larger volume than the liquid, there is the transition to the stratified flow pattern. In this regime, part of the liquid moves along the bottom channel wall in the form of a film co-flowing with the gas stream. This regime is observed only in minichannels of a nonround shape, since the film in round tubes with small diameters is closed into a complete circle and thus establishes annular flow [11]. Figure 4a-d shows the time dependence of the liquid thickness distribution in the channel restored from an intensity of the light from fluorophore. In the range of high superficial liquid velocities (U_{SL} > 0.02 m/s) the liquid pulsation can be observed in the lateral parts of the channel, fig 4d. By increasing superficial liquid velocity, liquid emissions occur from the lateral sides of the channel and one forms the film on the upper wall of the channel. There is the transition to the annular flow regime. This transition is monitored by digital video camera. The film in this pattern moves at the top and bottom walls of the channel. In the central part, the gas stream with the drops forms a core of the two-phase flow, fig 4c. Gas occupies most of the channel cross section, while the pulsating liquid movement occurs near with the lateral walls of the channel.

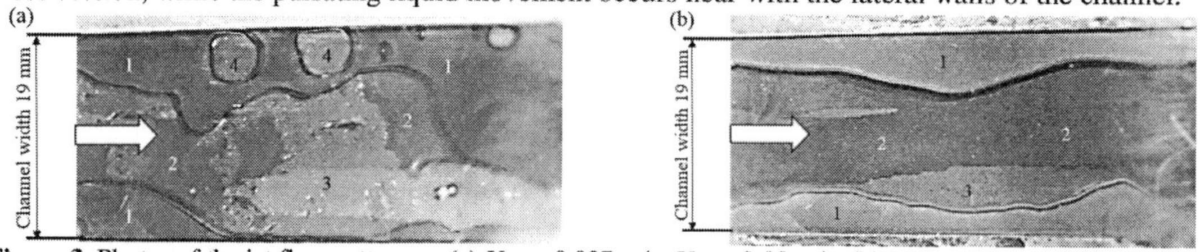

Figure 3. Photos of the jet flow pattern at: (a) U_{SG} = 0.027 m/s, U_{SL} = 0.03 m/s; (b) U_{SG} = 1.818 m/s, U_{SG} = 0.12 m/s; 1 - liquid, 2 - liquid film on the bottom channel wall, 3 - not wetted bottom channel wall, 4 - gas bubbles.

At high superficial liquid velocities and low superficial gas velocities, the bubble flow regime is observed. In this flow pattern, the liquid moves through the channel, containing many small gas bubbles, fig 4a. The size and amount of the bubbles are changed depending on the liquid and gas flow rate, but the bubble sizes are always smaller than the channel width. With increasing superficial liquid and gas velocities, the bubble frequency increases. By increasing the superficial gas velocity, the breaks in the bridges begin to appear, the transition to the churn flow regime takes place, fig 4a. This flow pattern is discussed in more details in [12].

Figure 4 shows comparison of two regime maps for the minichannels cross-section of 1×19 mm and 1×10 mm and the typical flow patterns in the channel 1×10 mm. The two-phase flow patterns in the channels do not differ, but the boundaries between them shift significantly. With an increase in channel width, the area of churn and jet regimes increase significantly, reducing the area of bubble regime. The areas of stratified and annular regimes remain practically unchanged.

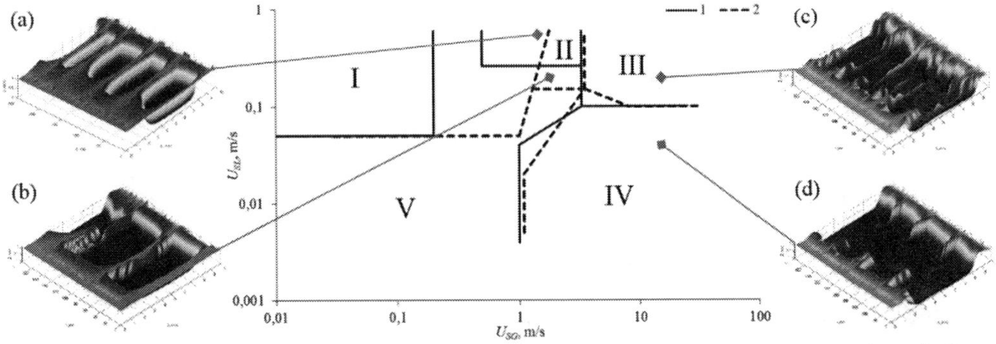

Figure 4. Comparison of the maps of flow regimes for the channels with cross sections (1) 1x19 mm^2, (2) 1x10 mm^2; flow regimes: (I) bubble, (II) churn, (III) annular, (IV) stratified and (V) jet. Flow patterns in the channel cross-section of 1×10 mm^2: (a) bubble, (b) churn, (c) annular, (d) stratified

In conclusion, it is should be noted that the features of the two-phase flow patterns have been defined in the channel with height of 1 mm and width of 19 mm. It is shown that the channel width has significant effect on the boundaries of two-phase flow regimes.

We gratefully acknowledge the support from the Ministry of Education and Science of Russia (№ 14.604.21.0053 Project identifier RFMEFI60414X0053).

References

1. V. V. Zamashchikov, A. A. Korzhavin, E. A. Chinnov, Combust. Explos. Shock Waves **50**, 381 (2014)
2. N. Shao, A. Gavriilidis, P. Angeli, Chemical Engineering Science **64**, 2749 (2009)
3. E. A. Chinnov, O. A. Kabov, High Temp. **44**, 773-791 (2006)
4. E. V. Rebrov, Theor. Found. Chem. Eng. **44**, 355 (2010)
5. V. Haverkamp, V. Hessel, H. Löwe et al., Chem. Eng. Technol. **29**, 1015 (2006)
6. O. A. Kabov, E. A. Chinnov, V. Cheverda, Microgravity sci. technol. **XIX-3/4**, 44-47 (2007)
7. E. A. Chinnov, V. V. Guzanov, O. A. Kabov, Tech. Phys. Lett. **35**, 653-656 (2009)
8. E. A. Chinnov, F. V. Ron'shin, O. A. Kabov, Int. J. Multiphase Flow **80**, 57-68 (2016)
9. J.R. Thome, A. Bar-Cohen, R. Revellin, I. Zun, Exp. Therm. Fluid Sci. **44**, 1 (2013)
10. E. A. Chinnov, Journal of Heat Transfer **138**, 7 (2016)
11. A. Barajas, R. L. Panton, Int. J. Multiphase Flow **19**, 337 (1993)
12. E. A. Chinnov, F. V. Ron'shin, V. V. Guzanov, D. M. Markovich, O. A. Kabov, High Temp. **52**, 681-687 (2014)

MATEC Web of Conferences **84**, 00034 (2016) DOI: 10.1051/matecconf/20168400034
International Symposium IPHT 2016

Features of two-phase flow patterns in horizontal rectangular microchannels of height 50 μm

Fedor Ron'shin[1,a], Vyacheslav Cheverda[1] and Oleg Kabov[1]

[1]*Kutateladze Institute of Thermophysics, 630090, Lavrentiev ave. 1, Novosibirsk, Russia*

Abstract. The horizontal microchannel with the height of 50 micrometres and width of 40 mm of a rectangular cross-section has been used to study two-phase flow. The classical patterns of two-phase flow in the channel (bubble, stratified, churn , jet, and annular) have been detected. Experimental information allows us to define the characteristics of the regimes and to determine precisely the boundaries between the patterns of the two-phase flows.

1 Introduction

There is a revolutionary development in pharmacy lab-on-chip systems where micro- and nanochannels more efficient than free-flowing liquid macrosystems [1] or ones with characteristic channels height larger than 1 mm [2]. The surface to volume ratio of the channel increases inversely proportional to its minimal transverse size with a decrease in the thicknesses of flat channels. Such systems are receiving increasingly wider application in pharmacy, microelectronics, aerospace industry, transport, and power engineering.

The reviews of the studies on two-phase flows in channels with different geometries can be found in [3-5]. In most publications relatively long channels are considered. In such channels the conditions of liquid and gas entry to the channel do not have a significant effect on the structure of two-phase flow. The length of the channels is two or more orders of magnitude higher than their transverse dimension in such systems. The short channels have broader prospects for using in technical applications, such as, biochips and cooling devices for microelectronics. Sufficiently small pressure drops can be achieved in the heat exchangers based on short microchannels. The number of publications on this subject is very limited. The gas-liquid flows in the short horizontal mini- and microchannels have been studied in [6-10]. It is shown that there are many factors affecting on the structure of the two-phase flow.

One way to solve the problems of microelectronic cooling is the use of short microchannels with a two-phase heat-transfer agent, where there is intense evaporation in the zone of active heat release with high velocities of subcooling gas-liquid flow under the conditions of relatively small pressure. The aim of this study is to investigate the regimes of two-phase patterns and boundaries between them in the short (length of 90 mm) horizontal channel with the height of 50 μm and width of 40 mm using the modern techniques.

[a] Corresponding author : f.ronshin@gmail.com

2 Experimental setup and methods of measurement

The experimental setup includes two computer-controlled circuits, which maintain the flow of liquid and gas phases. Figure 2 presents the schematic diagram of experimental setup with microchannel (2) and shows the arrangement of instruments for the applied measurement techniques.

The main part of the test section is a stainless steel plate with the length of 160 mm and width of 55 mm. The plate is covered with an optical glass. The height of the microchannel is 50 microns. The microchannel is created by placing two parallel constantan strips with height 50 μm, width 7.5 mm and length 160 mm between a stainless steel plate and glass window so the rectangular microchannel with the size of 50 μm × 40 mm × 90 mm is created.

The liquid is driven by a high precision syringe pump and injected into the microchannel by liquid nozzle (1). The gas have been supplied into the microchannel throw the gas nozzle (2) from the gas chamber (8). The distance between gas and liquid nozzle is about 70 mm. The liquid–gas interaction in the channel have been visualised in region (12) by the digital video and photo cameras in the Schlieren photography mode using the light beam reflected from the gas–liquid interface. The Schlieren method have been used to register deformation and visualize the liquid film surface. Light from a source (7) through a diffuser, a lens (6), a prism (10) and an optical glass cover (11) penetrates into the channel with the gas-liquid flow. Light, reflected from the gas-liquid interface was transmitted through separation prism (10), lens (6) and filter to camera objective (4). A knife (5), moved by a microscrew, cut the central part of the light beam. As a result, the video camera recorded the intensity of light reflected from the surface separating gas and liquid depending on the angle of this surface inclination from the axis of reference radiation.

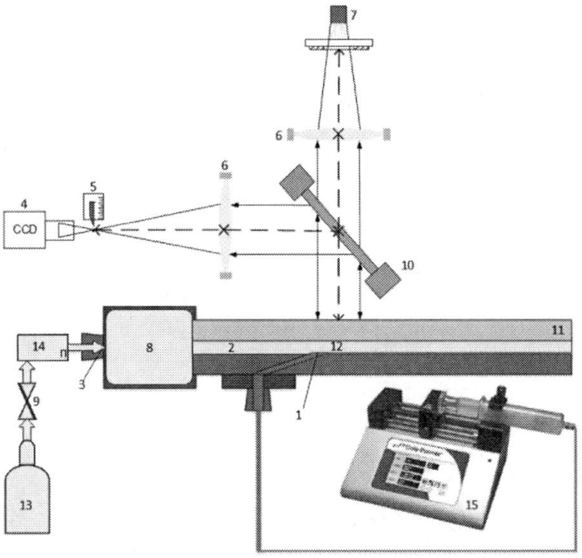

Figure 1. Schematic diagram of the experimental arrangement: (1) flat nozzle for the injection of liquid; (2) microchannel; (3) gas inlet; (4) digital camera; (5) optical knife; (6) lenses; (7) light source; (8) gas chamber; (9) pressure valve; (10) beam splitter; (11) optical window; (12) area of measurement; (13) gas balloon; (14) flow control and measurement gauge; (15) syringe liquid pump.

3 Experimental results and discussion

The basic two-phase flow patterns have been studied experimentally; the boundaries between various patterns have been detected. The map of flow regimes of the process has been created. Figure 2 shows the map of two-phase flow patterns in the 40-mm-wide horizontal channels with the height of 50 μm. The superficial velocities of gas U_{SG} and liquid U_{SL} (determined as the volume flow rates of gas or liquid divided by the cross-sectional area of the channel) have been used as the coordinates in this figure. The solid lines in the figure show the boundaries of the flow patterns for this channel. The following flow patterns have been determined: bubble, jet, stratified, churn, and annular ones.

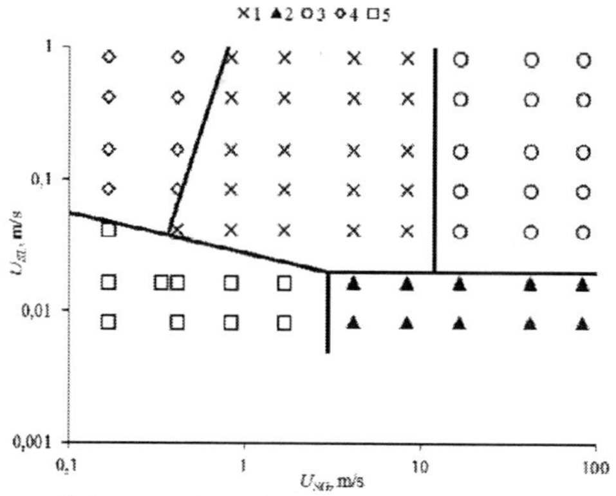

Figure 2. Map of the flow patterns of the two-phase flow in the channel with a cross section of 0.05×40 mm^2. Flow patterns: 1 – churn, 2 – stratified, 3 – annular, 4 – bubble, 5 – jet.

The bubble flow regime is observed in the range of high superficial velocities of liquid and low velocities of gas. The main part of the channel is occupied by liquid, where the gas bubbles move (Fig. 3a). The gas bubbles with the diameter from 1 to 10 mm have been observed. Sometimes bubbles coalesce, forming large bubbles. The liquid moves near the side walls of the channel. The rate of nucleation increases and the bubble sizes remain approximately the same with an increase in the gas velocity. For high superficial velocities of liquid at an increase in the superficial velocity of gas, the bubbles coalesce and bridges between the bubbles are broken, forming the churn flow regime. For the low superficial velocities of liquid at an increase in the superficial velocity of gas, the bubbles coalesce, forming a gas jet and the flow turns to the jet one.

In the range of low superficial velocities of liquid and gas from U_{SL} = 0.008 m/s, U_{SG} = 0.1 m/s to U_{SL} = 0.04 m/s, U_{SG} = 2 m/s, the jet flow pattern with many small drops is observed. This regime is characteristic for the channels with the height below 200 μm. The gas-liquid jet with predominance of gas moves along the channel. The feature of this regime in comparison with the higher channels is the presence of many drops in the gas jet. Two particular regimes that can be distinguished represent formation of immobile drops on the channel wall due to liquid film or bridge breakage and appearance of mobile drops because of the two-phase flow instability. The characteristic drop diameter in this flow is observed from several microns to several millimetres. If the upper part of the channel is dry, the drops do not move and exist long until their washing by a continuous liquid flow.

The churn flow pattern is observed at high superficial liquid velocities. The churn flow occupies a considerable part in the map. This pattern is considered in detail in the paper [10]. A transition from the bubble (with continuous liquid regions) to the churn regime is determined by the appearance of breaks in the liquid. On the contrary, a transition from the jet to the churn flow pattern is manifested by formation of filled stable regions, the number of which is greater than one. The churn flow pattern relates to development of instability in the jet flow pattern, which leads to increasing frequency of gas interaction with liquid supplied from the lateral walls of the channel. The continuous liquid bridges disappear at transition to the annular flow. The destruction of bridges contributes to the appearance of stationary droplets on the walls of the channel which exist until washing by a continuous liquid flow.

In the range of low superficial liquid velocities, the transition from the jet to the stratified flow pattern occurs at an increase in the superficial gas velocity. In this flow, a part of liquid moves along the bottom wall of the channel in the form of a film entrained by the gas flow. Under the stratified flow pattern, gas occupies almost the entire width of the channel. Liquid fills its full height only in a narrow area along the lateral walls of the channel.

The transition to the annular pattern occurs with an increasing superficial velocity of liquid. In the annular flow pattern, liquid moves along the channel walls as a film, and gas with droplets forms the flow core in the central part of the channel. As compares to the stratified flow, under the annular flow regime the film in the upper part of the channel occupies a larger area. In the side parts, there are the contacts with the lower film, resulting in formation of the annular flow. The transition from the stratified flow pattern to the annular one is manifested by a significant increase in the area of the film at the top of the channel. The film on the upper channel wall is formed immediately after the liquid input to the channel (as it occurs in the channel height below of 0.1 mm, [7]).

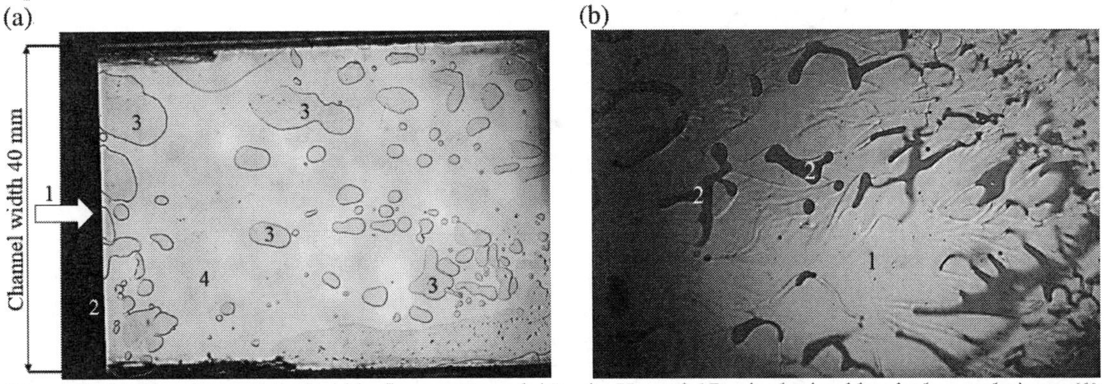

Figure 3. (a) The photo of the bubble flow at U_{SL} = 0.17 m/s, U_{SG} = 0.17 m/s obtained by shadow technique: (1) gas input to the channel, (2) slot in the lower channel wall for liquid input to the channel, (3) gas bubbles, (4) liquid; (b) the photo of the mobile drops in the annular pattern obtained by schlieren technique: (1) the liquid film on the upper channel wall, (2) mobile drops.

The feature of the channels height below of 200 μm in comparison with the higher channels is the presence of many drops in the gas jet. Drops appear for almost all two-phase flow patterns in the channel height of 50 μm. Two particular regimes that can be distinguished represent the formation of immobile drops on the channel wall due to liquid film or bridge breakage and appearance of mobile drops because of the two-phase flow instability. The second case is illustrated by the Schlieren image Figure 3b, where the dark regions correspond to liquid filling the channel from the top to the bottom and the bright regions represent the liquid on the inner surface of the top cover. After the ejection of liquid towards the center (caused by perturbation, [8]) and its subsequent return, a fraction of ejected liquid separated in the form of a drop. The diameters of these drops exceed the channel thickness and are typically from 50 μm to several millimetres, while their shapes are nearly circular due to the action of capillary forces.

We gratefully acknowledge the support from the Ministry of Education and Science of Russia (№ 14.604.21.0053 Project identifier RFMEFI60414X0053).

References

1. E. A. Chinnov, Journal of Heat Transfer **138**, 9 (2016).
2. V. V. Zamashchikov, A. A. Korzhavin, E. A. Chinnov, Combust. Explos. Shock Waves **50**, 381 (2014).
3. N. Shao, A. Gavriilidis, P. Angeli, Chemical Engineering Science **64**, 2749 (2009)
4. E. V. Rebrov, Theor. Found. Chem. Eng. **44**, 355 (2010).
5. E. A. Chinnov, O. A. Kabov, High Temp. **44**, 773 (2006).
6. O. A. Kabov, E.A. Chinnov, V. Cheverda, Microgravity sci. technol. **XIX**-3/4, 44 (2007).
7. E.A. Chinnov, O.A. Kabov, Doklady Physics **57**, 42 (2012).
8. E. A. Chinnov, F. V. Ron'shin, O. A. Kabov, Int. J. Multiphase Flow **80**, 57 (2016).
9. V. Haverkamp, V. Hessel, H. Löwe et al., Chem. Eng. Technol. **29**, 1015 (2006).
10. E. A. Chinnov, F. V. Ron'shin, V. V. Guzanov, D. M. Markovich, O. A. Kabov, High Temp. **52**, 681 (2014).

MATEC Web of Conferences **84**, 00035 (2016) DOI: 10.1051/matecconf/20168400035
International Symposium IPHT 2016

Evaporation of an intensively heated sessile droplet into the open atmosphere

Andrey Semenov[1,2,a], Geniy Kuznetsov[3], Dmitry Zaitsev[1]

[1]*Kutateladze Institute of Thermophysics, 630090, Lavrentiev ave. 1, Novosibirsk, Russia*
[2]*Novosibirsk State University, 630090, Pirogova str. 2, Novosibirsk, Russia*
[3]*National Research Tomsk Polytechnic University, 30, Lenin Ave, 634050, Tomsk, Russia*

Abstract. Evaporation of a heated sessile water drop was studied experimentally at a temperature difference between the substrate and surrounding atmosphere from 30 to 60^{0}C. The studies were performed on the substrates with micro- and nanocoatings with different wettability. The features of evaporation were studied for the pinned, partially pinned and depinned contact line (solid-liquid-gas interface). It is found that during the evaporation process the specific evaporation rate (mass loss per unit of the drop surface area) increases, particularly at the last stage of the drop lifetime.

1 Introduction

Among the effective solutions for removing high heat flux densities, there are the systems of spray cooling [1] as well as the systems using stratified two-phase flow in a microchannel [2]. Under the certain conditions, heat in these systems is removed due to intensive evaporation of liquid droplets that are in contact with hot surface. The process of liquid drop evaporation has been actively investigated in the last decade (see [3-6] and references therein). In most experimental and theoretical studies found in the literature, the isothermal evaporation of water droplets was studied. Research on nonisothermal evaporation of liquid droplets with a sufficiently high temperature difference between the substrate and the atmosphere is practically absent.

In this paper we study the process of evaporation of relatively large liquid droplets (drop volume varies from 100 to 400 microliters) to open atmosphere. A liquid drop is heated from below with the quasi-stationary heat transfer between the solid substrate and the drop. Ultra-pure water (by Milli-Q Company) is used as the working fluid. The use of water having a relatively high boiling point as a working fluid allows us to investigate the influence of the temperature difference between the substrate and the surrounding atmosphere on the rate of evaporation and heat transfer. The temperature difference between the substrate and the surrounding atmosphere ranges from 30 to 60 C. Three types of evaporation of the liquid droplets were studied: 1) constant contact angle (moving contact line), 2) constant surface area of contact (pinned contact line) 3) mixed type (partially pinned contact line).

[a] Corresponding author: semenov.itp@gmail.com

2 Experimental setup and technique of measurement

The experiments were accomplished on the experimental stand, the scheme of which is shown in Fig. 1. The test cell is an aluminum plate 50 mm thick, on the underside of which the heater was attached. As a heater we used Peltier element whose size is 40x40 mm^2. On the upper side of the aluminum plate we fixed a substrate, which was a disc 52 mm in diameter and 3.2 mm thick. The substrate surface temperature was measured with the thermocouple fixed at a distance of 1 - 2 mm from the contact line of a liquid droplet. The temperature under the substrate was measured by three thermocouples of K-type. Thermocouples were located radially at different distances from the center of the disc and allow to measure the temperature distribution along the radius of the substrate. We used three different substrates. Two substrates were made of a float glass plate with two different coatings: 1) the "Spin Teflon" coating, obtained by centrifugation of molten Teflon drops (with the contact angle hysteresis defined as the difference between the advancing and the receding contact angles, 122.5° - 112.7° = 9.8°), 2) the "Spray Teflon" coating, obtained by spraying of Teflon particles (contact angle hysteresis, 119.9° - 90.6° = 29.3°). The third substrate was made of stainless steel and had a developed surface structure that was obtained by bombarding the surface with calibrated aluminum oxide particles. The roughness of this surface was about 5 microns. Relative humidity in the room during the experiments was 20-30%. The room temperature was about 25°C.

Two optical techniques were used. The first one was the shadow method, which used a plane-parallel light source and a camera. The spatial resolution of the system was 6 microns per pixel and shooting frequency ranged from 3 to 8 frames per second. The shadowgraphs of the liquid drops were processed by the software package Drop Shape Analysis by KRUSS company. The second optical technique consisted of a digital video camera, which was placed atop the test section and allow to visualize the process of evaporation from the top and control the symmetry of the drop. In the event that a drop lost symmetry, the experiment was repeated.

The experiment started with a pretreatment of the studied substrate, which was thoroughly clean of the working surface; then, the entire substrate was placed into a tank containing distilled water for a day. Thereafter, in order to remove residual water from the surface, the substrate was purged with a special clean compressed air intended for cleaning optical components. A predetermined volume of a liquid drop placed on the surface, heated to the desired temperature, using an electronic syringe. The optical systems were focused, and then there was periodic shooting with the step of 10 seconds until complete evaporation of a liquid drop.

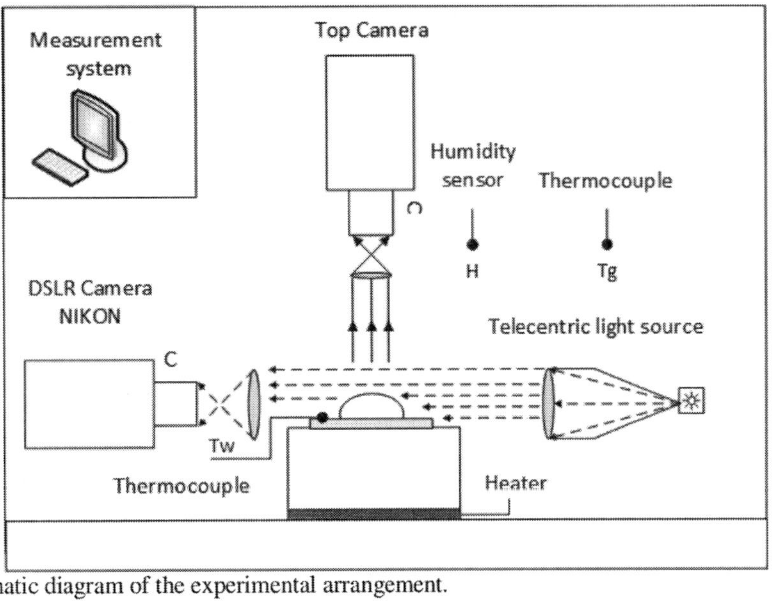

Figure 1. Schematic diagram of the experimental arrangement.

3 Results and discussion

Figure 2 shows evolution of droplets during evaporation on different surfaces. The temperature of the surface for each experiment is 65^{0}C. In the case of the Spin Teflon the evaporation occurs with constant contact angle (moving contact line), in the case of the Stainless Steel – with constant wetted diameter (pinned contact line), in the case of the Spray Teflon - mixed type (partially pinned contact line).

t, s	Spin Teflon	Spray Teflon	Stainless Steel
0		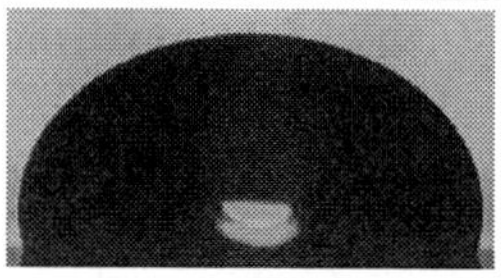	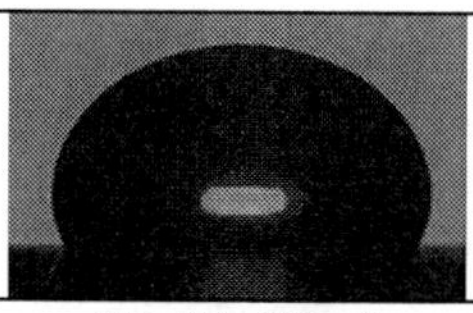
	V=96.65μl	V=100.67ul	V=85.72μl
200		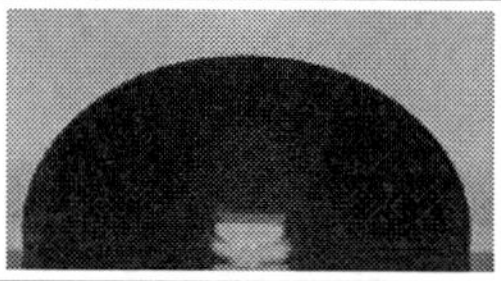	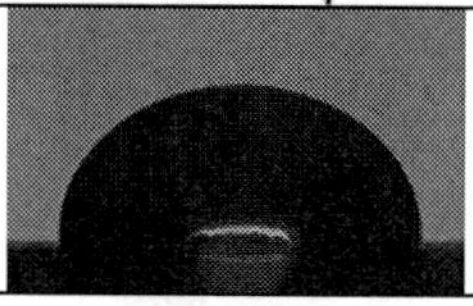
	V=76.47μl	V=82.77μl	V=53.88μl
400		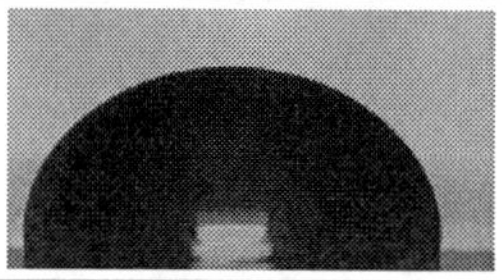	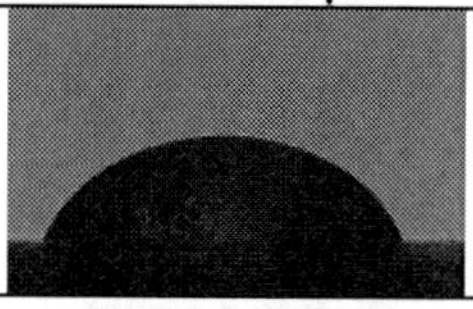
	V= 57.03μl	V=66.18μl	V=30.77μl
600	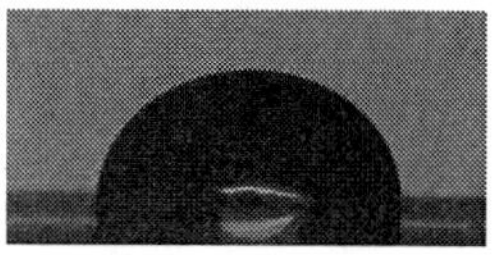		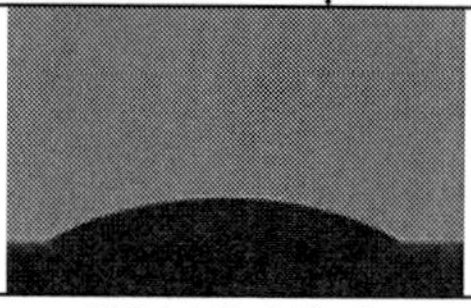
	V= 44.50μl	V=49.14μl	V=10.35μl
1000		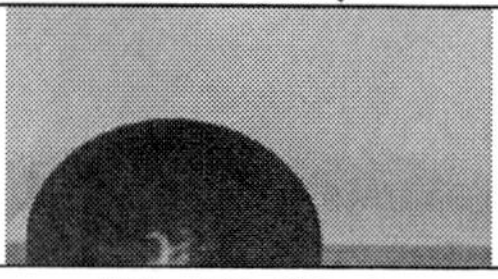	-
	V= 17.90μl	V=22.52μl	

Figure 2. Evolution of droplets during evaporation on different surfaces. T_w=65^{0}C.

Dependence of the specific evaporation rate of the liquid on time for different substrates is presented in Fig. 3. The rate of evaporation is calculated as the loss of the drop mass per unit drop surface area per unit time. It is seen that the specific evaporation rate increases with time (with decrease of drop volume) and at the last step of evaporation may be several times the initial value. The results are in good agreement with the findings of [7] where a comparison of the experimental and theoretical data on the dependence of the specific evaporation rate on time is presented. In contrast to [7], where the studies were carried out at a surface temperature T_w = 64^{0}C, in this paper we obtain new data for the surface temperatures range from 55 to 85^{0}C. The increase in the specific rate of evaporation is observed throughout the whole range of temperatures studied.

DOI: 10.1051/mateconf/20168400035

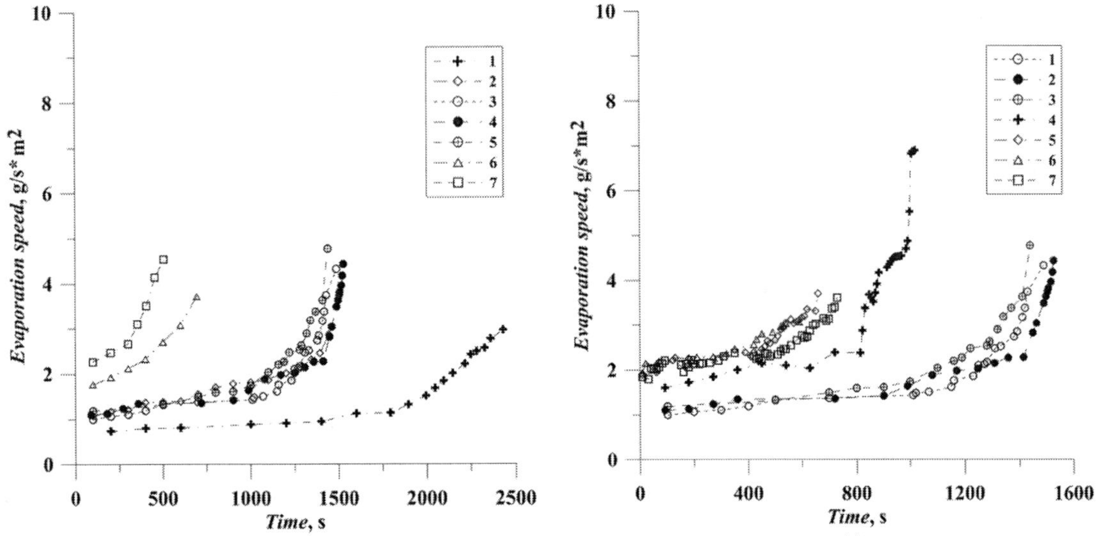

a) b)

Figure 3. Specific evaporation rate vs. time. a) - on the substrate with Spray Teflon coating: substrate temperature $T_w = 55^0C$ (1), 65^0C (2-5) (series of four runs), 75^0C (6), 85^0C (7); b) - different coatings at surface temperature $T_w = 65\ ^0C$: surfaces with Spray Teflon coating (1-3), Spin Teflon coating (4), stainless steel (5-7).

Acknowledgments

Authors gratefully acknowledge support of this work by the Russian Science Foundation (Agreement No. 14-19-01755).

References

1. Kim J. International Journal of Heat and Fluid Flow, **28**, 753 (2006).
2. Kabov O.A. and Zaitsev D.V. Multiphase Science and Technology, **21**, 249 (2009).
3. Ajaev V.S. J. Fluid Mech., **528**, 279 (2005).
4. Kuznetsov G.V., Feoktistov D.V., Orlova E.G. Journal of Engineering Physics and Thermophysics, **89**, 317 (2016).
5. Kuznetsov G.V., Feoktistov D.V., Orlova E.G. Thermophysics and Aeromechanics, **23**, 17 (2016).
6. Kabov O.A. and Zaitsev D.V. Doklady Physics, **58**, 292 (2013).
7. Gatapova E.Ya., Semenov A.A., Zaitsev D.V., Kabov O.A., Colloids and Surfaces A: Physicochemical and Engineering Aspects, **441**, 776 (2014).

The waves amplitudes increase due to interacting with thermocapillary structures

Evgeniy Shatskiy, Yuliya Koemets, Evgeny Chinnov

Kutateladze Institute of Thermophysics SB RAS, Lavrent'ev av. 1, Novosibirsk, 630090, Russia
Novosibirsk State University, 630090 Novosibirsk, Russia

Abstract. The analysis of experimental data on the simultaneous measurement of the thickness and temperature fields in the flowing heated liquid film at Re = 15 is presented. It is shown that increasing heat flux results in more expressed rivulet formation in the lower part of the heater (differential thicknesses in the rivulet and interrivulet area increases from 0.04 mm to 0.11 mm). It was found that the wave amplitude increase in the region where waves interact with the thermocapillary structures.

1 Introduction

Fluid film flows are widely used in cooling of the heat-generating surfaces in many technological applications: heat exchangers, condensers, evaporators. The main characteristics of these flows are their wavy amplitudes and structures. It is known that two-dimensional hydrodynamic waves in an isothermal liquid film are unstable to three-dimensional disturbances. It was found that the wavelength of the instability to transverse three-dimensional perturbations decreases with increasing Reynolds number. During breakdown of two-dimensional into three-dimensional waves in [1, 2] it was shown that there are isolated synchronous wave when there is no phase shift in the transverse direction of waves and the subharmonic waves, when the phase shift occurs.

When the liquid film flow over the heated surface except hydrodynamic instability that leads to the development of three-dimensional waves, there is also a thermocapillary instability which results in the appearance on the film surface of the stationary three-dimensional structures in the form of a number of rivulets with a thin film between them. Various regimes of the structures formation in a heated falling liquid film were discovered in [3–5]. Depending on the heating conditions and the temperature gradient on the film surface there are two types of thermo-capillary structures. Two mechanisms of growth of the relative amplitudes of the waves when heated falling liquid film are known in the literature: in the residual layer of the liquid film because of its intensive evaporation [6] and the film thickness decreases in interjet region by the action of thermocapillary forces [7].

2 Experimental setup

The setup is a closed circulation loop containing a reservoir with a pump, test section, filters, flow meters, pipes and valves. The test section consists of a carrier plate, a filmformer, a heat stabilizer, and a heater. The working fluid (water with dissolved rhodamine 6G as a dye) supplies by a pump in filmformer that included the collection chamber and the dispenser with a calibrated nozzle. The liquid flows down on the plate and then goes back into the reservoir. For the investigated conditions of the flow of a film of liquid, the formation of rivulets and propagation of developed threedimensional waves over the surface along the heater was observed. It has been shown in [8] that, for the isothermal case, the transition from a twodimensional wave motion to a threedimensional one on vertically falling films of liquid is accompanied by a redistribution of liquid in the horizontal direction. In a film of liquid being heated, the redistribution of liquid increases considerably due to the influence of thermocapillary forces. The formation of rivulets and of a thin film between them occurs [9]. The parameters of hydrodynamic waves are significantly varied when they pass along the heater.

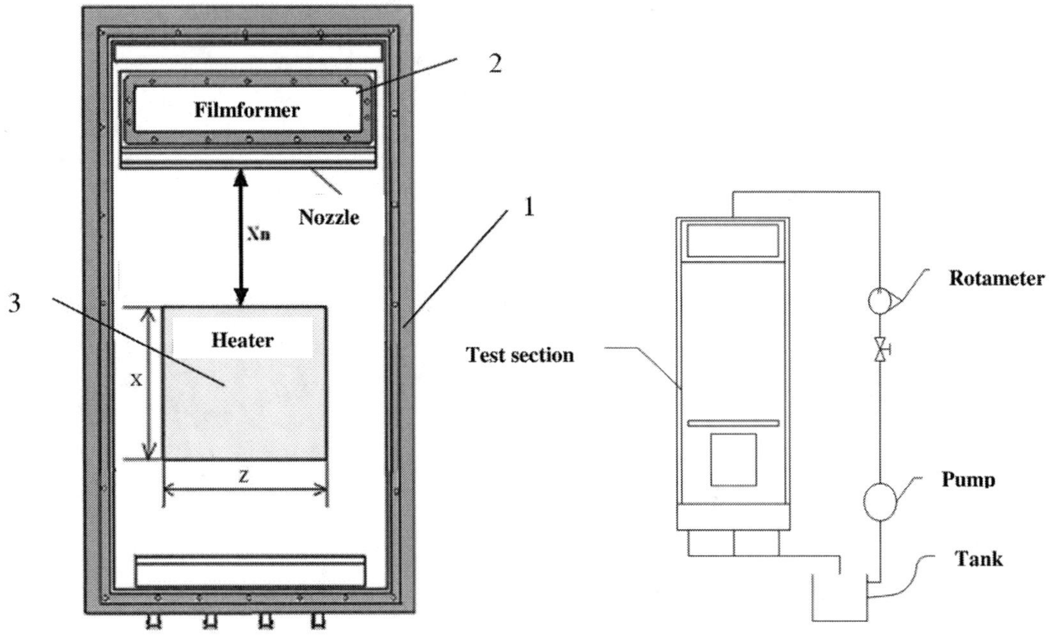

Figure 1: Experimental setup: 1 — Test section; 2 — filmformer; 3 — heater.

The filmformer locates in the upper part of the test section. The distance from the bottom edge of the nozzle to the upper edge of the heater is 263 mm, which corresponds to the area of transition from two-dimensional to three-dimensional waves. The heating element is a flat copper heat exchanger 150 mm wide and 100 mm long. Inside the heater, there are rectangular channels through which the heating liquid is pumping. The heat flux from the heater surface is calculated from the temperature difference of pumping fluid on the inlet and outlet of the heat exchanger for a given mass flow rate. The temperature of the heater surface facing the film was measured by three thermocouples located on the vertical axis of symmetry. The boundary condition on the surface is close to T = const providing high value of temperature gradients near the upper edge of the heater. The heat flux varied in the range 0.5-6 W/cm^2. The initial water temperature at the nozzle exit was 28 °C. The film Reynolds number equals to 15. For determination of the instantaneous field of film thickness and wave characteristics, a modified fluorescent method is used. A distinctive feature of the layout of the measuring system in the experiment is the location of the radiation source and the receiver from the free surface of the film. To energise the fluorophore the laser diode pumping is used (RLM-532-2000). The laser continuously illuminating an area with the size of 120x120 mm. Re-emmited light is recording with a digital camera PCO 1200h, with a frequency of 500 Hz in full frame (1280x1024 pix.). Before the camera, there is a red filter to cut off the reflected laser light. When measured in the area of 100x100 mm, the system provides a spatial resolution of 0.1 mm.

3. Results and discussion

Figure 1 shows the dependence of the amplitude of the incoming waves to the heater along the rivulet from the path length for several values of the heat flux. It is seen that with increasing of the heat flux there is a local maximum of the thickness in the initial part of the heater. This trend is related to the fact that when the heat flux reaches the threshold value at the upper edge of the heater in the residual layer there are thermocapillary periodic structures. These structures arise because of the high temperature gradients at the leading edge of the heater (up to 10 K/mm), resulting in occurrence of thermocapillary forces directed against the flow. These structures make a disturbance in the following wave front, leading to an increase in the amplitude of the waves. It was found that the amplitude of hydrodynamic waves in the upper part of the heater (5 mm < X <30 mm) can be increased by more than half when interacting with the thermocapillary instability of type A ($q= 2.55\ W/cm^2$). While moving along the heater the wave amplitude decreases, although the value is substantially higher than at lower heat flux ($q = 0.98\ W/cm^2$) when thermocapillary structure type A are not yet formed. Also at high heat fluxes there is a transverse movement of the rivulets and in combination with an amplitude increase it results in an increase of the stability of the heated liquid film to rupture.

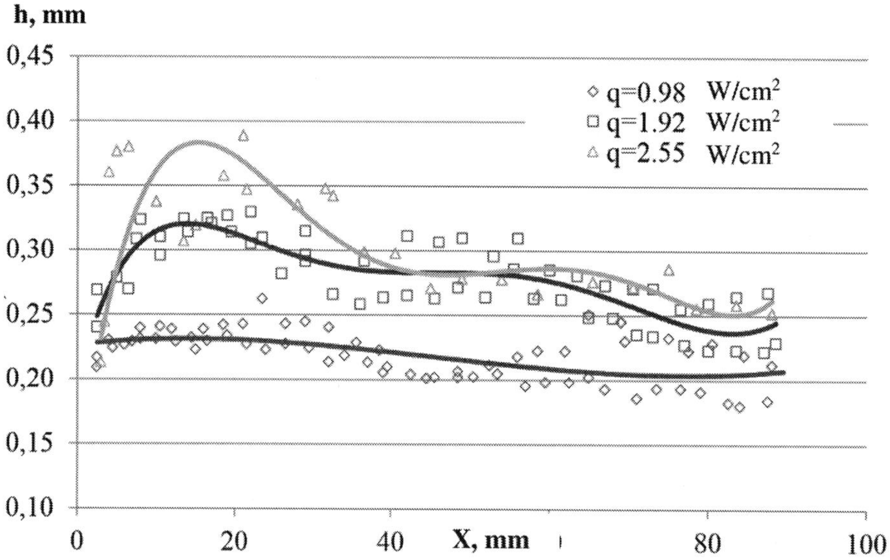

Figure 2: Dependence of the amplitude of the wave from the path length of the heater

This work was supported by a grant from the Russian Science Foundation (project Nr. 15-19-30038).

References

1. Joo S. W., Davis S.H. J. Fluid Mech. **242**. 529-547 (1992).
2. Liu J., Schneider J. B., Golub J.P. Phys. Fluids. **7**. 55–67 (1995).
3. O.A. Kabov, Heat Transfer Research, **27**, 221-226 (1996)
4. O.A. Kabov, I.V. Marchuk, V.M. Chupin, Russ. J. Eng. Thermophys. **6** (2) 104–138 (1996).
5. O.A. Kabov, E.A. Chinnov, Russ. J. Eng. Thermophys. **7** (1/2) 1–34 (1997).
6. Pavlenko A. N. et. al. Journal of Engineering Thermophysics. **11**, 321 – 333 (2002)

7. Chinnov E. A. International Journal of Heat and Mass Transfer, **71**, 106-116.
8. S. V. Alekseenko, V. V. Guzanov, D. M. Markovich, and S. M. Kharlamov. Technical Physics Letters, **38**, 739-742.
9. Chinnov E.A., Nazarov A.D., Saprykina A.V., Zhukovskaya O.V., Serov A.F. High Temperature, **45**, 657-664.

Cooling of high power LED chip by boiling on the longitudinal finned surfaces

Evgeniy Shatskiy[a]

Kutateladze Institute of Thermophysics SB RAS, Lavrent'ev av. 1, Novosibirsk, 630090, Russia
Novosibirsk State University, 630090 Novosibirsk, Russia

Abstract. In this work, we have created a new experimental setup for studying heat transfer from a local heat source. The boiling heat transfer on the local heaters with a diameter of 5 and 1 mm has been investigated. It is shown that on the finned surfaces the overheating relative to the saturation temperature in comparison with a smooth surface decreases up to three times for the heater with a diameter of 5 mm. There is up to two times increase in heat transfer coefficient on finned surfaces as compared to the smooth ones. For finned surfaces on the heater with a diameter of 1 mm the surface overheating relative to the saturation temperature decreases in four times. More than three times increase is observed for the heat transfer coefficient on finned surfaces as compared to the smooth ones.

1 Introduction

The future of a number of sectors of the economy is linked with the technology of solid-state light sources and one of the main characteristics of uptime is the removal of heat. Study of the evaporation mechanism on the surfaces of small size was started in [1]. The main objective was to study the temperature fields of the heating surface and liquid, and to study conditions for the formation and existence of a thin film of liquid at the base of the vapor bubble. In this work, as in the follow-up, dedicated to the study of a single center of vaporization [2], the heat flux limited with value at which there is a steady generation of a single center of evaporation, that is relatively modest values. A review of studies on heat transfer during vaporization at the surface of the small size is contained in [3].

Cooling of heat-emitting surfaces in different technological applications is an important problem, which can be solved by such processes as film flows [4, 5], etc. In our days the problem of creating an effective compact cooling system for powerful LEDs whose efficiency and durability are highly dependent on the efficiency of the power removal remains unsolved.

Study of evaporation mechanism on the surfaces of small size (SSS) was started in [6]. Heat transfer at boiling of dielectric liquids on the heaters of small size in the sunk vertical channels was studied in [7]. A review of works on heat transfer at evaporation on SSS may be found in [3]. In the literature, there is a dependencies for the critical heat flux density for each type of SSS, having the general form:

[a] Corresponding author : shatskiy.itp@gmail.com

$$q_{cr}/q_{cr\infty} = k(d_s/l_\sigma)^n \tag{1}$$

but differing in correlation multipliers; and d_s is the diameter of the heating surface. Here

$$q_{cr\infty} = 0.13 r_{lv} \rho_v \left(\frac{\sigma(\rho - \rho_v) g}{\rho_v^2} \right)^{0.25} \tag{2}$$

r_{lv} is the latent heat of vaporization, ρ_v is the vapor density, ρ is the fluid density, σ is the surface tension, and g is the gravity acceleration.

For locally heated horizontal plates k=3.4, n=-0.5 [3], for horizontal cylinders k=1.4 [8], and for spheres k=2.5 [8]. The reason for such significant effect of the size of SSS on q_{cr} is connected with the feature of hydrodynamic situation that occurs around the surface of small size (SSS).

The aim of the present work is studing of pool boiling of water on the surfaces with different geometry.

2 Experimental setup

To study the heat transfer processes that occur at boiling on the longitudinal finned surfaces, a new experimental setup was created. Schematic diagram of the experimental setup is shown in Fig. 1.

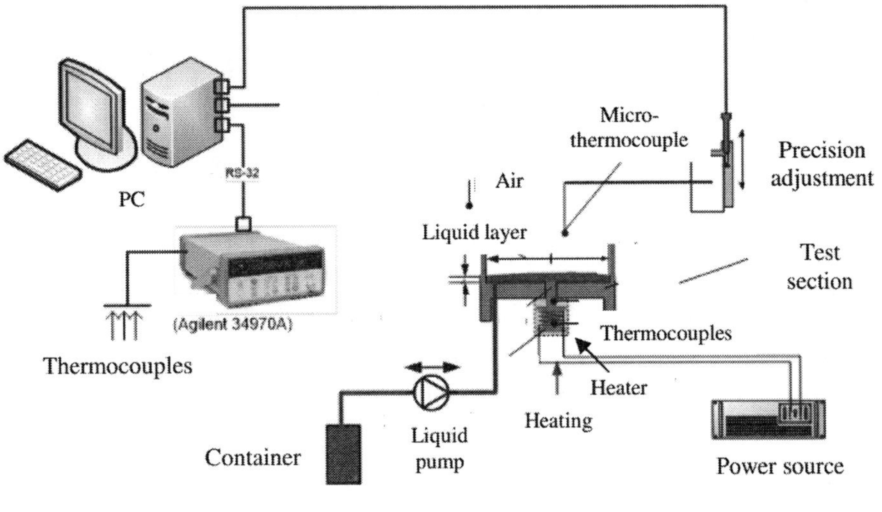

Figure 1. – Scheme of experimental setup.

The test section is a PTFE cilindeer with diameter 28mm. The working fluid (distilled water) is supplied to the test section using a peristaltic pump, where a horizontal layer of liquid is formed. The thickness of the liquid layer is controled by the amount of liquid (the geometry of test section is known) and varies depending on the operating parameters of the experiment (20-50 mm). In the center of the test section there is a local heating performed by a heater. The heating element is made of brass with a round head with a diameter of 1 mm and 5 mm. The tip of the core tightly inserted into the base. The source of heat is a nichrome tape wound on the shank of the core. In order to minimize heat losses, the heating element was carefully insulated. The fiberglass was used as a heat insulator, which was wound in several layers around the core of the heating element. It was estimated that the whole incoming power transferred to the liquid and heat losses did not exceed 20% [9].

To intensify the heat transfer from the heater surface to the liquid the surface of the heater D = 5 mm was milled with radial grooves (Fig. 3). For heater with diameter 1 mm special headers have been developed which also have been milled with grooves (Fig. 2). Heaters parameters are shown in Table 1, and the photographs in Figs. 2, 3.

Table 1. Parameters of the heaters.

Diameter of the heater, mm	Number of fins	Surface area, mm^2
1	0 (smooth surface)	78,5
1	20 (small finns)	277,4
1	30 (large finns)	360
5	0 (smooth surface)	19,6
5	16	248,6

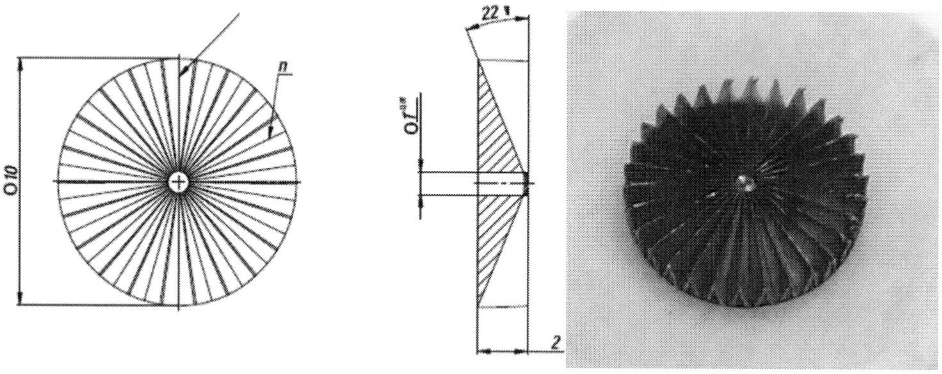

A B

Figure 2. Drawing of the heater header for D = 1 mm (A) and photo (B).

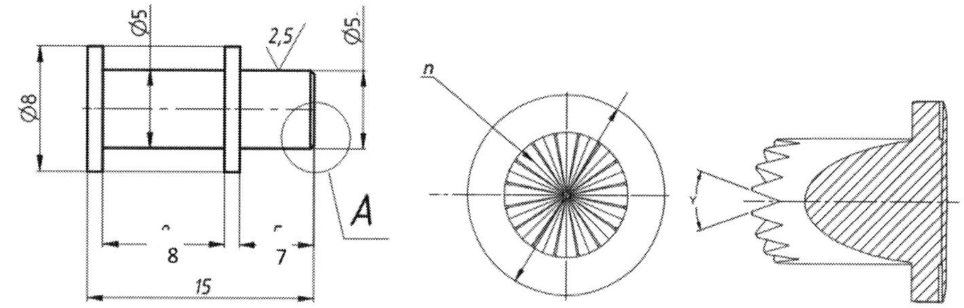

A B

C

Figure 3. Drawing the heater D = 1 mm (A) smooth surface (B) finned surface, (C) photo.

The power of the heating element is controlled by the power source. The temperature of the heater (at the interface with the liquid) and the fluid is measured by a number of thermocouples. The data from thermocouples is collected using data acquisition system, and the temperature measurement precision is 0.1 ^{0}C.

3 Results and discussion

Curves of boiling for heaters with D=5 and 1 mm on a smooth and finned surfaces are presented (Figs. 4 and 5). It was found that the boiling curves for the finned surfaces are higher than smooth. Experiments on the heaters D=5 mm with a smooth and finned surfaces showed that overheating on the finned surfaces is three times smaller (Fig. 4). And the heat transfer coefficient increases up to two times on the finned surface compared with the smooth.

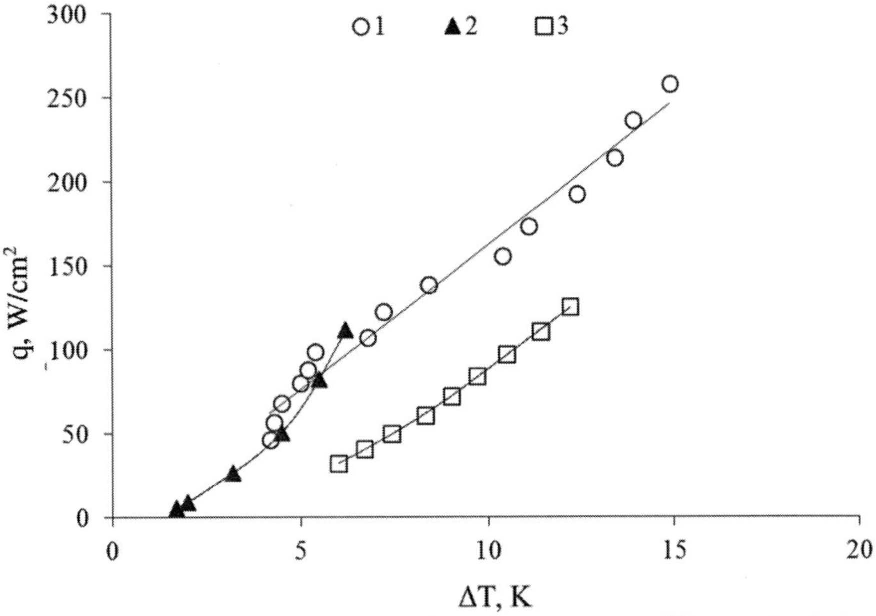

Figure 4. The dependence of the heat flux from the temperature difference for the heater for D = 5 mm. 1 - large fins, 2 - small fins 3 - smooth surface.

Registration of heat transfer crisis was carried out by the thermocouples at the moment of instantaneous growth of temperature on the heater. The crisis of heat transfer has been registered on the heater with large fins (Fig. 4) and amounted to 256 W/cm^2.

The studies of heat transfer during boiling on a heater with a diameter of 1 mm are shown on Fig. 5. It should be noted that the heat flux density was calculated using the area of the heater excluding the header. On the smooth surface overheating temperature reached about 200 C. The heat flux density before the heat transfer crisis reaches 900 W/cm^2.

DOI: 10.1051/matecconf/20168400037

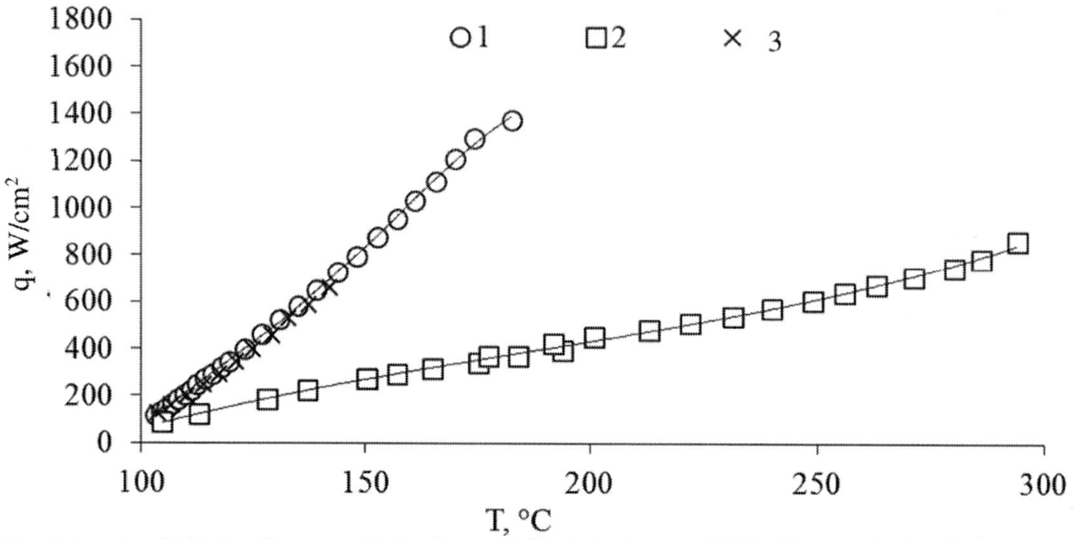

Figure 5. Dependence of the heat flux density from the temperature of the heater with a diameter D = 1 mm. 1 - head with large fins, 2 - smooth surface, 3 - head with small fins.

On the finned surfaces overheating relative to saturation temperature decreases up to four times. More than three times increases the heat transfer coefficient on the finned surface compared with the smooth. It should be noted that the type of finning has no significant effect on the boiling curve. As a result of the experimental work the density of heat flux reached 1400 W/cm^2.

In this study, we carried out experiments for the conditions of a large volume of fluid (thickness of the layer > 10 mm) and maintaining the temperature of the fluid at saturation temperature. It was found that radial fining of the heater surface may lead to significant increase in critical heat flux density compared to the smooth one.

The work was supported by the RF Ministry of Education and Science (Agreement No. 14.604.21.0053, project ID RFMEFI60414X0053).

References

1. Moor F. D., Mesler R. B. AIChE Journal **7** (1961)
2. Borisov V. Z., Kirillov P. L. Journal of Engineering Physics and Thermophysics **28,** 910 (1970)
3. Tolubinskiy, V.I., Antonenko, V.A., Kudritskiy, G.R., Ostrovskiy, Yu.N Industrial Heat Engineering **9** (1987)
4. Chinnov E. A. Journal of Heat Transfer **138** (2016)
5. Kabov O.A., Zaitsev D.V., Cheverda V.V. and Bar-Cohen A. Experimental Thermal and Fluid Science **35**, 825 (2011)
6. Moor F. D., Mesler R. B. AIChE Journal **7** (1961)
7. Chinnov E.A., Dyatlov A.V., Kravchenko D.N. Siberian Physico-Technical Journal **5** (1992)
8. Linard D., Dir B. Teploperedacha C **2** (1973).
9. Yu. V. Lyulin, S. E. Spesivtsev, I. V. Marchuk and O. A. Kabov. Technical Physics Letters **41** (2015)

MATEC Web of Conferences **84**, 00038 (2016) DOI: 10.1051/matecconf/20168400038
International Symposium IPHT 2016

Thermocapillary breakdown of a horizontal spot-heated liquid layer

Serafim Spesivtsev [1,2,a], Yuriy Lyulin[1,3]

[1] *Institute of Thermophysics, 1, Lavrentiev Ave., 630090, Novosibirsk, Russia*
[2] *Novosibirsk State University, 2, Pirogov Ave., 630090, Novosibirsk, Russia*
[3] *Tomsk Polytechnic University, 30, Lenina Ave, Tomsk, 634050, Russia*

Abstract. Thermocapillary breakdown of thin (0.3 – 0.7 mm) horizontal layers of liquid (ethanol) when heated from a localized hot spot was investigated experimentally. The effects of layer thickness and the surface properties of the substrates on the breakdown dynamics were studied. Visualization and control of the liquid layer were carried out using schlieren and shadowgraphy techniques. Main steps of the breakdown process were determined and average velocity of the dry spot formation was measured. Significant influence of the substrate properties on breakdown dynamics has been found. It was shown that one of the main factors affecting the dynamics of layer breakdown and the formation of dry spots in the heating area besides the thermocapillary effect is evaporation.

1 Introduction

Study of heat transfer from a local heat source becomes one of the most important problems in thermophysics. The problem is closely connected to the cooling of microelectronic equipment [1]. Permanent consumption of electricity by the device inevitably leads to increasing the microprocessor temperature and degradation of its performance and reliability. Continuous development and complication of microchip structure produce nonuniform heat flux distribution on the chip surface. Heat flux density in some regions is much higher than the chip average [2], of the order of 1 kW/cm^2. These specific regions called "hot spots" could have size from several hundred microns to 1-2 millimeters. However, using special localized cooling it is possible to produce large performance gains in microprocessors. Nowadays, there are several effective techniques for cooling of local hot spots such as spray cooling, boiling in microchannels, thermoelectric coolers. One of the promising methods for removing such high heat fluxes from a spotted heat source is technology based on evaporation of a thin liquid layer. Dynamics of evaporation essentially depends on the conditions in the layer [3]. In particular, the breakdown of liquid layer leads to dramatic decreasing of heat transfer from a spotted heat source [4]. Processes of liquid layer rupture are actively studied experimentally [4-6] and theoretically in the present time [7-9]. The goal of the present work is to study using schlieren technique the breakdown dynamics of a horizontal evaporating liquid layer when heated from a localized hot spot.

[a] Corresponding author: simafir94@gmail.com

2 Experimental setup

Experiments were conducted on the setup shown in Fig. 1. The working fluid is supplied to the test cell with the help of the syringe pump and a horizontal liquid layer is formed. The thickness of liquid layer is varied from 300 to 700 μm. The layer of liquid is opened to the atmosphere and maintained on the surface of the working area using sharp edge on the substrate perimeter. Spot heating of the horizontal liquid layer takes place in the center of the substrate. The test cell consists of caprolon base, metallic substrate and the heating element. The caprolon base has a special cut on the upper side for installation of the substrate and a central through hole with a diameter of 1.6 mm. The substrate is made of stainless steel and has a diameter of 50 mm and a thickness of 1 mm. In the center of the substrate is a closed hole with a diameter of 1.6 mm and a height of 0.8 mm. The heating element is made of brass and has a round tip with a diameter of 1.6 mm and a height of 3 mm. It is tightly inserted into the closed hole of the substrate through the caprolon base. Thermal paste is used for better thermal contact between the heater tip and the substrate. The distance between the tip and the upper side of substrate is 0.2 mm. The power of heating element is controlled by the power supply. Insulating material is located on the underside of the heater to minimize heat losses. Temperature in the test cell is measured by thermocouples (type K) with an accuracy of 0.1°C. Location of the thermocouples is shown in Fig. 1. Relative humidity and atmosphere temperature are measured using the thermohygrometer Testo 645 with an accuracy of 2% and 0.1°C, respectively.

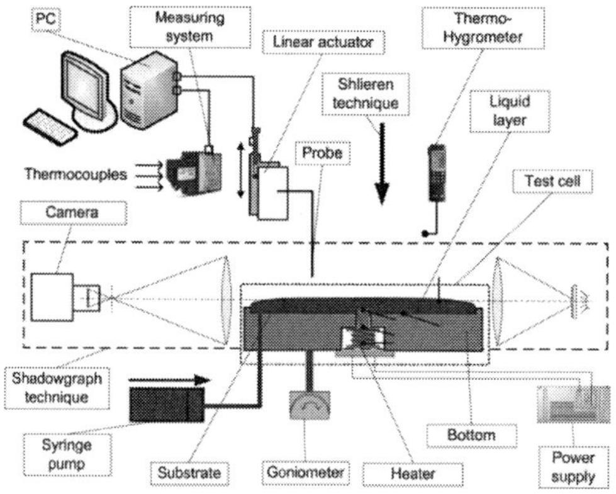

Figure 1. Scheme of the experimental setup.

The heat flux density is determined by measuring the temperature difference between two different sections along the heater tip. The height of the horizontal liquid layer is maintained in constant position during the all experiment. The probe with a diameter of 100 μm is installed at a required distance from the substrate surface. With the help of high-precision syringe pump the flow rate is chosen for providing a constant layer thickness, taking into account the evaporation. In the experiment the layer surface is maintained at a distance of 3-5 μm below the probe. If the height of the liquid layer is increased upper the required one the liquid meniscus is formed due to the contact the probe and liquid surface. The event of the contact is instantly visualized on the camera. Precise linear actuator and shadowgraph technique are used in order to determine the position of the probe. The probe moves in range of 5 mm by steps of 1 μm. Linear actuator is controlled from personal computer and by special software. Observation of the layer surface and the probe is carried out using video camera of shadowgraph technique with a resolution 2592x1944 pixels and a field of view of 4.3x5.8 mm. For visualization the deformation of the surface and recording the layer breakdown dynamics schlieren technique with Phantom 3798M high-speed camera was used (at 2500 frames per second the resolution is 256x256 pixels and the field of view is 7x7 mm). The test cell was installed in a horizontal position with the help of two-axis goniometer. Equilibrium contact angle of the substrate surface was defined by Young-Laplace method [10] at room temperature of 25 ± 2 ° C. The values of the contact angle is $\theta_1 = 8 \pm 1$ ° for smooth steel substrate and $\theta_2 = 6 \pm 1$ ° for rough steel substrate. Profilometer "Micro Measure 3D station" was used for measuring the roughness of substrates.

Average values of the roughness are R_{a1} = 0.037 μm and R_{a2} = 0.327 μm for smooth and rough steel substrate, respectively.

3 Results and discussion

Experiments were conducted at atmospheric pressure, temperature and relative humidity of 28±2°C and 25±3%, respectively. Layer thickness was ranged from 300 to 700 μm. Ethanol (95%) was used as a working fluid. Heater temperature range was 20 - 105°C. Heat flux density was varied from 0 to 95 W/cm². Injection liquid flow rate was up to 200 μl/min. During the experiment heat flux is increased up to a critical value at which the liquid layer ruptures. At this moment heating process is stopped to prevent the failure of heating element.

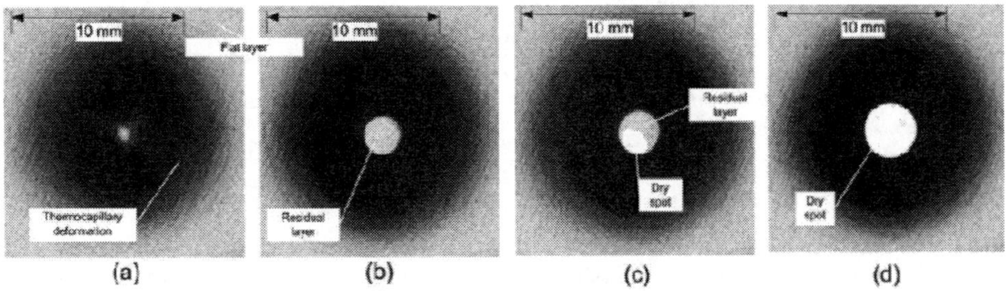

Figure 2. Visualization of breakdown dynamics and formation of the dry spots.

It was found that breakdown process consists of several steps. At the beginning, thinning of the liquid layer over the heating area (Fig. 2a) due to the effect of thermocapillary forces [11, 12] and evaporation is observed. Further thinning leads to the formation of residual liquid layer in the area of the local heating, Fig. 2b. Then the residual layer evaporates until its thickness reaches the critical and breakdown of the liquid layer occurs, Fig. 2c, [4, 5]. After the breakdown the whole area of the local heating rapidly dries and quite symmetrical circular dry spot is formed, Fig. 2d.

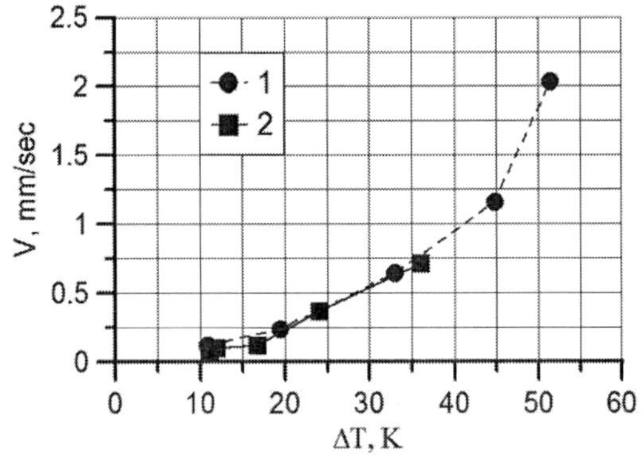

Figure 3. Dependence of the dry spot velocity on the radial temperature drop. 1- smooth substrate, 2- rough substrate.

Dependence of the characteristic velocity of dry spot formation on the radial temperature drop is shown in Fig. 3. The difference ΔT is taken between the heater temperature and the temperature at a distance of 15 mm from the center. The temperature values were measured at the moment of layer breakdown for different layer thickness (from 300 to 700 μm) for each substrate. The average propagation velocity of dry spot was determined as the ratio of the dry spot diameter to the time of dry spot formation. The time of dry spot formation is calculated from the moment of breakdown to the complete evaporation of residual liquid layer. For rough steel substrate the time of dry spot formation varies from 1.39 to 7.85 seconds depending on the layer thickness. For smooth substrate the time range is from 0.67 to 7.88 seconds. It was found that the dry spots velocity increases with the radial temperature drop. Temperature in the center of substrate increases and more intense evaporation of residual layer takes place. Thus, one of the main factors influencing the residual layer breakdown and the heating region draining is evaporation.

Effect of the surface properties on dynamics of layer breakdown was found. Dry spots were arisen in various points simultaneously and dewetting process was observed on the rough substrate, Fig. 4a. In case of the smooth substrate the residual layer breakdown occurs in one point, Fig. 4b.

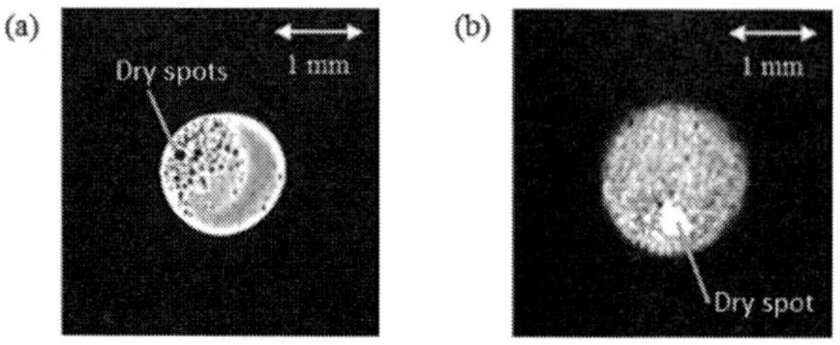

Figure 4. Effect of the surface properties on breakdown dynamics (layer thickness 700 μm): a – rough substrate, $R_a = 0.327$ μm; b – smooth substrate, $R_a = 0.037$ μm.

4 Conclusions

Influences of the layer thickness, intensity of heating and type of substrate on the breakdown dynamics of liquid layer were studied. The existence of residual liquid layer over the heating area before the breakdown has been proved. Velocity of dry spots formation increased when the local heating intensity is increased. One of the main factors that influence the breakdown of residual liquid layer and dry spot formation is evaporation. Significant influence of the surface properties on breakdown dynamics of liquid layer has been found. In case of the rough substrate dry spots were arisen in various points simultaneously and dewetting process was observed. For the smooth surface the residual layer ruptures in one place.

The study was financially supported by the Russian Science Foundation (Project 14-19-01755).

References

1. A. Bar-Cohen, P. Wang, J. Heat Transfer **134**, 5 (2012)
2. R. Mahajan, C. Chiu, G. Chrysler, Proceedings of the IEEE **94**, 8 (2006)
3. A.A. Fedorets, I.V. Marchuk, O.A. Kabov, Interfacial Phenomena and Heat Transfer **1**, 1 (2013)
4. Yu.V. Lyulin, S.E. Spesivtsev, I.V. Marchuk, O.A. Kabov, Technical Physics Letters **41**, 11, 1034-1037 (2015)
5. D.V. Zaitsev, D.A. Rodionov, O.A. Kabov, Microgravity Science and Technology, **19**, 100 (2007)
6. D.V. Zaitsev, O.A. Kabov, Microgravity Science and Technology, **19**, 174 (2007)
7. V. S. Ajaev, Interfacial Phenomena and Heat Transfer **1**, 1 (2013)
8. M. B. Williams and S. H. Davis, J. Colloid Interface Sci. **90** (1982)
9. J. P. Burelbach, S. G. Bankoff and S. H. Davis, Phys. Fluids A **2**, 322–333 (1990)
10. I.V. Marchuk, V.V. Cheverda, P.A. Strizhak, and O.A. Kabov, Thermophysics and Aeromechanics **22**, 3 (2015)
11. I. V. Marchuk, Journal Engineering of Thermophysics **18** (2009)
12. I.V. Marchuk, Journal Engineering of Thermophysics **24** (2015)

Cavitating flow control through continuous tangential mass injection on a 2D hydrofoil at a small attack angle

Mikhail V. Timoshevskiy[1,2], Ivan I. Zapryagaev[1,2], Konstantin S. Pervunin[1,2,a] and Dmitriy M. Markovich[1,2,3]

[1]*Kutateladze Institute of Thermophysics, Siberian Branch of the Russian Academy of Sciences, 1, Lavrentyev Ave., Novosibirsk, 630090, Russia*

[2]*Department of Physics, Novosibirsk State University, 2, Pirogov Str., Novosibirsk, 630090, Russia*

[3]*Institute of Power Engineering, Tomsk Polytechnic University, 30, Lenin Ave., Tomsk, 634050, Russia*

Abstract. We studied cavitating flow over the suction side of a symmetric 2D foil – a scaled-down model of high-pressure hydroturbine guide vanes (GV) – in different cavitation regimes at the attack angle of 3°. High-speed imaging was used to analyze spatial patterns and time dynamics of the gas-vapour cavities. A hydroacoustic pressure transducer was employed to register time-spectra of pressure fluctuations nearby the hydrofoil. A PIV technique was applied to measure the velocity fields and its fluctuations. The active flow control was implemented by means of a continuous liquid supply with different flow rates through a slot channel located in the GV surface. It was found that the active mass injection does not influence the primary flow upstream of the slot channel position. For the cavitation-free and cavitation inception cases, the injection was shown to make the turbulent wake past the GV section more intense. However, at the developed cavitation regimes the active flow management made it possible to reduce substantially the amplitude or even totally suppress the periodic cavity length oscillations and pressure pulsations associated with them.

1 Introduction

Unsteady cavitation phenomena occurring within the ducts of real hydropower systems are known to be one of the main sources of instabilities that cause significant vibrations, overloads and, consequently, fatigue damages. These might lead to unstable operating conditions, reduced efficiency and safety problems. Presently, two main classes of partial cavity instabilities are distinguished [1]: intrinsic and system. In order to hinder the development of the unsteadiness a number of methods are applied to manipulate the flow dynamics. Probably, the most simple in implementation and effective of them is continuous tangential mass injection along a test body surface to introduce an additional momentum to overcome the adverse pressure gradient [2].

The paper aims at the experimental study of partial cavity flows around a scaled-down model of guide vanes (GV) of a high-pressure turbine at a small attack angle, different cavitation numbers and control of the cavitation by means of continuous mass injection through a spanwise slot channel in the foil surface. We report here on high-speed visualization of spatial structure and dynamics of attached

[a] Corresponding author: pervunin@itp.nsc.ru

and cloud cavities and their integral characteristics, a comparison of the frequency spectra of pressure pulsations near the profile made by a hydrophone and PIV measurements of the spatial distributions of the mean velocity and its turbulent fluctuations around the cavitating GV model.

2 Test foil and experiment description

The test foil of C = 100 mm chord length was a scaled-down model of a Francis turbine guide vane (GV), which is a two-dimensional body, with a blunt trailing edge of 1.31 mm height (Fig. 1). It was made of brass with the mean surface roughness of 1.5 μm. The maximum width of GV is H_{max} = 0.2206C measured at the distance of x_{max} = 0.4396C from its leading edge. The aspect ratio (the ratio of the foil span to its chord) was AR = 0.8 that is very close to the one in some practical full-scale conditions.

In order to provide flow control, the hydrofoil was manufactured with a spanwise slot channel in its surface (Fig. 1) through which the liquid was supplied from a bypass circuit directly to the test section along the GV surface. The slot nozzle exit of 0.6 mm height and 70 mm width was located at the distance of 0.6C from the foil leading edge so that the liquid was injected across the whole foil span, except for the end caps of 5 mm width. The nozzle chamber inside the GV profile had a helical geometry with a contraction towards the exit (Fig. 1a). The maximum exit velocity of the slot wall jet was U_{in} = 10.6 m/s. The measurement uncertainty of the injection flow rate did not exceed 2%.

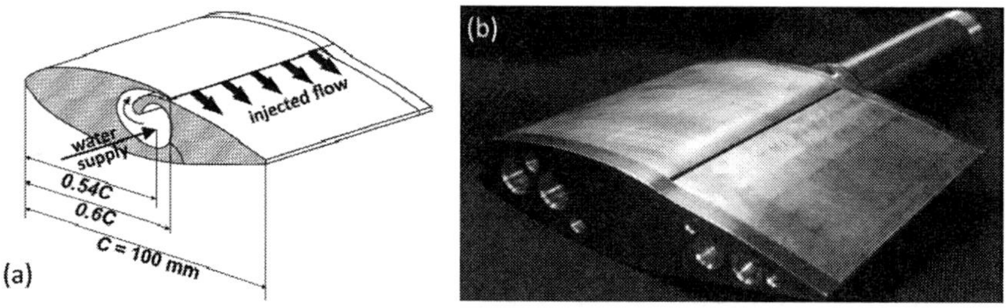

Figure 1. (a) A 3D scheme and (b) photograph of the fabricated GV profile.

The experiments were carried out in the Cavitation tunnel in Kutateladze Institute of Thermophysics SB RAS. Its description as well as the details on experimental conditions and applied measurement techniques can be found in [3]. The foil attack angle α was 3°. Pressure measurements were carried out close to the test foil by means of a hydroacoustic pressure transducer – a hydrophone B&K Type 8103 that was flush-mounted into the sidewall of the test section. It was vertically located at the centre of the test section and 80 mm shifted behind the GV trailing edge.

3 Results

3.1 Visualization and pressure measurements

Cavitation is initiated right behind the leading edge on the suction side of the GV profile in form of travelling bubbles (Fig. 2-1.a). At σ = 0.95, initially transitional cavity that is characterized by relatively small clouds (compared to cloud cavitation regimes) shedding from the cavity closure region is transformed to a typical travelling bubble cavitation as soon as the injection through the slot is started (Fig. 2-1.b). This leads to a decrease in the dimensions of the cavitation area on the GV surface. However, pressure fluctuations for both cases are almost identical and reveal no effect of the injection (Fig. 2-1.c).

This influence detected visually is probably linked with the existence of a shear layer past the backward facing step produced by the slot and its wall, which leads to an abrupt reduction of the shear stresses once the injection begins. Weakened shear stresses cannot anymore produce strong spanwise vortices (like in the case without injection) in the centres of which an intense evaporation occurs and, thus, cavitation is suppressed.

When the cavitation number is slightly decreased to 0.89, the partial cavity on the GV section becomes unsteady and starts to oscillate periodically in the unforced flow case (Fig. 2-2.a). This is also proved by the pressure measurements. The main maximum that is clearly seen in Fig. 2-2.c at St $= fC/U_0 = 0.21$ corresponds to the frequency of the attached cavity pulsations. At this regime, the liquid injection results in stabilization of the attached cavity dynamics (Fig. 2-2.b), changing the partial cavity regime to transitional one like that shown in Fig. 2-1.a.

At the same time, the peak in the pressure fluctuations spectrum at St = 0.21 vanishes completely (Fig. 2-2.c). When the cavitation number is further reduced to $\sigma = 0.84$, the active mass injection does not already change the cavity unsteady behaviour (Figs. 2-3.a and 2-3.b) but the amplitude of the pressure pulsations diminishes about three times (cf. the peak heights in Fig. 2-3.c).

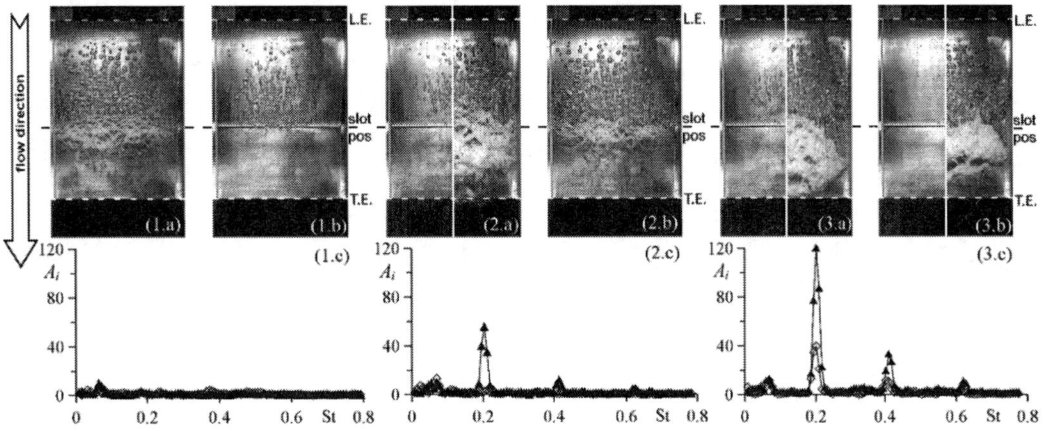

Figure 2. Instantaneous images of partial cavities (top view) on the suction side of the GV model at $\alpha = 3°$ (a) without and (b) with mass injection through the slot channel and (c) comparison of the frequency spectra of pressure pulsations, where A_i denotes fast-Fourier-transform coefficients (◇ – with injection, ▲ – w/o injection), when: (1) $\sigma = 0.95$, $U_{in}/U_0 = 0.79$ (transitional cavitation/travelling bubbles), (2) $\sigma = 0.89$, $U_{in}/U_0 = 0.74$ (unsteady cavity/ transitional cavitation) and (3) $\sigma = 0.84$, $U_{in}/U_0 = 0.8$ (unsteady cavitation in both cases). For unsteady regimes, half-images indicating the phases of the attached cavity evolution when it is of (left) the shortest (just before its detachment) and (right) maximal lengths are shown together.

3.2 Velocity distributions

Fig. 3 display selected profiles of the streamwise component of the time-averaged velocity (plotted as the normalized velocity defect, $(U - U_0)/U_0$) and the rms of the streamwise component of its turbulent fluctuations $\tilde{u}$ at several locations along the GV model. The reference point (the origin of coordinate axes) coincides with the leading edge of the GV section in the measurement plane at $\alpha = 0°$.

As seen, the injection does not affect the profiles of the mean and fluctuating velocities upstream of the slot channel position at all (cf. the profiles at sections $x/C = 0.1$ and 0.4 in Fig. 3). Moreover, the injection flow rate corresponding to the range of the injection velocities from $U_{in}/U_0 = 0$ to 0.76 was found not to practically influence the distributions of these characteristics (not shown in the images) so that the global difference is only between the unforced and forced flow conditions. Thus, below only these two cases are considered, without taking into account the values of U_{in}/U_0.

Besides the aforementioned modification of the cavitation pattern from transitional cavitation to travelling bubbles (Figs. 3-1.a and 3-2.a), the mass injection also alters noticeably the distributions of the mean velocity and turbulence intensity within the boundary layer, decreasing the boundary layer thickness by half. Thus, at these regimes the active control system allows a favourable and efficient flow manipulation.

When transition to unsteady regimes occurs (Figs. 3-1.b and 3-1.c), the mass injection does not, however, influence the transversal size of the boundary layer but somewhat reduces the mean flow velocity inside it. In addition, under the forced conditions, the turbulent fluctuations are in general lower (typically by roughly 30%) in comparison with the unforced flow conditions (Figs. 3-2.b and 3-2.c).

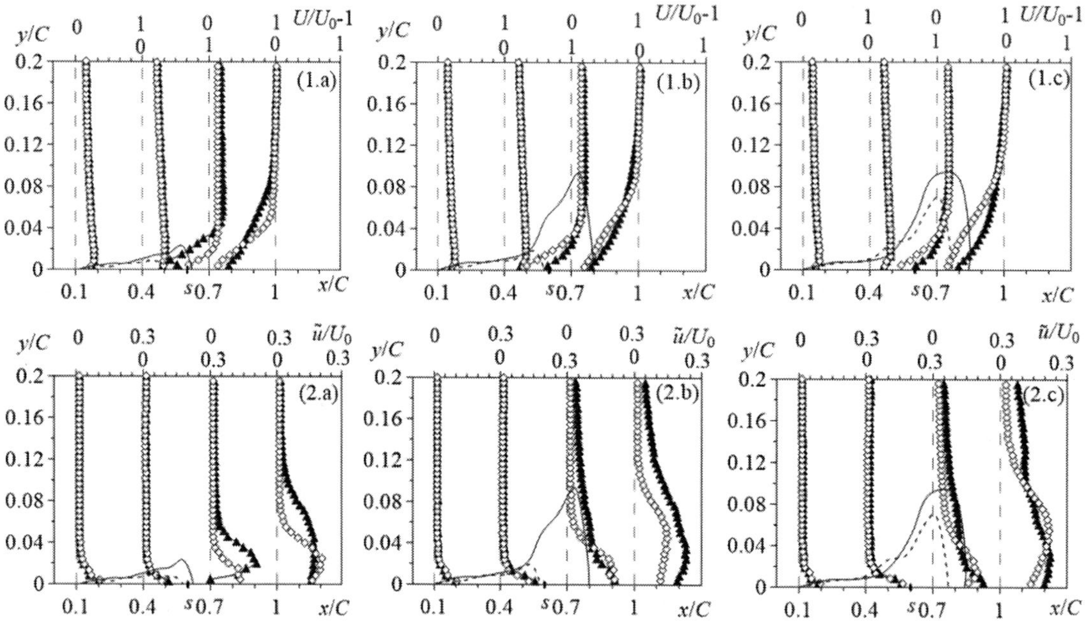

Figure 3. Downstream evolution of the streamwise component of the (1) mean velocity and (2) turbulence intensity (rms values) over the suction side of the GV model for (a) transitional cavitation/traveling bubbles at σ = 0.95, U_{in}/U_0 = 0.66; (b) unsteady cavity/transitional cavitation at σ = 0.89, U_{in}/U_0 = 0.74 and (c) unsteady cavitation in both cases at σ = 0.84, U_{in}/U_0 = 0.76. α = 3°. ▲ – w/o injection, ◇ – with injection. Solid and dashed lines represent interfaces (extracted from visual data) of the attached cavities of maximum size on the GV section for the unforced and forced flow conditions. *s* denotes the position of the slot channel in the foil surface. The flow direction is from the left.

4 Conclusions

The continuous liquid supply through a slot channel in the GV section surface allows a favourable and efficient flow manipulation, especially at the regimes with developed cavitation. In particular, the flow management makes it possible to reduce substantially the amplitude or totally suppress the periodic cavity length oscillations and pressure pulsations associated with them. However, the physics of the regarded effects remains unclear, which, therefore, requires supplementary experiments for a broader range of the governing parameters.

Acknowledgments. The research was funded by a grant from the Russian Scientific Fund (Project No. 16-19-00138) through IT SB RAS.

References

1. M. Callenaere, J.-P. Franc, J.-M. Michel and M. Riondet, J. Fluid Mech. **444**, 223–256 (2001)
2. P.G. Kozhukharov, V.H. Hadjimikhalev, V.I. Mikuta, L.I. Maltzev, ASME Fluids Eng. Div. (Int. Symp. Jets and Cavities, ASME Winter Ann. Meeting, Miami Beach, FL, USA) **31**, 67–74 (1985)
3. A.Yu. Kravtsova, D.M. Markovich, K.S. Pervunin, M.V. Timoshevskiy and K. Hanjalić, Int. J. Multiphase Flow **60**, 119–134 (2014)

MATEC Web of Conferences **84**, 00040 (2016) DOI: 10.1051/matecconf/20168400040
International Symposium IPHT 2016

Interfacial phenomena in films of monoliquids and binary mixtures in vacuum

Yuri Vyazov, Victor Prihodko, Igor Yarygin and Vyacheslav Yarygin[a]

Kutateladze Institute of Thermophysics SB RAS, Rarefied Gases Laboratory, 630090, Lavrentieva ave. 1, Novosibirsk, Russia

Abstract. The problem of joint gas outflow with near-wall liquid film from an axisymmetric cylindrical channel into vacuum is studied experimentally. It is shown that liquid film at the exit edge of the channel not only disintegrates into droplets, but also emerges on the external surface of the channel and starts moving in the opposite direction to the gravitational force. The phenomenon revealed is strongly dependent on physical properties of liquid and pressure in the surrounding space. Temperature of a film formed on the external surface of the channel is measured. It is shown that due to evaporation in vacuum liquid film is cooled to temperature at which the saturated vapor pressure is close to pressure in vacuum chamber.

Introduction

Recently both experimental and theoretical studies on gas jet expansion into the vacuum attracted a lot of attention. At the same time the problem of liquids expansion into the vacuum, including outflow of near-wall films combined with gas flow has not been studied thoroughly. From the practical point of view an interest in studying such flows is associated primarily with space applications. A case in point is ejection of liquids from orbital stations, including propellant refueling systems, orientation thrusters in which the fuel film is used for cooling combustion chamber and nozzle walls [1, 2]. Since the saturated vapor pressure of liquids used usually is by several orders of magnitude higher than pressure in the outer space, liquid, being ejected into the vacuum, becomes "instantly overheated". This causes liquid boiling-up, disintegration into droplets, and phase transitions on the surface and inside droplets. Co-current gas flow has a significative effect on the behavior of near-wall liquid film at the exit edge of the nozzle under ejection into the vacuum. The limiting angle of jet divergence under gas outflow from a nozzle into the vacuum exceeds 90° relative to the nozzle axis. Gas flows at angles over 90° are usually called backflows. It was shown [3] that under action of backflow near-wall liquid film entering vacuum at the exit edge not only disintegrates into droplets, but also emerges on the external surface of the channel and starts moving in the opposite direction even against gravitational force. Due to this effect operation of orientation thrusters of space vehicles is accompanied by backflows of propellant components droplets, which are usually harmful and should be minimized. More detailed information about the features of the near-wall liquid films expansion into vacuum and accompanying physical processes in application to modeling operation of space vehicles and orbital stations orientation thrusters, including the International Space Station (ISS), can be found in [4, 5]. It should

[a] Corresponding author : yarygin@itp.nsc.ru

be noted that in these studies ethanol was used as a model liquid because its main physical properties (density, saturated vapor pressure, heat of vaporization, viscosity, surface tension) are close to unsymmetrical dimethylhydrazine, which is currently used as a fuel for ISS orientation thrusters. We can conclude that the observed phenomenon of near-wall film emergence on the external surface, its rise and further evolution, including phase transitions (evaporation, cooling down up to freezing) depend mainly on the physical properties of the liquid (saturated vapor pressure and heat of vaporization) and therefore represent independent scientific interest. The current paper is devoted to experimental study on the effect of physical properties of monoliquids and binary mixtures on the process of near-wall film expansion with co-current gas flow from cylindrical channel into the vacuum.

Experimental setup and measurement techniques

Experiments were carried out on a VIKING vacuum gas-dynamic setup of Kutateladze Institute of Thermophysics. The large volume of the working chamber (150 m^3) allowed operation in pulse mode with high flow rates of gas and liquid. Cylindrical channel with 5 mm diameter, 20 mm length and 1 mm wall thickness was mounted vertically in the vacuum chamber, with exit part facing downwards. The liquid entered through a circular gap of 0.1 mm width into the stagnation chamber of the channel (nozzle) and moved down the nozzle walls as a film. Simultaneously, gas was blown through the nozzle. At the nozzle exit, a freely expanding gas jet with dispersed liquid droplets was formed. Time interval of gas and liquid feeding was equal several seconds, and it was sufficient for quasi-steady flow conditions settlement. Initially both gas and liquid were at room temperature, initial pressure in vacuum chamber for majority of experiments was equal to 1 Pa, and the final one could be 1-2 orders of magnitude higher. Snapshots of the liquid film outflow process into the vacuum were taken in experiments with the help of a digital camera, and liquid film temperature at the external surface of the channel was measured by Texas Instruments LM50 temperature sensors.

We used purified air as a working gas, and ethanol, dodecane, water and water-ethanol mixture as working liquids. Their physical properties are shown in Table 1. Dodecane was chosen as a liquid with many similar physical properties but with much lower saturated vapor pressure as compared to ethanol. Water, which has high heat of evaporation, was chosen as a working liquid because of a number of important applications in the field of aerospace technologies. Although a number of papers studied water injection into vacuum from tubes accompanied by droplet formation and freezing [6-7] we found no studies on ejection of water in the form of near wall film. As liquid film behaviors of pure ethanol and pure water are drastically different under ejection into vacuum, we also used water-ethanol mixture to achieve better understanding of the ejection process.

Table 1. Physical properties of working liquids.

Parameter	Units	Ethanol C_2H_6O	Dodecane $C_{12}H_{26}$	Water H_2O	Water–eth. 50% (vol.)
Saturated vapor pressure at 20°C	kPa	5.86	0.0097	2.31	3.3
Dynamic viscosity at 20°C	Pa·s, 10^{-3}	1.2	1.492	1.002	2.77
Density at 20°C	kg/m^3, 10^3	0.79	0.75	1	0.927
Surface tension at 20°C	N/m, 10^{-3}	22.8	25.5	72.8	27
Melting point at 100kPa	°C	-114.5	-9.55	0	-32.5
Boiling point at 100kPa	°C	78.3	216.3	100	83
Heat of vaporization at 100kPa	kJ/kg	840	360	2256	1859

Results

Snapshots of near-wall film ejection process into the vacuum chamber with chamber pressure of 1 Pa for dodecane, ethanol and water are shown in Fig. 1. Flow patterns during ejection into the vacuum for all working liquids show emergence of near-wall liquid film on the external surface of the channel and backward motion even against gravity. This is drastically different from the ejection into the atmosphere when usual gas-droplet flow takes place behind the nozzle edge.

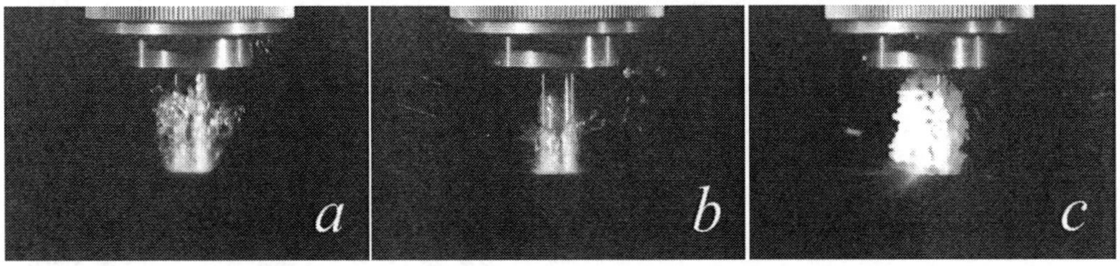

Figure 1. Outflow of liquid films into the vacuum. *a* – dodecane, *b* – ethanol, *c* – water.

Snapshots in Fig. 1 show the flow at the moment of maximum film rise. One can see, that rise of dodecane film is the highest, probably because dodecane has the smallest saturated vapor pressure. For the same reason dodecane film entering the vacuum chamber doesn't go through explosive disintegration into droplets as shown in Fig. 1a. Rise of ethanol film is the smallest and ethanol film entering the vacuum chamber shows significant disintegration of liquid film into droplets under its motion over the external surface of the nozzle. Arguably this behavior is due to high saturated vapor pressure of ethanol at room temperature that leads to instant overheating of ethanol film. The process of intensive droplet detachment from external surface in experiments with ethanol is shown in Fig. 1*b*. A special feature of water near-wall film ejection into the vacuum is formation of ice layer on the external surface of the nozzle. Although this phenomenon was expected in experiment, the rise of the ice layer to the full height of the tube shown in Fig. 1*c* was unanticipated. As for liquid film of water-ethanol mixture, its behavior during the experiment was close to the one of ethanol film. However upon discontinuation of liquid feeding remaining liquid film on the external surface started to freeze. Presumably it is caused by faster evaporation of ethanol into the vacuum and increase in concentration of water component in mixture.

Liquid film temperature measurement results for water, ethanol, and water-ethanol mixture with 50% volumetric concentration under ejection into the vacuum chamber with initial pressure of 1 Pa are shown in Figure 2.

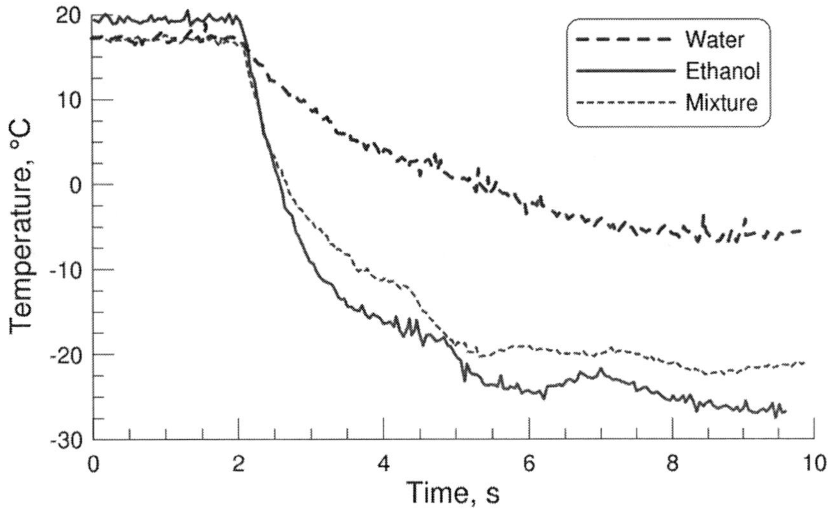

Figure 2. Liquid film temperature on the external surface of the nozzle near its exit edge.

Gas feeding was started in experiments at the moment of time t = 0.9s and stopped at t = 4.9s, while liquid feeding was started at t = 1.0s and stopped at t = 4.2s. One can see that it takes about 1s for liquid film to reach the nozzle exit edge. At the exit edge liquid film emerges on external surface and begins to cool intensively. Temperature results allowed us to better understand how liquid films are cooled under considered conditions at different absolute pressures in vacuum chamber (from 1 to 1300 Pa). It was shown that the final temperature of liquid film depends on the vacuum chamber pressure p_c. For example for ethanol, which does not freeze in the vacuum, at p_c = 1300 Pa liquid film is cooled to the temperature of about –4 °C, and at p_c = 130 Pa the film temperature decreases to –28 °C. This result has a clear explanation, assuming that the film of ethanol is cooled to the temperature, at which the pressure of saturated ethanol vapors becomes close to the pressure in the vacuum chamber. In other words, the lower the pressure in the vacuum chamber, the lower the temperature of the liquid film. For water the situation is different due to the fact that water undergoes phase transition in our experiment. According to the measurement results, at p_c = 1300 Pa the water film is cooled to $T \approx 1.5$°C, and at p_k = 130 Pa it freezes and the temperature drops to $T \approx -5.5$ °C. From Fig. 2 it is possible to see, that temperature of water-ethanol mixture is higher than for pure ethanol and lower than for pure water, but further detailed study on effect of composition ratio on film temperature is required. As for dodecane, it was shown that due to low saturated vapor pressure it is almost not subject to evaporation, therefore its film temperature is nearly steady in experiments.

Conclusion

A series of experimental studies on jet expansion of gas from a cylindrical channel with near-wall liquid film into the vacuum was carried out. It was shown that under expansion into the vacuum physical properties of liquids (primarily saturated vapor pressure and heat of vaporization) and pressure in the surrounding space (vacuum chamber) determine film behavior at the exit edge of the channel (backward turn, emergence on the external surface, and rise even against gravity) and its further evolution (disintegration into droplets, cooling down to freezing). Our results expand existing knowledge on the features of liquids ejection into the vacuum in the form of thin films, and could be relevant for a number of space applications.

Acknowledgments

The study was supported by the Russian Foundation for Basic Research, research project No. 16-08-00436.

References

1. H. Trinks, AIAA Paper No. 87-1607 (1987).
2. H. Trinks, I. Kaelsch, AIAA Paper No. 87-1603 (1987).
3. V.N. Yarygin, V.G. Prikhodko, I.V. Yarygin, Yu.N. Vyazov, Vacuum, **109**, 401-404 (2014).
4. V.N. Yarygin, Yu.I. Gerasimov, A.N. Krylov, L.V. Mishina, V.G. Prikhodko, and I.V. Yarygin, Thermophys. Aeromech., **18**, 333-358 (2011).
5. V. Yarygin, Yu. Gerasimov, A. Krylov, V. Prikhodko, and I. Yarygin, Microgravity Sci. Technol., **23**, 15-23 (2011).
6. H. Fuchs, H. Legge, Acta Astronaut., **6**, 1213-1226 (1979).
7. J.W. Alred, N.L. Smith, K.C. Wang, F.E. Lumpkin, and S.M. Fitzgerald, TFAWS (1997).

Interphase interaction of near-wall liquid film with co-current gas flow inside a nozzle

Yuri Vyazov, Victor Prihodko, Igor Yarygin and Vyacheslav Yarygin[a]

Kutateladze Institute of Thermophysics SB RAS, Rarefied Gases Laboratory, 630090, Lavrentieva ave. 1, Novosibirsk, Russia

Abstract. The results of experimental study on high-velocity co-current gas flow interaction with near-wall liquid film are presented in the paper. Local parameters of near-wall liquid film are measured with the help of capacity-type probes. It is shown that co-current gas flow has strong influence on near-wall liquid film, leading to intensive wave formation, detachment of droplets from the film surface and their entrainment by the gas flow. A model for the film motion with co-current gas flow, linking together thickness and velocity of the film with value of shearing stress at gas-liquid boundary, is suggested.

Introduction

Flow of liquid in the form of thin films (with thickness less than 1 mm) is widely used in various heat-and-mass exchangers that involve condensation of stationary and moving vapor, absorption or desorption of gas, drying, distillation, fractioning, chemical processes, etc. Another important application of film flows is protection of walls from the high-temperature gas flow, for example, protection of liquid-propellant thrusters combustion chamber and supersonic nozzle using the near-wall fuel film.

Thin film flows attracted a great deal of attention recently. The most thorough and extended studies are concerned with gravitation films. Important role in such flows is played by the mechanisms of instability leading to evolution and interaction of waves of various character. Thermocapillary instability effect and phase transformation can also be significant. In such phenomena condensation has a stabilizing effect, and evaporation destabilizes the flow. The results of these studies can be found for example in monographs [1, 2].

The stress film flows are studied extensively in experiments. However, these experiments are carried out at relatively low velocities of the co-current flow of about 10 m/s. Interaction of the near-wall film with the co-current gas flow is mainly a function of the flow parameters, such as velocity, density, and static pressure. Even at low flow velocities, liquid film loses its stability, and capillary waves are formed on its surface, whereas, without the co-current flow, the film flow is waveless. An increase in velocity of the co-current gas flow facilitates interfacial interaction and leads to formation of 2D and 3D waves on the film surface. Beginning with some values of gas flow velocity and density (Weber number), droplets get detached from the film surface and entrained by the co-current flow [3].

Another special feature of such two-phase flows is interaction of the near-wall film with the co-current gradient flow, for example inside nozzles. A significant drop in static pressure in the

[a] Corresponding author : yarygin@itp.nsc.ru

flow (approximately, by the factor of 2 for Mach number 1, and by the factor of 30 for Mach number 3) can cause the situation, when the static pressure in the flow above the film becomes lower than the liquid saturated vapor pressure. This leads to film boiling-up and evaporation, as well as formation of a transverse vapor flow from the film surface to the boundary layer. Similar to injection through a permeable wall into the boundary layer, this vapor flow can essentially affect the value of interface friction causing a decrease in the shearing stress on the gas-liquid interface in comparison with the modes of gas-film interaction without evaporation. Currently there are no complete studies with accurate and reliable data on the values of shearing stress on the gas-liquid interface under the condition of high relative velocities of gas flow. Existing experimental studies are focused on the measurement of flow integral characteristics and averaged parameters of waves without taking their non-stationary character into consideration [4, 5].

Our early studies were devoted to solving the problem of the International Space Station (ISS) external contamination by the jets from orientation thrusters, in which the fuel film is used for nozzle walls cooling. Experimental studies on modeling operation of orientation thrusters showed that behind the nozzle exit complicated flow structure of droplet phase takes place, and droplets are dispersed at angles up to 180° relative to jet axis. Such flows of droplets are the reason of the ISS external surface contamination [6]. This paper presents the results of experimental studies on the interaction of co-current gas flow with near-wall liquid film for supersonic nozzle in a wider range as compared to the earlier modeling experiments of the flow parameters on the magnitude of gas flow pressure above near-wall liquid film. Dodecane, as well as ethanol were used as working liquids, which allowed us to carry experiments for liquids with significantly different values of saturated vapor pressure.

Experimental setup and measurement techniques

Measurements of local parameters of a near-wall liquid film (film thickness and film velocity) in the present study were carried out with the help of capacity-type probes [7]. Film thickness and velocity were measured near the exit edge of the supersonic nozzle (geometrical Mach number M = 3) with 10 mm throat diameter, 20 mm exit diameter and 20 mm nozzle length (Fig. 1). Four probes were mounted around the nozzle for film thickness measurements (results were avaraged by four probes), and two consequently mounted probes were used for film velocity measurements.

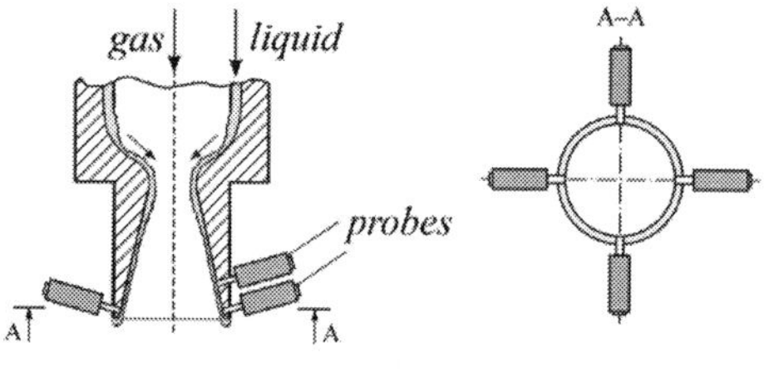

Figure 1. Scheme of test section.

Purified air was used as a working gas, and ethanol and dodecane were used as working liquids. Initially both gas (flow rate from 2 to 20 g/s) and liquid (flow rate from 0.5 to 2 g/s) were at the room temperature. Gas flow regime was turbulent. All experiments were carried out at the large-scale vacuum gas-dynamical setup VIKING of Kutateladze Institute of Thermophysics SB RAS (volume of vacuum chamber is about 150 m^3). Ethanol and dodecane were used in experiments because their main physical properties (density, specific evaporation heat, viscosity, surface tension) are similar to properties of the rocket fuel. At the same time they differ significantly in saturated vapor pressure. Ethanol has rather high vapor pressure equal to 5.86 kPa at 20°C and evaporates easily in vacuum. Dodecane has much lower saturated vapor pressure 0.0097 kPa at 20°C and almost does not evaporate.

A great deal of attention in our research was paid to calibration of the capacity-type probes. We developed the method with use of dielectric plugs, which were tightly inserted into the channel and had a groove of certain depth over the probe, filled with liquid. The method allowed us to obtain

DOI: 10.1051/mateconf/20168400041

calibration curves for probes taking into account their placement on the channel surface, and to have liquid film thicknesses measurements margin of error less than 10%.

Results

Ethanol film thickness diagram, measured by capacity-type probes is shown in Fig. 2. Solid line corresponds to the case when the film was not boiling. Stagnation pressure of gas in the nozzle was equal p_0 = 100 kPa, and pressure of gas over the liquid film near the nozzle exit was about 3.3 kPa. Dashed line corresponds to the case when the film was boiling. Stagnation pressure in these experiments was equal to p_0 = 13 kPa and pressure over the liquid film was 0.43 kPa, respectively. Reynolds numbers in both cases were Re_l = 14. One can see that evaporation and boiling have strong influence on the behavior of the film, causing large fluctuations and increase of its mean thickness.

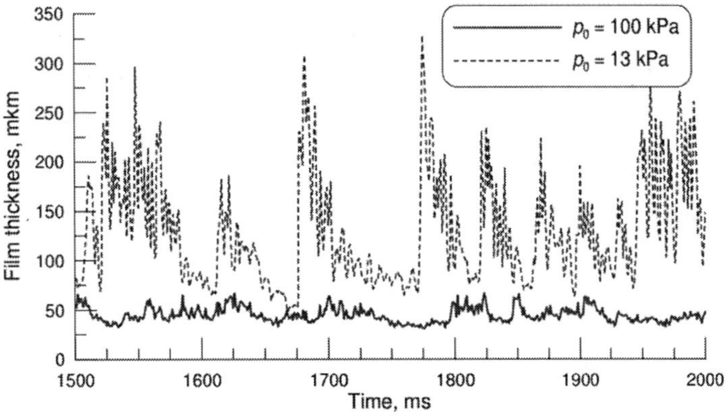

Figure 2. Ethanol film thickness without boiling (p_0 = 100 kPa) and with boiling (p_0 = 13 kPa).

Experiments allowed us to obtain mean film thickness and velocity inside the channel. It is shown that film velocity both for ethanol and dodecane increases from 0.2 m/s for $Re_{gas} = 3.5 \cdot 10^4$ up to 0.8 m/s for $Re_{gas} = 3.3 \cdot 10^5$, where Reynolds number Re_{gas} is calculated by the critical diameter of the nozzle. Measured data for ethanol and dodecane film thickness are shown in Fig. 3.

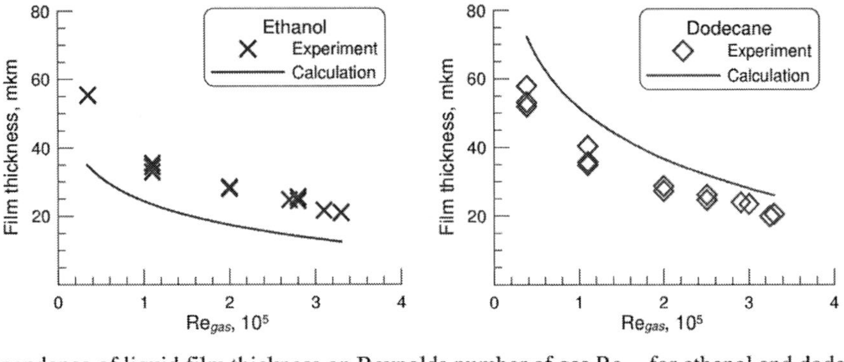

Figure 3. Dependence of liquid film thickness on Reynolds number of gas Re_{gas} for ethanol and dodecane.

The solid lines in these figures indicate mean film thicknesses δ calculated for the experimental conditions with the help of the model [8]:

$$\delta = \sqrt{\frac{m\mu}{\pi R \rho \tau}}, \tag{1}$$

where m, μ, ρ - liquid flow rate, viscosity and density, respectively, R – channel radius, τ - shearing stress at interface border. For these calculations it is necessary to know the value of shearing stress τ.

In our model the value of shearing stress τ at interface border is assumed as shearing stress τ_w on the nozzle wall without the film. Corresponding dependencies for turbulent boundary layer with pressure gradient in the external flow and absence of laminar region were taken from the reference-book, and they were as follows:

$$C_f = \frac{2\tau}{\rho V^2} = 0{,}0131\mathrm{Re}^{**-1/6}, \text{ where } \mathrm{Re}^{**} \text{ is determined as } \mathrm{Re}^{**7/6} = \frac{0{,}0076}{\upsilon V^{2{,}75}} \int_0^x V^{3{,}75} dx. \quad (2)$$

Our model despite its simplicity allowed us to estimate the film thickness and velocity under the motion inside the nozzle with the co-current gas flow. At the same time one can see that calculation results for dodecane are above experimental data and calculation results for ethanol are below experimental data. We believe that this can be explained by the influence of evaporation and wave formation that reduce and increase shearing stress respectively. For dodecane under considered conditions the main interaction processes between the co-current flow and the film are wave formation and droplet detachment, and the effect of film boiling is insignificant. Both wave formation and droplet detachment should increase τ at the interface in comparison with τ set in calculations on a smooth solid wall. For ethanol under considered conditions boiling plays a significant role. It follows from comparison of calculation and experimental data, shown in Fig. 3, that at film boiling τ should decrease, i.e., change in the opposite direction in comparison with a change in τ at wave formation and droplet detachment. These arguments are plausible, if we accept the existence of the analogy between the processes of evaporation and substance injection through a permeable wall.

Conclusion

Experimental study on interaction of high-velocity co-current gas flow with near-wall liquid film in supersonic nozzle was carried out. Local parameters of near-wall films – wave characteristics, thickness, velocity with high time resolution of 1 kHz (gas flow velocity up to 540 m/s) were measured with the help of capacity-type probes. We have demonstrated that character of interaction (wave dispersion, evaporation, detachment of droplets from the film surface) is mainly determined by gas flow parameters (primarily, velocity and density) and physical properties of liquid (viscosity and saturated vapor pressure). The model of gas flow interaction with near-wall film was suggested. This model allows estimation of the shearing stress increase at the interface border under conditions of wave formation compared to thin laminar film with the help of measured values of thickness and velocity.

Acknowledgments

The study was supported by the Russian Foundation for Basic Research, research project No. 16-08-00436.

References

1. S.V. Alekseenko, V.E. Nakoryakov V.E., and B.G. Pokusaev, *Wave Flow of Liquid Films* (1994).
2. S. Kalliadasis, C. Ruyer-Quil, B. Scheid, and M. G. Velarde, *Falling Liquid Films*, (2012).
3. I.I. Gogonin, J. Eng. Phys. Thermoph., **77**, 454-470 (2004).
4. H. Han, Z. Zhu, K. Gabriel, Nucl. Engin. Des., **236**, 2580–2588 (2006).
5. D.A. McNeil, A.D. Stuart, Int. J. of Heat and Fluid Flow, **25**, 58–73 (2004).
6. V. Yarygin, Yu. Gerasimov, A. Krylov, V. Prikhodko, and I. Yarygin, Microgravity Sci. Technol., **23**, 15-23 (2011).
7. V. Prikhodko, I. Yarygin and V. Yarygin, Microgravity Sci. Technol., **21**, 277-281 (2009).
8. I.V. Yarygin, V.F. Levchenko, Proc. ICMAR, **4**, 314-318 (2004).

Velocity field measurements in an evaporating sessile droplet by means of micro-PIV technique

Anna Yagodnitsyna[1,2,a], Artur Bilsky[1,2], Mina Roudgar[3], Joël De Coninck[3] and Oleg Kabov[1,4]

[1]*Kutateladze Institute of Thermophysics SB RAS, Novosibirsk, Russia*
[2]*Novosibirsk State University, Novosibirsk, Russia*
[3]*Laboratory of Surface and Interfacial Physics. Université de Mons, Belgium*
[4]*National Research Tomsk Polytechnic University, Tomsk, Russia*

Abstract. Velocity fields are measured in evaporating sessile droplets on two substrates with different contact angles and contact angle hysteresis using micro resolution particle image velocimetry technique. Different flow patterns are observed in different stages of droplet evaporation: a flow with vortices and a radial flow. Flow structure is found to be similar for droplets on different substrates.

1 Introduction

The interest towards droplet evaporation is connected with number of scientific and technological processes such as DNA molecules visualization, ink-jet printers, spray cooling and so on [1-3]. Flow structure in evaporating droplets can dramatically influence the heat and mass transfer processes and particle deposition on a substrate. Hence, experimental data of velocity fields inside evaporating droplets are highly demanded not only for technological processes design and optimization but also for numerical code verification. A number of works have been focused on experimental study of flow structure in evaporating sessile droplets. The authors of the work [4] used the ray tracing method in order to correct optical disturbances during PIV measurements from the side view of the droplet. In the work [5] micro-PIV technique was used for velocity field measurements in the small region near the contact line of evaporating water droplet on a glass substrate. In the present work in order to visualize flow structure in evaporating droplets on the substrates with different contact angles micro-PIV technique is applied.

2 Experiment details

The scheme of experimental setup is shown in Figure 1. Micro-PIV system is used for velocity field measurements in an evaporating sessile droplet. The system is made of Zeiss Axio Observer.Z1 microscope with 10x/0.25 lens and motorized stage, double-pulsed Nd:YAG green laser, double exposure CCD camera with 11 Mpix resolution and a personal computer with ActualFlow software for data acquisition, processing and visualization. The motorized stage is able to move in three directions with 1 micrometer resolution. It is used for accuretly moving the focal plane. The working

[a] Corresponding author : yagodnitsinaaa@gmail.com

 DOI: 10.1051/mateconf/20168400042

fluid is an ultra-pure water seeded with 3.2 μm fluorescent polystyrene particles. Droplets are created by high precision micropipette. The droplet volume is equal to 1.5 μl with 1% accuracy. During experiments the environment temperature and the humidity has been controlled. Evaporation time for each droplet has been also measured as a time between droplet pipetting and its dry-out The standard deviation of evaporation time is not exceed 12% for each substrate.

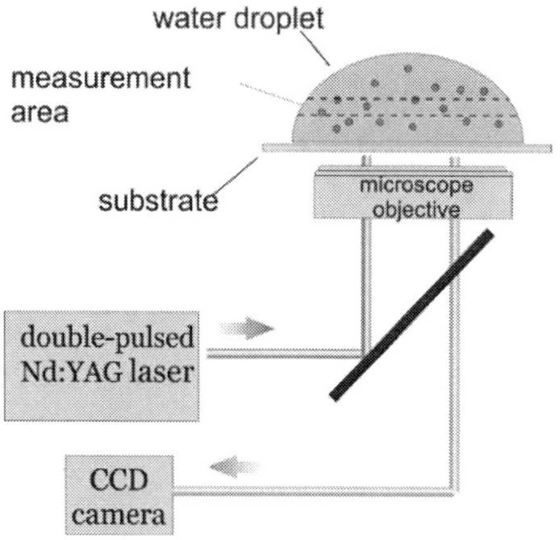

Figure 1. The scheme of experimental setup.

Experiments have been performed for two transparent substrates with different receding contact angles (RCA) and contact angle hysteresis (CAH): the substrate covered by octadecyltrichlorosilane monolayer (OTS) and the substrate covered by SU-8 photoresist. The contact angle and contact angle hysteresis measured on DSA 100 (Kruss) are presented in the Table 1.

Table 1. The properties of the used substrates.

Substrate	**RCA**	**CAH**
OTS	95°±5°	10-15°
SU-8	65°±5°	30°-35 °

Droplets are illuminated and observed from the flat bottom side, so optical distortions are not contribute to the measurement results. Velocity field measurements are performed in 9 cross-sections parallel to the substrate plane located at different distances from the substrate ranging from 13 μm to 465 μm. The depth of correlation is equal to 55 μm. For each cross-section several measurements for different droplets are done in order to obtain sufficient statistical information. The measurements are carried from the beginning of droplet evaporation up to the end.

In-house software package "ActualFlow" is used to process the acquired data. Image preprocessing is included masking and minimum intensity subtraction calculated for every 20 particle images. These procedures allow eliminating background noise and decreasing out-of-focus particles influence. Instantaneous velocity fields are calculated using iterative cross-correlation algorithm. Final spatial resolution of measured instantaneous velocity fields is equal to 30 μm per velocity vector.

3 Results

Particle trajectories visualization and instantaneous velocity fields at different time starting from the beginning of droplet evaporation at the distance 93 μm from the SU-8 substrate plane are presented in Fig. 2 and Fig. 3. One can see that at the beginning of evaporation flow structure does not present typical radial flow [5]. Instead, there are two symmetrical vortices in the droplet, which presumably are connected with some local depinning of the contact line. The maximum velocity near the edges are not exceed 6 μm/s. When the height of the droplet becomes smaller the flow turns into radial with higher velocities near the edges of the droplet. In this evaporation stage maximum velocity decreases significantly down to 1.5 μm/s. The similar flow in an evaporating droplet is found in three-dimensional direct numerical simulations in [6]. Analysis of velocity fields in evaporating droplet on the OTS substrate shows the same behaviour of the flow as in evaporating droplet on SU-8 one. During the time between 78 and 220 seconds from the beginning of evaporation for OTS substrate the droplet cross-section radius has been decreased by approximately 30%. In comparison for SU-8 substrate the droplet cross-section radius reduction is equal to 15% which is consistent with the statement that the large-contact-angle hysteresis droplets can be pinned [7].

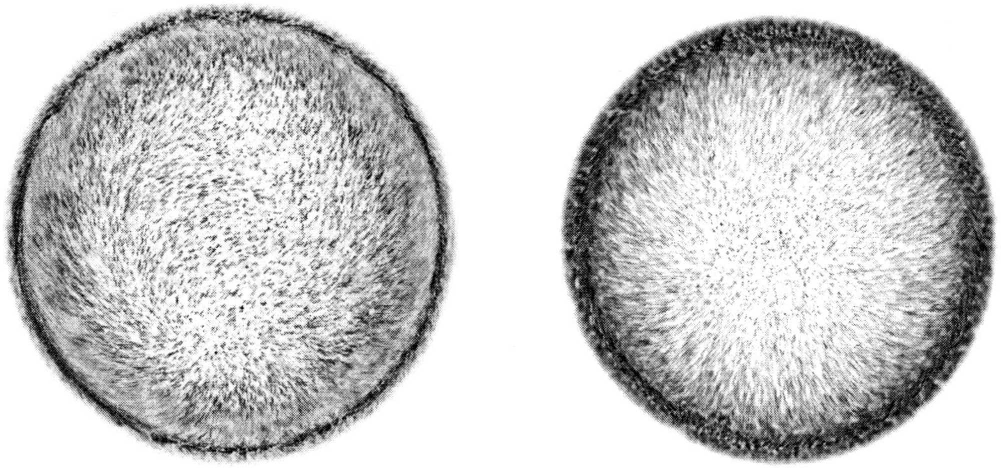

Figure 2. Trajectories of seeding particles in the droplet at the distance 93 μm from the SU-8 substrate. Left: 78s past the beginning of evaporation. Right: 220 s past the beginning of evaporation.

4 Conclusion

In order to visualize flow structure in evaporating sessile droplets on substrates with different properties micro-PIV technique has been used. Instantaneous velocity fields are measured at different distances from substrate plane. Analysis of instantaneous velocity fields shows that flow structure in evaporating droplet changed significantly during the evaporation process. At the first stage of evaporation, two vortices have been found in the droplet. Similar vortices have been obtained for droplets on both substrates. At the second stage of evaporation droplet height decreases and velocity fields become radial with higher velocities near the edges of the droplets.

Acknowledgments

This work was supported by a grant from the Russian Science Foundation (project 14-19-01755).

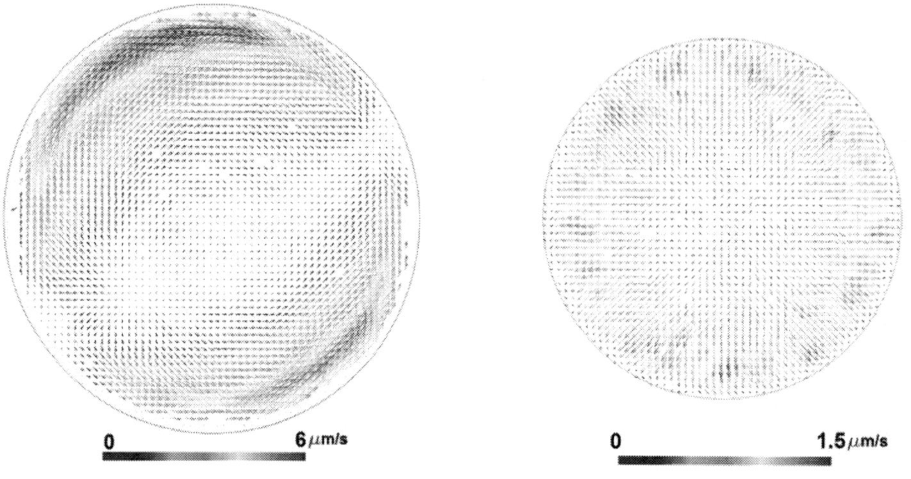

Figure 3. Instantaneous velocity fields in a droplet cross-section located at the distance 93 μm from the substrate surface made of SU-8. Left: 78s after the beginning of evaporation. Right: 220 s after the beginning of evaporation. Full evaporation time is equal to 558 s.

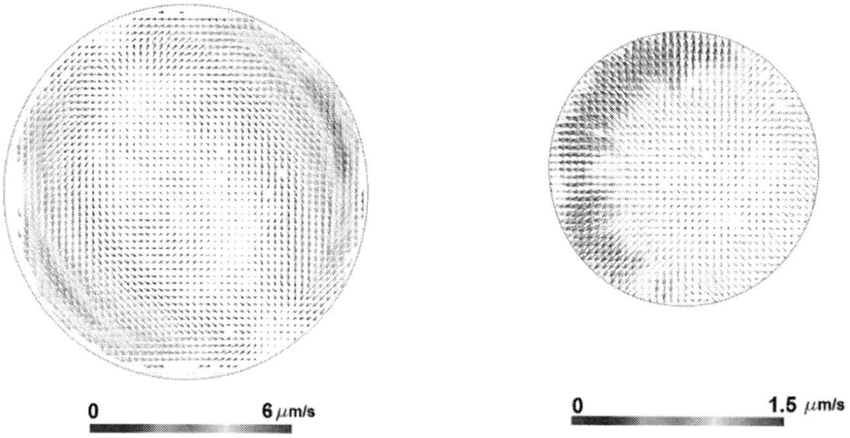

Figure 4. Instantaneous velocity fields in a droplet cross-section located at the distance 93 μm from the substrate surface made of OTS. Left: 78s after the beginning of evaporation. Right: 220 s after the beginning of evaporation. Full evaporation time was equal to 418 s.

References

1. H. Erbil, Adv. Colloid Interface Sci., **170** (2012)
2. A.L. Karchevsky, I.V. Marchuk, O.A. Kabov, Appl. Math. Model., **40 (2),** 1029-1037 (2016)
3. E.Ya. Gatapova., A.A. Semenov, D.V. Zaitsev, O.A. Kabov, Colloids and Surfaces A: Physicochemical and Engineering Aspects, **441,** 776 (2014)
4. K. Hyoung Kang, S. Joon Lee, C. Mook Lee, I. Seok Kang, Meas. Sci. Technol., **15** (2004)
5. H. K. Dhavaleswarapu , C. P. Migliaccio, S. V. Garimella , J.Y. Murth y, Langmuir, **26(2)** (2010)
6. P. J. Sáenz, K. Sefiane, J. Kim, O.K. Matar, P. Valluri, J.Fluid Mech., **772** (2015)
7. O.A. Kabov, D.V. Zaitsev, Doklady Physics, **58**, 292 (2013)

The effect of the channel height on the critical heat flux in a shear-driven liquid film

D. Zaitsev[1,a], O. Kabov[1]

[1] *Institute of Thermophysics SB RAS, 1, Lavrentiev Ave, 630090, Novosibirsk, Russia*

Abstract. The fast development in semiconductor technology has led to increasingly higher chip power dissipation and greater non-uniformity of the heat dissipation with the result of localized hot spots, often exceeding 1kW/cm^2 in heat flux. As was shown in previous works of the authors, thin and very thin liquid films driven by a forced gas/vapor flow (stratified or annular flows), i.e. shear-driven liquid films in a narrow channel are promising candidate for an innovative cooling technique optimizing the tradeoffs between performance and cost. The goal of the present work is to study the effect of the channel height on the critical heat flux (CHF) in a locally heated liquid film driven by the shear stress of gas in a channel.

1 Introduction

One of the most important problems of thermal physics today is cooling of microelectronic equipment. Currently, the production of processors changes over from the 14 nm manufacturing process technology to the 7 nm technology. New IBM processors use 7 nm transistors. This has been made possible through the use of a silicon-germanium alloy in the transistor design instead of pure silicon, the traditional material for the manufacture of microchips. A normal-size processor can accommodate more than 20 billion transistors having a size of 7 nm. In addition to reducing the size of the microprocessor unit, the use of the silicon-germanium alloy has improved the switching speed of transistors and lowered the power consumption. The average heat flux density on the chips of commercially available computers and other electronic devices is up to 200-300 W/cm^2. In local areas of 100 μm^2 to a few cubic millimeter, the heat flux density reaches values of 1 kW/cm^2 or more (IBM). Devices of the next generation may have higher heat fluxes and pulsed loads [1], whose combination requires a new thermal management level. Two-phase flows are of particular interest for use in cooling systems because of high heat-transfer coefficients. The latent heat of evaporation in two-phase flow usually provides much greater cooling capacity than single-phase flow, allowing the use of lower mass flow rates.

The use of two-phase systems is not limited to microelectronics. Recently, there has been an increased interest in transport processes in mini- and microsystems, caused by the rapid development of medicine, as well as the miniaturization of devices in various fields of technology, for example, in the automotive, aerospace, and transportation industries. Thus, there has been a breakthrough development of heat exchange systems with micro- and nanodimensions, and these systems are much more energy efficient than macrochannels with channel dimensions of 3-100 mm.

---

[a] Corresponding author : zaitsev@itp.nsc.ru

As shown in [2], thin and very thin liquid films driven by a forced gas/vapor flow (stratified or annular flows), i.e. shear-driven liquid films in a narrow channel are promising candidate for an innovative cooling technique optimizing the tradeoffs between performance and cost. Works [3-7] are devoted to experimental and analytical study of hydrodynamics, heat and mass transfer and crisis phenomena during joint motion of intensively evaporating liquid film and gas flow in the horizontal channel with a local heater (simulating an electronic device). The goal of the present work is to experimentally study the effect of the channel height on the critical heat flux in such a cooling system in the range of the channel heights from 0.17 to 2.00 mm.

2 Experimental setup

Schematic of the test section is presented in figure 1. The main part of the test section is a 4 mm thick stainless steel plate with a flush-mounted copper rod. At the working surface the rod has a 1×1 cm square head emulating surface of a computer chip. The rod is heated by a heating spiral coiled around its lower part (not shown in figure 1). Such construction of the heater provides the condition at the heater surface T = const (which is confirmed by thermocouples measurements). The test section is covered with a transparent cover made of optical glass so that a flat channel is formed (the cover is not shown in figure 1). The channel height, H, varies from 170 to 2000 μm. Gas, pumped by a compressor, flows through the channel and passes to the atmosphere at the end of the test section. The liquid is supplied from a thermostat, gets into the channel through a liquid nozzle and flows under the friction of the gas along the stainless steel plate as a film. The liquid is accumulating at the bottom part of the test section and is returned into the thermostat. Several thermocouples are embedded in the stainless steel plate and in the copper rod, allowing determination of the working surface temperature. The temperature of the heater surface is calculated taking into account the depth at which the thermocouples are embedded (2 mm). All the thermocouples are individually calibrated to an accuracy of 0.1°C. The test section is oriented horizontally.

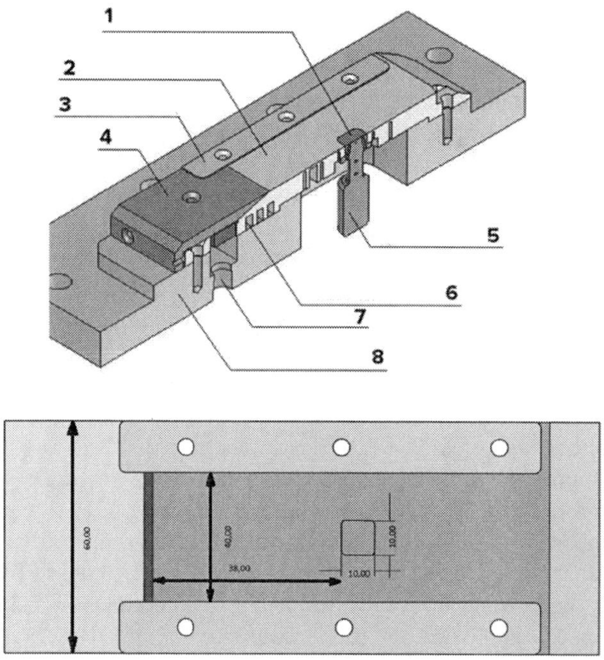

Figure 1. Design of test section: 1- source of the local heating; 2- stainless steel plate; 3- channel height control; 4- knife; 5- copper rod; 6- thermostabilizer; 7- liquid inlet; 8- textollite substrate.

The working surface of the test section was mechanically polished. The morphology of the surface was analyzed using scanning electron microscope (JEOL JSM6700F) and atomic force microscope (Solver Pro NT MDT). The root mean square (RMS) surface roughness, measured on a 22 μm² area of the heater surface, was found to be 790 nm.

Ultra-pure water (Merck Millipore) with initial temperature of 24°C is used as the working liquid. Air with temperature of 24-27 °C and relative humidity of 15-30% is used as the working gas. The experiments are carried out under stationary conditions. Superficial gas velocity (volumetric gas flow rate divided by cross-sectional area of the channel) U_{Sg} varies from 2.5 to 50 m/s. Liquid Reynolds number Re_l varies from 18 to 70.

The heat flux q is determined by the electrical power dissipated on the heating spiral. Thermal conductivity of copper is 400 W/mK which is almost 30 times higher than that of stainless steel (15 W/mK). This provides moderate heat spreading from the heater to the stainless steel plate: about 15% at q>200 W/cm², about 20% at q=100 W/cm² and up to 30% for smaller heat fluxes (according to estimation using measurements of thermocouples embedded into the stainless steel plate). Heat spreading into the atmosphere has been calculated as the ratio of the heat flux between two thermocouples embedded into the copper rod (estimated by measurements of thermocouples) and the total heat flux determined by the electrical power. In order to minimize heat spreading from the heating spiral into the atmosphere, the heating spiral is wrapped with a layer of heat insulation material (mineral wool). Heat loses into the atmosphere don't exceed 10% at q>200 W/cm². Thus, the heat loses into the atmosphere and heat spreading into the steel plate in total don't exceed 25% at heat fluxes higher than 200 W/cm². All data below are presented without taking into account the heat losses and heat spreading.

3 Experimental results

Figures 2 and 3 show the flow in the channel prior occurrence of the heat transfer crisis for the channel height H=250 and 2000 μm, respectively. For H=250 μm the heater is covered with nonstable, evaporating liquid film at both superficial gas velocities U_{Sg}=5 and 30 m/s (at U_{Sg}=5 m/s the liquid is touching the glass cover). However, for H=2000 μm the heater is covered with nonstable liquid film only at relatively high gas velocity (U_{Sg}=33 m/s), while for smaller gas velocity (U_{Sg}=2.5 m/s) the first dry patch propagates all over the heater (the liquid is not touching the glass cover in this case), causing dryout and thus the occurrence of the heat transfer crisis.

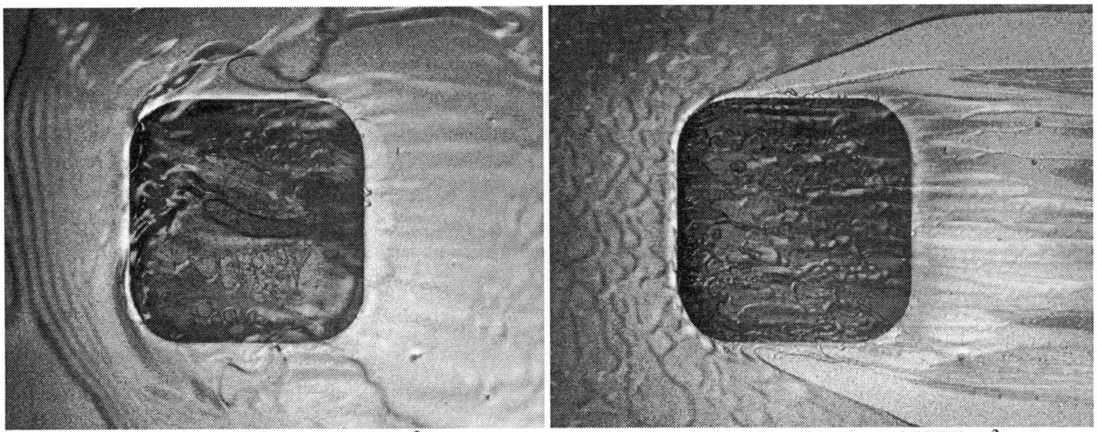

a) U_{Sg}=5m/s, q=210W/cm² b) U_{Sg}=30m/s, q=250W/cm²

Figure 2. Photograph of the flow prior to the critical heat flux for H=250 μm, Re_l=18. The flow directed from left to right.

DOI: 10.1051/mateccconf/20168400043

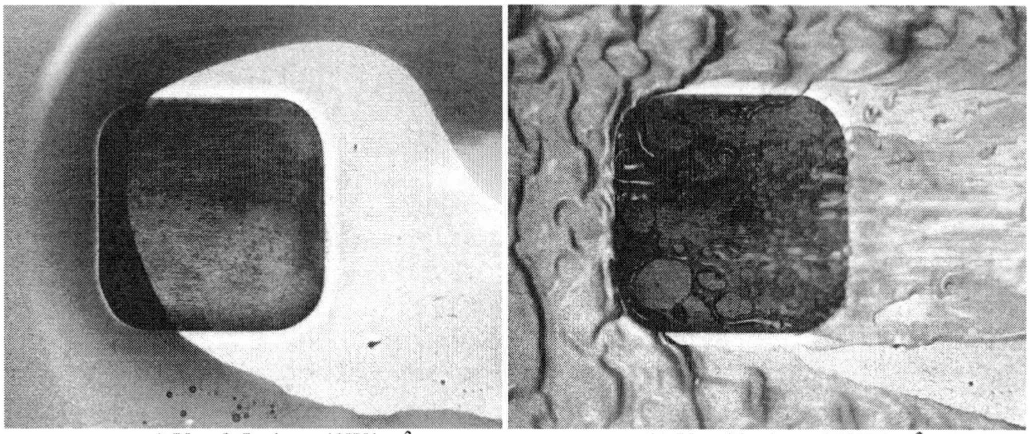

a) U_{Sg}=2.5m/s, q=40W/cm^2 b) U_{Sg}=33m/s, q=250W/cm^2

Figure 3. Photograph of the flow prior to the critical heat flux for H=2000 μm, Re_l=18. The flow directed from left to right.

Figure 4 shows the effect of the channel height on the critical heat flux for different superficial gas velocities. It is seen that for H=170-250 μm the critical heat flux is weakly dependent on the superficial gas velocity, while for H=1-2 mm the critical heat flux strongly decreases with decrease of the superficial gas velocity. For relatively small superficial gas velocities the critical heat flux sharply decreases with the channel height, while for relatively high superficial gas velocities the critical heat flux slightly increases with the channel height.

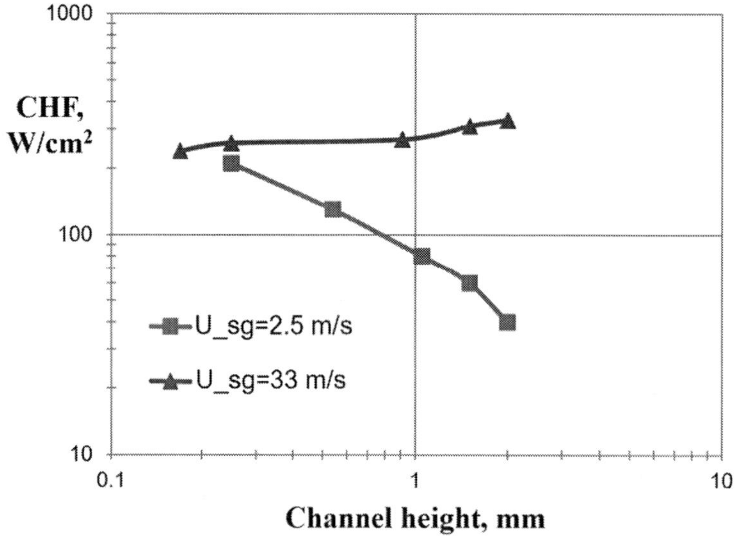

Figure 4. Critical heat flux vs. channel height for different superficial gas velocities, Re_l=18.

4 Conclusions

For a locally heated liquid film shear-driven in a microchannel with the height of H=170-250 mm the critical heat flux was found to be weakly dependent on the superficial gas velocity, while for the liquid film shear-driven in a minichannel with the height of H=1-2 mm the critical heat flux was found to strongly decrease with the decrease of the superficial gas velocity. Thus we conclude that cooling systems with shear-driven liquid films in *micro*channels are more reliable than those with shear-driven liquid films in *mini*channels.

This work was supported by the Ministry of Education and Science of Russia (Agreement No 14.604.21.0053, project identifier RFMEFI60414X0053).

References

1. Bar-Cohen A, Wang P. ASME. J. Heat Transfer, **134**(5), 051017 (2012).
2. Kabov O.A., Lyulin Yu.V., Marchuk I.V. and Zaitsev D.V. International Journal of Heat and Fluid Flow **28** 103-112 (2007).
3. Zaitsev D.V., Rodionov D.A. and Kabov O.A. Technical Physics Letters **35** (7) 680-682 (2009).
4. Kabov O.A. and Zaitsev D.V. Multiphase Science and Technology **21** (3) 249-266 (2009).
5. Kabova Y.O. Technical Physics Letters, **39** (10) 902-905 (2013).
6. Kabova Y., Kabov O., Gambaryan-Roisman T., Stephan P., Kuznetsov V.V. Int. Journal of Heat and Mass Transfer, **68**, 527-541 (2014)
7. Gatapova E.Ya., Graur I.A., Sharipov F., Kabov O.A. Int. Journal of Heat and Mass Transfer **83** 235–243 (2015).

MATEC Web of Conferences **84**, 00044 (2016) DOI: 10.1051/matecconf/20168400044
International Symposium IPHT 2016

Micro-droplet cluster dynamics in a heated layer of liquid at dry spots formation

D. Zaitsev[1,a], D. Kirichenko[1,2]

[1]*Institute of Thermophysics SB RAS, 630090, Novosibirsk, Russia*

[2]*Novosibirsk State University, 630090, Novosibirsk, Russia*

Abstract. Organization of a cloud of micro-droplets of condensate into an ordered hexagonal structure levitating over a heated layer of liquid (the so called micro-droplet cluster) has been observed in several recent experimental works, however the nature of this phenomenon is still not completely understood. In the present paper the micro-droplet cluster is studied on a test section with relatively large heating area, under the condition when the liquid layer ruptures and dry spots form on the heater. It has been found, that the micro-droplets can levitate not only over the heated liquid surface, but also over a dry surface of the heater, following the rupture of the liquid layer. The micro-droplets can move over the dry surface as far as 3 mm away from the contact line, and can exist for up to 3.5 seconds after the rupture.

1 Introduction

A micro-droplet cluster is a spatially ordered structure of micro-droplets, discovered in [1] over the surface of liquid layer spot-heated from below. The cluster consists of several of hundreds of drops packed into one layer, forming a hexagonal structure. The drop diameter is on the order of 10 microns; the distance between the drops is of several drop diameters, while the drops levitation height above the liquid surface is comparable to the drop diameter. The possible mechanism of drops levitation is the Stokes force acting onto a drop from the steam flow arising from the heated liquid surface [2,3].

Among the effective solutions for removing high heat fluxes, there are systems of spray cooling [4] as well as the systems using stratified two-phase flows in microchannels [5,6]. Under the certain conditions, heat in these systems is removed due to evaporation of liquid microdroplets that are in contact with the surface [7,8]. However, the phenomenon of droplet levitation above a heated surface may have a significant effect on operation efficiency of the two-phase cooling systems, since the evaporation rate of levitated droplets is much lower than that of droplets in contact with the heated surface.

In this paper the micro-droplet cluster is studied at a relatively large heater of size 1×1 cm^2, (in works [1-3] the size of the heater was around 1 mm) under the condition when the liquid layer ruptures and the micro-droplets move onto the dry surface of the heater.

[a] Corresponding author : zaitsev@itp.nsc.ru

2 Experimental setup

Schematic and photograph of the experimental setup is presented in Fig. 1. The main part of the test section is a 3 mm thick stainless steel plate with a flush-mounted copper rod. At the working surface the rod has a 1×1 cm^2 square head. The rod is heated by a heating spiral coiled around its lower part. A layer of insulation material (mineral wool) is superimposed over the spiral. The heater surface temperature, T_w, and surface temperature of stainless steel plate are measured at several points by thermocouples. The heat flux, q, is determined by the electric power released at the spiral. Ultra-pure degassed distilled water with the initial temperature of 25°C is used as the working liquid. The test section is open to the atmosphere; the temperature of ambient air is 23±2°C. The test section is installed horizontally, and a square cooper cooling circuit is mounted on the working surface, so that the heater is in its centre. The inner size of the circuit is 40×40 mm^2. Water with the temperature of 25°C is pumped through the circuit. Before conducting experiment, a given volume of liquid is put onto the working surface by means of a syringe. The initial thickness of the liquid layer is h_0=0.40 mm. To register dynamics of cluster formation, the high-speed camera FASTCAM SA1.1 (5400 fps, resolution of 1024x1024 pixel, optical resolution of up to 1 micron/pixel) is used.

The working surface of the test section was mechanically polished. The morphology of the surface was analysed using scanning electron microscope (JEOL JSM6700F) and atomic force microscope (Solver Pro NT MDT). The root mean square (RMS) surface roughness, measured on a 22 μm^2 area of the heater surface, was found to be 790 nm.

The equilibrium contact angle on the working surface at different points was measured by the method of a sessile drop at the room temperature of 23 ± 2°C. To obtain the profile of the drop surface, the shadow method was used with a collimated light source and digital camera Nikon D800 with resolution of 6 micron/pixel. The resulting images were processed using the software by KRUSS. The advanced static contact angle measured at different points over the surface of the copper heater was 74±9°.

During the experiment the heat flux increased with small steps up to a threshold value of q=27 W/cm^2, when a dry spot was formed on the heater. Several runs were made. At the moment of the layer rupture the temperature of the heater surface was T_w=85±3°C.

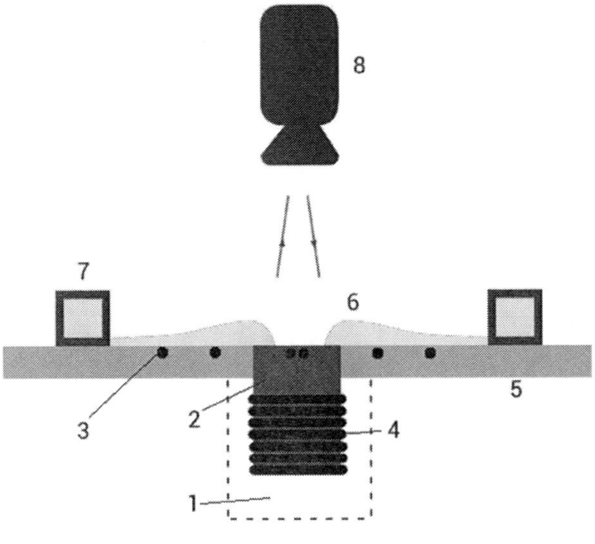

Figure 1. Schematic and photograph of the experimental setup. 1 – heat-insulating material, 2 – copper rod, 3 – thermocouple, 4 – nichrome spiral, 5 – stainless steel plate, 6 – liquid, 7 – cooling circuit, 8 – high-speed camera.

3 Experimental results

Increase of the heat flux induces formation of the cluster of micro-drops levitating over the heated layer (Fig. 2a). Usually, the cluster was of an elliptical shape. The average size of the cluster (defined as the mean between minor and major diameters of the ellipse inscribed in the array of micro-droplets) was found to be approximately from 500 to 4000 μm. Figure 3 represents variation of the average size of the cluster with time for 5 different runs (zero time corresponds to the beginning of recording, the last point in each run corresponds to the moment when the cluster coalescences with liquid; for mechanism of the coalescence refer to [3]). No cluster with an average size higher than approximately 4000 μm was observed. In most of previous studies [1-3] the authors use the term 'droplet cluster', meaning that the droplets form a localized one-layer array. However, the cluster in their experiments is localized due to the use of localized heating (the size of the heater is around 1 mm), and as such the size of the cluster is roughly coincides with the size of the heater. In contrast, we use a heater having size much larger than typical size of the cluster, which allows us to obtain larger arrays of droplets and measure their size.

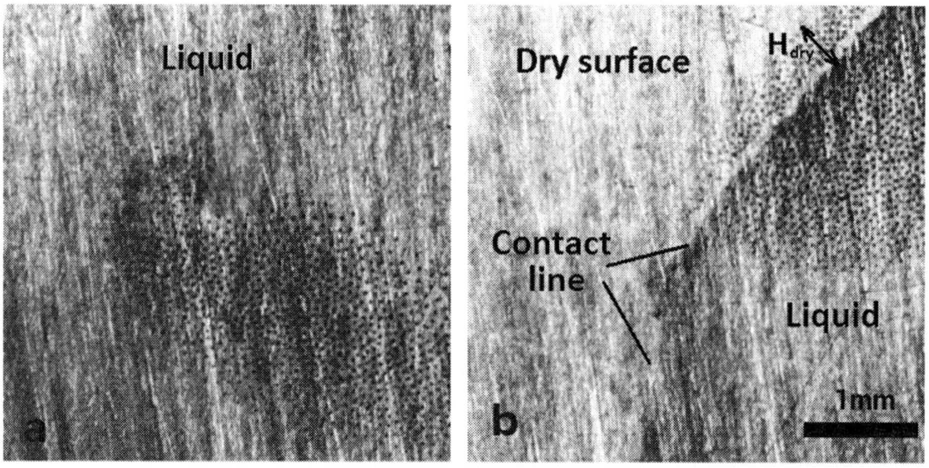

Figure 2: Array of micro-drops levitating over a water layer heated from below (the heat flux is 27.0 W/cm^2, the heater surface temperature is 83±3°C). Central area of the heater is shown. a) Before the rupture of liquid layer; b) just after the rupture.

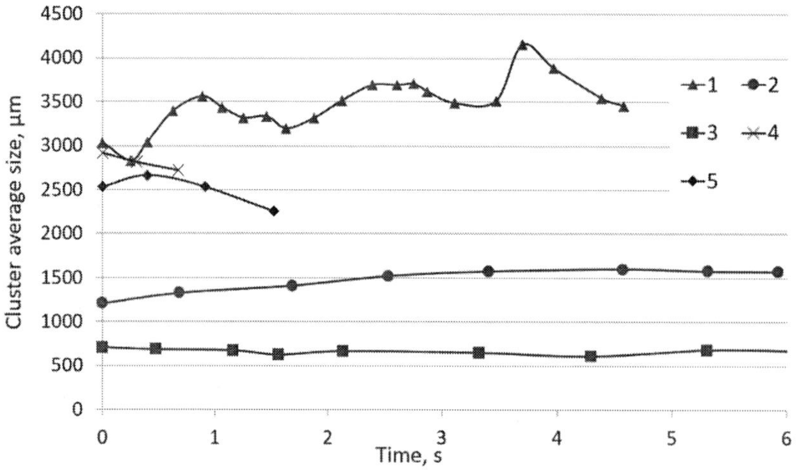

Figure 3: Time variation of the average size of the micro-drop cluster over heated layer of liquid for 5 different runs. The heat flux is 27.0 W/cm^2, the heater surface temperature is 70-83°C.

Figure 4 show the average distance between centre s of micro-drops in the cluster for two runs presented in Fig. 3 (runs 1 and 3). It is seen that in spite the fact that the size of the cluster differs roughly 7 times for these runs, the average distance between the micro-drops is close (around 175-200 μm, which is in good agreement with results of [1-2]).

Upon the rupture of liquid layer (at a threshold heat flux) the levitating micro-drops move to the dry spot. It has been found out that usually micro-droplets do not touch the heater surface, but levitate over the surface at some distance, Fig. 2b. Due to evaporation, the droplet size decreases with time. The drops can move over the dry surface as far as 3 mm away from the contact line, and can exist for up to 3.5 seconds after the rupture, Fig. 5. It should be noted, that the evaporation rate of levitating micro-droplets is much less as compared to the micro-droplets that fall down the heater surface. As soon as a micro-droplet touches the heater surface, it disappears within a fraction of a second (less than 0.1 s). Thus, the phenomenon of micro-droplet levitation may have a significant effect on operation efficiency of the two-phase cooling systems.

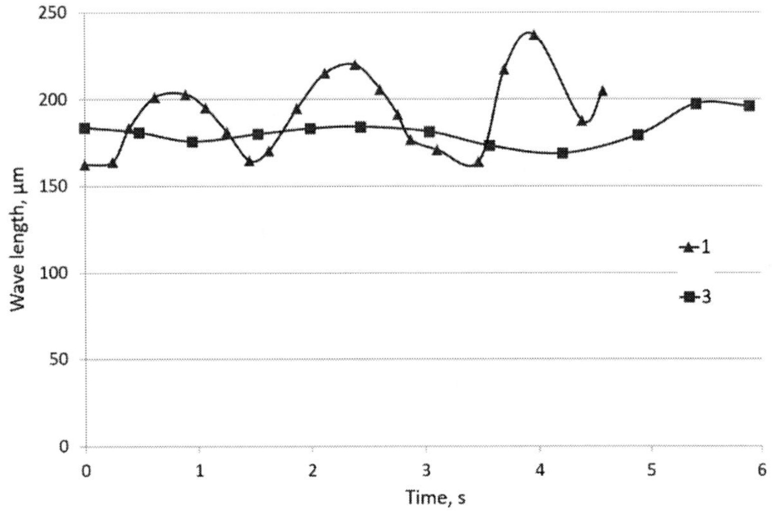

Figure 4: The average distance between centres of micro-drops in the cluster for runs 1 and 3 from Fig. 3.

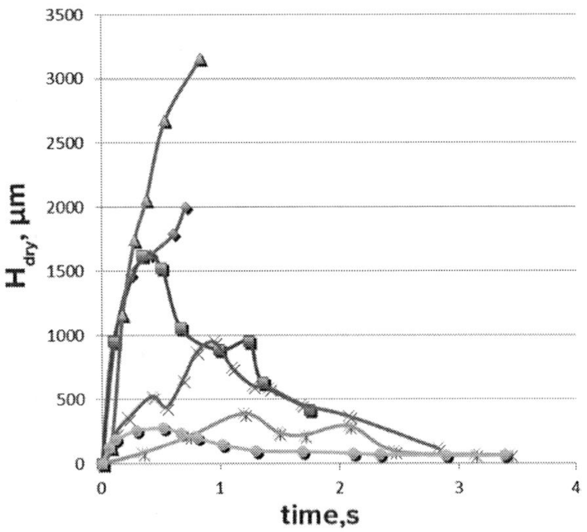

Figure 5: Distance from the contact line to the boarder of the array of micro-drops levitating over the dry surface of the heater (see Fig. 2b), versus time. Data are prasented for 6 runs. Zero time corresponds to the moment of the layer rupture. The heat flux is 27.0 W/cm^2, the heater surface temperature is 83±3°C.

4 Conclusions

The micro-droplet cluster has been studied on the heater having size greater than typical size of the cluster. The average size of the cluster was found to vary from approximately 500 to 4000 μm, with the distance between micro-drops being around 175-200 μm. It has been found, that the micro-droplets can levitate not only over the heated liquid surface, but also over a dry surface of the heater, following the rupture of the liquid layer. The micro-droplets can move over the dry surface as far as 3 mm away from the contact line, and can exist for up to 3.5 seconds after the rupture.

This work was supported by the Ministry of Education and Science of Russia (Agreement No 14.604.21.0053, project identifier RFMEFI60414X0053).

References

1. A. A. Fedorets, JETP Lett. **79**, 372 (2004).
2. A.A. Fedorets, I.V. Marchuk and O.A. Kabov, Technical Physics Letters, **37** (2), 116 (2011).
3. A.A. Fedorets, I.V. Marchuk and O.A. Kabov, JETP Lett. **99**, 266 (2014).
4. Jungho Kim. Int. Journal of Heat and Fluid Flow, **28** (4), 753-767 (2006).
5. Kabov O.A., Lyulin Yu.V., Marchuk I.V. and Zaitsev D.V. Int. Journal of Heat and Fluid Flow, **28**, 103-112 (2007).
6. Zaitsev D.V., Rodionov D.A. and Kabov O.A. Technical Physics letters, **35** (7), 680-682 (2009).
7. Semenov A.A., Feoktistov D.V., Zaitsev D.V., Kuznetsov G.V., and Kabov O.A. Thermophysics and Aeromechanics, **22** (6), 771-774 (2015).
8. Kuznetsov G.V., Feoktistov D.V., Orlova E.G. Thermophysics and Aeromechanics, **23** (1), 17-22 (2016).

MATEC Web of Conferences **84**, 00045 (2016) DOI: 10.1051/matecconf/20168400045
International Symposium IPHT 2016

Numerical investigation of heat and mass transfer processes in a spherical layer of viscous incompressible liquid with free boundaries

Alla Zakurdaeva[1,2,a]

[1]*Altai State University, Faculty of Mathematics and Information Technologies, Russia*
[2]*Institute of Thermophysics SB RAS, Laboratory of Enhancement of Heat Transfer. Russia*

Abstract. The results of mathematical modelling of the dynamics of a mixture of the viscous incompressible liquid and gas, which fills a spherical layer with free boundaries and contains a gas bubble within itself, are presented in this paper. Spherical symmetry is assumed, and it is considered that the dynamics of the layer is determined by thermal, diffusive and inertial factors. On the basis of constructed numerical algorithm the studies of the formation of the liquid glass layers, which contain the carbon dioxide gas within themselves, have been conducted. The impact of the external thermal regime, external pressure and the density of gas in the bubble at the initial time on the dynamics of the layer, diffusion and heat-and-mass processes inside it is investigated. The results of numerical investigation of the full and simplified thermal problem statement, without consideration of gas diffusion, are compared.

1 Introduction

The study of spherical liquid layers is a task of current interest in connection with the investigation of properties of some new materials, such as spheroplast or sensitizers for emulsion explosives [1, 2]. The mathematical and numerical modeling of the formation of spherical microballoons is held in [3-7].

The dynamics of a spherical layer with free boundaries, consisting of the viscous incompressible liquid and gas and containing a gas bubble within itself [3-5], is investigated numerically in this paper. It is assumed that the gas is insoluble in the liquid and all transfer coefficients depend on the temperature. The problem is considered in a short-term weightlessness, that involves a spherical symmetry of processes, i.e. only radial component of the velocity of the fluid is nonzero. All physical quantities depend on the distance from the origin and vary with time. The Navier-Stokes equations, as well as the equations of the heat transfer and diffusion of the gas form the basis of mathematical model, describing the processes within the liquid layer. Inside the gas bubble the pressure, density and absolute temperature of the gas satisfy the ideal gas law.

[a] Corresponding author : alla2300@bk.ru

2 Statement of the problem

Let the domain $R_1(t) < r < R_2(t)$ define a spherical layer, where $r = R_1(t)$ and $r = R_2(t)$ are its internal and external boundaries. The consequence of the original mathematical model is the following system of equations in the dimensionless form [4]:

$$\frac{dV}{dt} = \frac{1}{2}V^2 \frac{(R_2^2 + R_1^2)(R_2 + R_1)}{R_2^3 R_1^3} + \frac{1}{\mathrm{Re}}\left[P_g S - P_{ex} S - 2\overline{Si}\sigma(T)\frac{R_2 + R_1}{R_2 R_1}\right]\frac{R_2 R_1}{R_2 - R_1} - \tag{1}$$

$$-\frac{4}{\mathrm{Re}}\nu(T)V\frac{R_2^2 + R_2 R_1 + R_1^2}{R_2^2 R_1^2}, \quad t > 0, \quad V(0) = V_0;$$

$$\frac{dR_1}{dt} = \frac{V}{R_1^2}, \quad t > 0, \quad R_1(0) = R_{10}; \tag{2}$$

$$\frac{d\rho_g}{dt} = -\frac{3V}{R_1^3}\rho_g + \frac{3}{Pe_d R_1}D(T)\frac{\partial C}{\partial r}, \quad t > 0, \quad \rho_g(0) = \rho_{g0}; \tag{3}$$

$$\frac{\partial T}{\partial t} + \frac{V}{r^2}\frac{\partial T}{\partial r} = \frac{1}{Pe}r^{-2}\frac{\partial}{\partial r}(r^2\chi(T)\frac{\partial T}{\partial r}), \quad t > 0, \quad T(0,r) = T_0(r); \tag{4}$$

$$\frac{\partial C}{\partial t} + \frac{V}{r^2}\frac{\partial C}{\partial r} = \frac{1}{Pe_d}r^{-2}\frac{\partial}{\partial r}(r^2 D(T)\frac{\partial C}{\partial r}) \quad t > 0, \quad C(0,r) = C_0(r). \tag{5}$$

Here V is the rate of change of the spherical layer's volume, $V = r^2 \upsilon$, where υ is the radial velocity of the fluid, ρ_g is the density of the gas in the bubble, T is the temperature, C is the gas concentration in the fluid layer, P_g, P_{ex} are the pressure in gas bubble and external pressure, $\nu(T)$, $\sigma(T)$ are the kinematic viscosity and surface tension coefficients, $\chi(T)$, $D(T)$ are the thermal diffusivity and diffusion coefficients. The following dimensionless parameters arose as a result of bringing the system of equations to dimensionless form: the Reynolds number $\mathrm{Re} = (\upsilon_* r_*)/\nu_*$, the Peclet number $Pe = (\upsilon_* r_*)/\chi_*$, the diffusive Peclet number $Pe_d = (\upsilon_* r_*)/D_*$ and also parameters $S = (r_* P_*)/(\rho_* \nu_* \upsilon_*)$, $Si = \sigma_*/(r_* P_*)$, $\overline{Si} = Si \cdot S$. Characteristic values are designated by index *. They are selected so that $r_* = t_* \upsilon_*$, $\chi_* = \kappa_*/(\rho c_1)$, ρ is the density of fluid (chosen as the characteristic value), $\kappa(T)$ is the thermal conductivity coefficient, c_1 is the heat capacity of the fluid.

Equations (1), (2) are consequences of the Navier-Stokes equations and kinematic conditions at the inner boundary of the fluid layer, the equation (3) determines the change in the mass of the gas bubble, the equations (4) and (5) represent the heat transfer and diffusion processes.

The statement of the problem is supplemented by the kinematic, dynamic conditions and Henry law that links the gas concentration at the boundaries with the pressure outside this area. On the internal free surface the energy balance equation and the condition of temperature continuity are also determined. On the external free boundary the heat exchange with the atmosphere is fulfilled [3].

Inside the gas bubble the ideal gas law is prescribed:

$$P_g = \tilde{R}\rho_g T_g. \tag{6}$$

Here $\tilde{R} = R\rho T_*/P_*$ is a dimensionless parameter, R is the universal gas constant [4].

3 Numerical algorithm

The numerical computations have been carried out according to the following steps:

• Solving the Cauchy problem for a system of the ordinary differential equations (1)-(3) to find the rate of change of the spherical layer's volume V, the density of the gas in the bubble ρ_g and the internal radius R_1 by the fourth-order Runge–Kutta method.

• Determination of the external radius of the layer R_2 with the help of volume conservation law $R_2^3(t) - R_1^3(t) = R_{20}^3 - R_{10}^3$.

• Transition from the domain with moving boundaries (r,t) to the fixed region (x,t) with the help of the new dimensionless spatial variable $x = (r^3 - R_1^3(t))/(R_{20}^3 - R_{10}^3)$.

• Computation of the temperature T within the fluid layer for finite-difference analogue of equation (4) by Thomas algorithm complicated by a parameter, which is the unknown value of temperature at the internal free boundary [6].

• Computation of the gas concentration C in the fluid layer for finite-difference analogue of equation (5) by means of the ordinary Thomas algorithm [7].

• Determination of the pressure inside the gas bubble P_g through the ideal gas law (6).

4 Calculation results

Numerical studies on the formation of the liquid glass layers, which contain the carbon dioxide gas within themselves, have been conducted. The initial state of the "gas-liquid" system is characterized by the following values: $R_{10} = 0.02$ cm, $R_{20} = 0.05$ cm. A parabolic distribution of the gas concentration $C_0(x)$ at the initial moment is assumed to be given. Initial distribution of the temperature is constant: $T_0(x) = 1171.1$ K. The external atmosphere heats up from $T_{ex1} = 1171.1$ K to $T_{ex2} = 1673$ K under the law:

$$\begin{cases} T_{ex} = T_{ex1} + \dfrac{(T_{ex2} - T_{ex1})}{t_2 - t_1}(t - t_1), t_1 \le t \le t_2; \\ T_{ex} = T_{ex2}, t > t_2. \end{cases}$$

Here $t_1 = 0$ s, $t_2 = 0.3$ s.

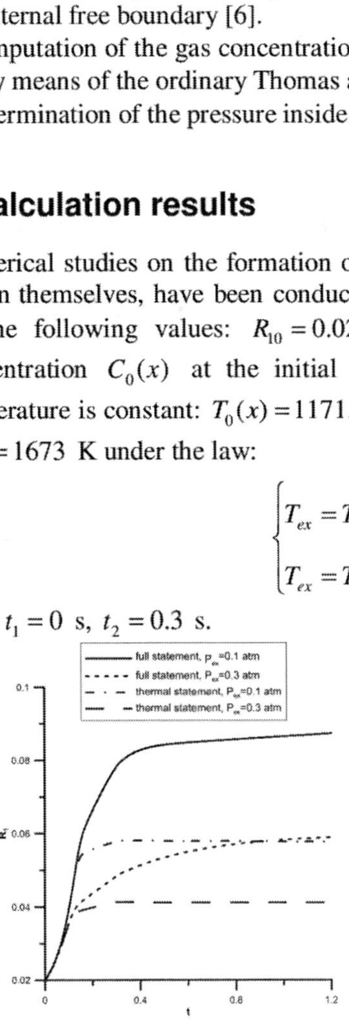

Figure 1. Comparison of results for internal radius change at different external pressure ($\rho_{g0} = 0.92 \cdot 10^{-3}$ g/cm^3).

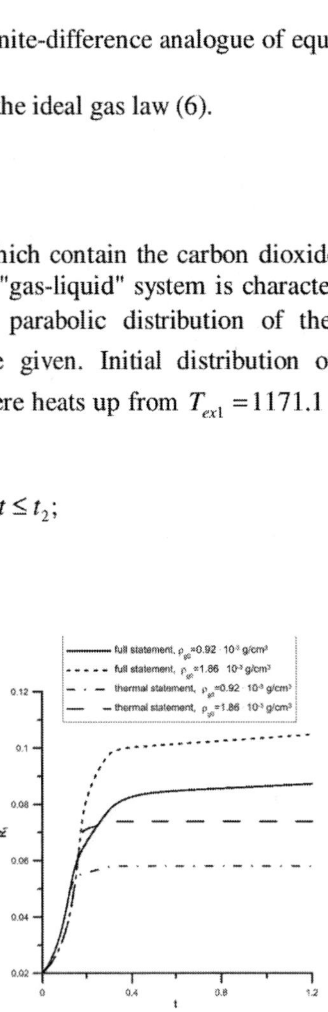

Figure 2. Comparison of results for internal radius change at different density of the gas in the bubble at initial time ($P_{ex} = 0.1$ atm).

The comparison of results for the full and simplified thermal problem statement, when the gas diffusion is not taken into account, at different external pressure (Fig. 1) and at different density of the gas in the bubble at initial time (Fig. 2) has been carried out. It is clear that for the full problem statement a more intense extension of the liquid layer is observed and a stationary regime is achieved later than in the case of the thermal statement. Also Fig. 1 demonstrates that the increase of external pressure P_{ex} vastly curbs the expansion of the layer while Fig. 2 shows that the increase of density of the gas in the bubble at initial time ρ_{g0} contribute to the process.

The comparison of results for the Dirichlet and Newton conditions for the temperature at the external free boundary has been performed (see Fig. 3).

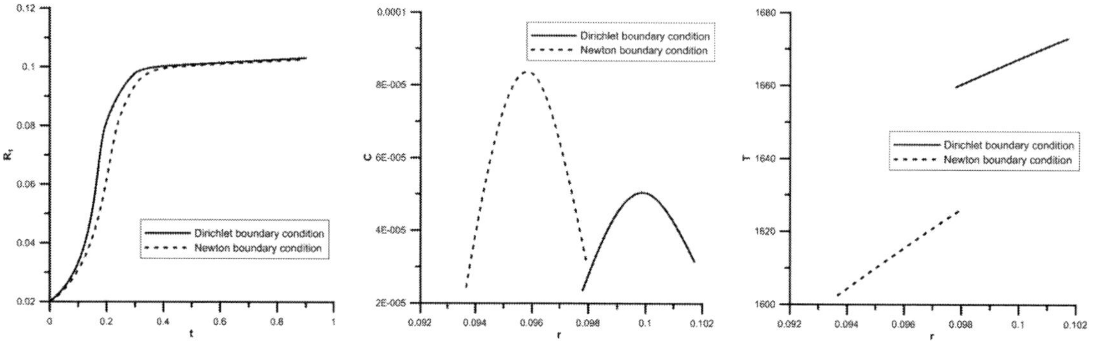

a) b) c)

Figure 3. Comparison of results for the Dirichlet and Newton boundary conditions for the temperature at the external free boundary ($\rho_{g0} = 1.86 \cdot 10^{-3}$ g/cm³, $P_{ex} = 0.1$ atm): a) Variation of internal radius R_1 , b) Distribution of gas concentration C at the moment $t_0 = 0.3$ s, c) Distribution of temperature T at the moment $t_0 = 0.3$ s.

5 Conclusions

The problem of the dynamics of a spherical layer of viscous incompressible liquid and heat and mass transfer processes inside it is being solved. The numerical algorithm for the problem solving is constructed. Numerical studies for the "liquid glass-carbon dioxide" system for different values of the density of the gas in the bubble at initial time and external pressure are conducted. The influence of external thermal regime on the dynamics of the layer, distribution of temperature and gas concentration inside it is investigated. The comparison of results for the Dirichlet and Newton boundary conditions for the temperature at external surface of the sphere is performed. Results for the full and simplified thermal problem statement (when the gas diffusion is not taken into account) are compared. It is determined that besides the external thermal regime a significant impact on the behaviour of the spherical layer is provided by diffusion of gas.

Acknowledgments. The study was supported by the Russian Ministry of Education and Science (project identifier RFMEFI61314X0011).

References

1. A. G. Anshits, N. N. Anshits, A. A. Deribas, S. M. Karakhanov, N. S. Kasatkina, A. V. Plastinin, A. Yu. Reshetnyak, V. V. Silvestrov, Fizika goreniya i vzryva **41** (**5**), 119-127 (2005).
2. E. V. Karpov, Prikladnaya mehanika i tehnicheskaya fizika **50** (**1**), 197-204 (2009).
3. O. N. Goncharova, Dinamika sploshnoy sredi **82**, 66-79 (1987)
4. O. N. Goncharova, V. V. Pukhnachev, Modelirovanie v mehanike **5**, 83-95 (1990)
5. O. N. Goncharova, Dinamika sploshnoy sredi **106**, 36-48 (1993)
6. A. V. Zakurdaeva, E. V. Rezanova, Omskiy nauchniy vestnik **3** (**143**), 312-315 (2015)
7. A. V. Zakurdaeva, E. V. Rezanova, Prikladnaya matematika i fundamentalnaya informatika **2**, 37-40 (2015)

MATEC Web of Conferences **84**, 00046 (2016) DOI: 10.1051/matecconf/20168400046
International Symposium IPHT 2016

Combustion of liquid fuel in rectangular mini and microchannels

Valery V. Zamashchikov[1,3,a], Aleksey A. Korzhavin[1], and Evgeny A. Chinnov[2,3]

[1]*Institute of Chemical Kinetics and Combustion SB RAS, 630090 Novosibirsk, Russia*
[2]*Kutateladze Institute of Thermophysics SB RAS, 630090 Novosibirsk, Russia*
[3]*Novosibirsk State University, 630090 Novosibirsk, Russia*

Abstract. It is shown that flame spread rate can be high and comparable with velocities of flame propagation in the stoichiometric homogeneous gas mixture. The flame spread rate depends on velocity of oxidizer. It can either increase or decrease with arise of oxidizer velocity, depending on the oxygen content. The flame surface is significantly distorted with increase in average flame spread rate. It is shown that the flame spread rate can be significant and comparable with the laminar burning velocity of the stoichiometric homogeneous gaseous mixture.

1 Introduction

The trend towards a decrease in characteristic sizes of devices in various technical fields determines development of studies on hydrodynamic and heat transfer in mini- and microchannels. At a decrease in the height of flat channels the ratio of surface area to channel volume increases inversely to its minimal transverse size, and this causes high heat transfer intensity in the microsystems. In application to the processes occurring in reacting systems, for instance, burning of combustible liquids in small channels, heat transfer intensification can both improve and deteriorate the situation. Therefore, investigation of combustion processes in such systems is important for development of new applications and improvement of energy efficiency of available technical solutions.

To implement the combustion process, which includes the co-current motion of combustible liquid and gas it is important to get information about the regimes of two-phase flow in the considered channels. For instance, it is necessary to know in advance the conditions suitable for the separate flow, common for combustion.

It is shown in the review on the studies of the two-phase flows [1], that the flows in channels of different cross-sections depend significantly on channel transverse dimensions. According to analysis of the maps of air-water flow patterns in horizontal rectangular channels with height 0.1-1 mm [2-3], at low liquid and gas velocities, the phases flow separately.

Flame spread over pools of liquid with the flash temperature higher than ambient temperature was investigated by Ross and Miller [4, 5]. It is commonly assumed [6, 7] that liquid in the pre-flame area is heated through convective heat transfer, occurring due to the gradient of surface tension The space flame spreading is impossible in the conditions of restricted without oxidizer supply [8]. The pulsating flame spread regime implementation and mean flame spread rate depend on opposed air flow

[a] Corresponding author : albor@kinetics.nsc.ru

velocities. Flame spread rates of an order of 0.5 cm/s were obtained. Increase in the flame spread rate oppose the oxidizing flow is possible by enriching the oxidizer with oxygen. Furthermore the flame spread rate depends on the channel height. In this connection the experiments were carried out in a narrow flat channel, the height of which may be varied, and with oxidizing flow that can be enriched with oxygen, n-butanol was used as combustible liquid.

In the present work, we have studied the flames propagating over the surface of n-butanol in micro- and minichannels. This experimental study was aimed at determination of the features of diffusion combustion of fuel film at the counter-current flow of oxidizer in a narrow flat channel.

2 Experimental setup

The flames, propagating above the surface of n-butanol in mini channels at $h>h_{cr}$, were studied. Here h_{cr} is quenching size for a slot. The channel height could vary, and it was an order less than its width. This was made to achieve the 2D structure of the flows and flame front, as far as possible.

The working section and scheme of equipment location are shown in Fig. 1. The working section consists of two parallel quartz plates, the distance between which (the channel height) h is set by two metal gaskets with the thickness from 1 to 0.5 mm. The oxidizing gas mixture was fed into the channel. The oxidizing mixture was prepared in the mixer of high pressure by partial pressures. The prepared mixture was fed to the flow controller El-Flow of Bronkhorst production. The accuracy of mixture preparation was 5%. The average velocity V of the oxidizer was determined as the ratio of the oxidizer flow rate to the cross-sectional area of the channel. The combustion process was registered from the top in the region between the liquid input and channel outlet by the digital video and photo cameras.

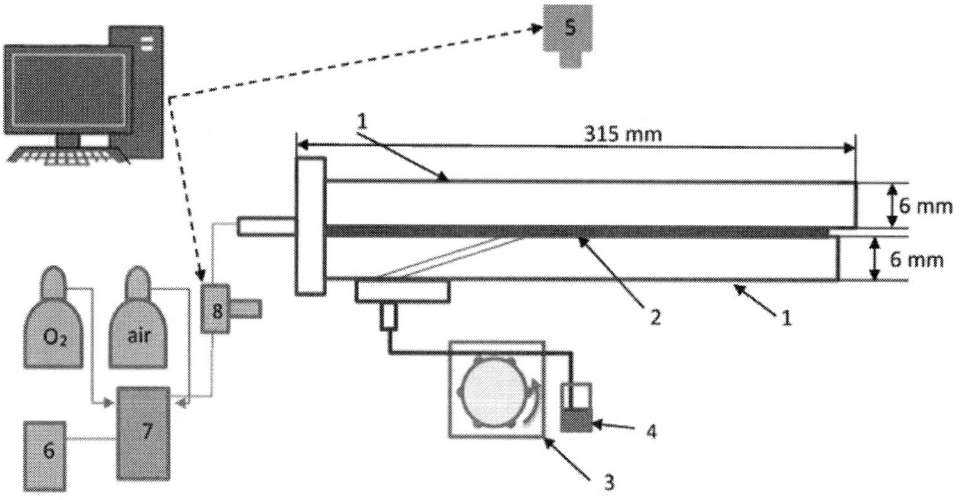

Figure 1. The scheme of setup. 1 – quartz plates, 2 – metal gasket, 3 – peristaltic pump, 4 – vessel with n-butanol, 5 – video camera, 6 – mixer of air and oxygen, 7 – gas control panel, 8 – gas flow controller El-Flow.

Ignition was performed by the naked flame at the channel outlet. After ignition we recorded the flame front coordinate and obtained the dependence of this coordinate on time. The flame velocity S was determined by inclination of this dependence. The accuracy of velocity measurements was 5%.

3 Results

Dependences of average flame propagation velocity on average gas velocity in the channels of 1 and 0.5-mm height at different oxygen content in the mixture are shown in Fig. 2. Average flame propagation velocity *S* determined on coordinate of the leading point of the flame in the beginning and final moments of flame propagation. As expected with increase in oxygen percentage in the oxidizer flame rate also increases. It is also seen in Fig. 2 that at addition of up to 60 % of oxygen into the oxidizer, dependences of flame velocity on average gas velocity are similar to those obtained in the higher channels [9]. However, for pure oxygen this dependence is changed to the opposite one (Fig. 2). Noteworthy also the average flame velocity in the oxygen flow in the channels with the height of 1 and 0.5 mm decreases with a decrease in the channel height, as it is expected.

Now as far as effect of environmental temperature on flame propagation velocity is concerned. Notice, according to [9] flame propagation velocity is dependent on environmental temperature. However flame propagation velocity is sensible to the temperature only at moderate oxidizer rates [9] and effect of temperature is weakened at increase in oxygen percentage in the oxidizer. Furthermore to avoid mistakes due to variation of environmental temperature the experiments were carried out at negligible variation of environmental temperature.

Complex structure of the flame front was observed in channel with 0.5 mm height (Fig. 3). Probably because of film nonuniformity, the presence of liquid on the upper plate and small average film thickness (about 0.1 mm), the flame front is asymmetrical. Notice, if the oxidizer is pure oxygen, in the channel of 1-mm height the average flame velocity becomes comparable with the laminar burning velocity of the flame of homogeneous stoichiometric hydrocarbon-oxygen mixture (3 m/s) [10]. At that the flame propagates with pulsations and flame velocity amounts to 7 m/s; i.e. these values are higher than the laminar burning velocity of homogeneous gas flame.

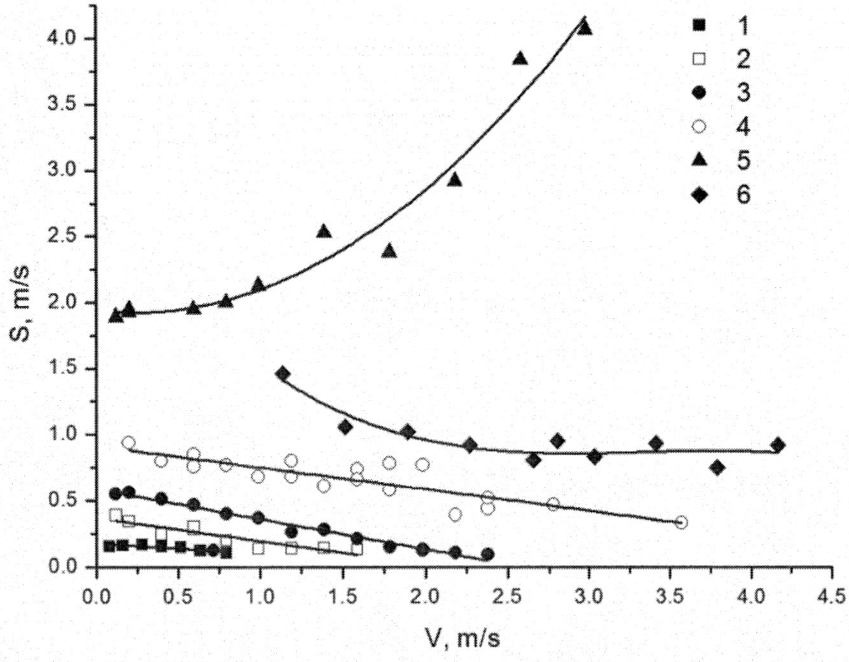

Figure 2. Average flame propagation velocity vs average gas velocity for different oxygen additive and channel heights. Channel height 1 mm (1-5), 0.5 mm – 6; ambient temperature 25.0-25.7°C. Oxygen additive in air: 1 - 24%, 2 - 30%, 3 - 40%, 4 - 60%, 5 – 6 pure oxygen (without air).

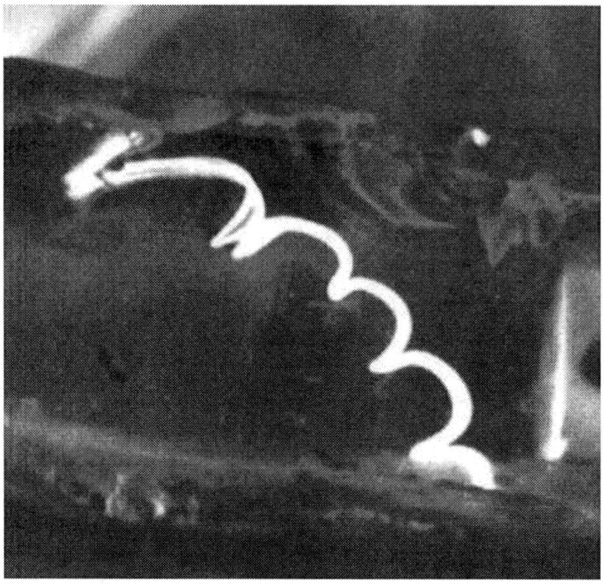

Figure 3. Flame in the channel of 0.5-mm height. Oxygen flow rate – $48 \cdot 10^{-6}$ m^3/s ($V = 2.3$ m/s).

4 Conclusions

Flame propagation against the oxidizer flow above the thin films of liquid fuel with vapor pressure below the lean limit in a narrow flat channel was studied experimentally.

It is shown that flame propagation velocity can be high and comparable with velocities of flame propagation in the stoichiometric homogeneous gas mixture.

The flame propagation velocity depends on velocity of oxidizer. It can either increase or decrease with arise of oxidizer velocity, depending on the oxygen content.

The flame surface is significantly distorted with increase in average flame propagation velocity.

Acknowledgments

This work was supported by the grant of Russian Science Foundation (Agreement No. 15-19-30038).

References

1. E.A. Chinnov, O.A. Kabov High Temperatures, **44**, 5 (2006).
2. E.A. Chinnov, F.V. Ron'shin, O.A. Kabov Thermophysics and Aeromechanics, **22**, 5, (2015).
3. E.A. Chinnov, F.V. Ron'shin, O.A. Kabov IJMF, **80** (2016).
4. H. D. Ross, F. J. Miller Proc. Combust. Inst., **26** (1996).
5. H. D. Ross, F. J. Miller Proc. Combust. Inst., **27** (1998).
6. P. L. Garcia-Ybarra, et al. Proc. Combust. Inst., **26** (1996).
7. W. A. Sirignano, I. Glassman Combust. Sci. Technol., **1** (1970).
8. V.V. Zamashchikov Combust. Sci. Technol., **181** (2009).
9. V.V. Zamashchikov, A. A. Korzhavin, E.A. Chinnov Combustion, Explosion and Shock Waves, **50**, 4 (2014).
10. E.S. Golovina, G.G. Fyodorov Proc. Combust. Inst., **6** (1957).

AUTHOR INDEX